Hard Disk Drive

Drive

Mechatronics and Control

AUTOMATION AND CONTROL ENGINEERING

A Series of Reference Books and Textbooks

Editor

FRANK L. LEWIS, Ph.D.

Professor
Applied Control Engineering
University of Texas at Arlington
Fort Worth, Texas

Hard Disk Drive

Mechatronics and Control

Abdullah Al Mamun
The National University of Singapore

GuoXiao Guo
Western Digital Technology
Lake Forest, California, U.S.A.

Chao Bi
Data Storage Institute
Singapore

CRC Press
Taylor & Francis Group
Boca Raton London New York

CRC Press is an imprint of the
Taylor & Francis Group, an **informa** business

CRC Press
Taylor & Francis Group
6000 Broken Sound Parkway NW, Suite 300
Boca Raton, FL 33487-2742

© 2007 by Taylor & Francis Group, LLC
CRC Press is an imprint of Taylor & Francis Group, an Informa business

No claim to original U.S. Government works

International Standard Book Number-10: 0-8493-7253-4 (Hardcover)
International Standard Book Number-13: 978-0-8493-7253-7 (Hardcover)

Visit the Taylor & Francis Web site at
http://www.taylorandfrancis.com

and the CRC Press Web site at
http://www.crcpress.com

To my parents
Late Md. Abdul Mazed
and
Mrs. Rabeya Khatun
- Abdullah Al Mamun

To: Ying, Tianrui and Tianxin
- Guoxiao Guo

To: Ruoling and Bi Hui
- Chao Bi

Authors

Abdullah Al-Mamun graduated in 1985 from IIT, Kharagpur, India, with a B.Tech. (Hons.) degree in Electronics and Electrical Communication Engineering, and received a Ph.D. in 1997 from the National University of Singapore (NUS).

Prior to joining NUS in 1999 as a member of the faculty in the department of Electrical and Computer Engineering, he worked in various capacities in Siemens (Bangladesh) Limited, Magnetic Technology Center (MTC), Singapore, and Maxtor Peripherals Ltd. His research interests include precision mechatronics, control and signal processing in disk drive servomechanism, nonlinear control and mobile robots. His work has so far contributed to more than 30 papers in reputable international journals and conference proceedings.

Guoxiao Guo received B.Eng. (1989) and M.Eng. (1994) degrees from the Department of Automation, Tsinghua University, Beijing, China, and a Ph.D. (1997) from the School of Electrical and Electronics Engineering, Nanyang Technological University, Singapore.

He joined the A*STAR-Data Storage Institute, Singapore in 1995 as a research engineer, and became the manager of the Mechatronics and Recording Channel Division before joining Western Digital, USA. While in DSI, he was an Adjunct Fellow and a Member of the Graduate School of Integrative Science and Engineering at National University of Singapore. His research interests include nonlinear and robust control, mechatronics and MEMs, vibration analysis and control with application to nano positioning systems. He has authored or co-authored over 40 journal publications in these areas.

Chao Bi obtained a B.Eng. degree from the Hefei University of Technology in 1982, an M.Eng. degree from the Xian Jiaotong University in 1984, and a Ph.D. from the National University of Singapore in 1994.

He is currently a research scientist in the A*STAR-Data Storage Institute, Singapore and an adjunct fellow in the National University of Singapore. His research interests include the design, control and testing of permanent-magnet AC motors, electromagnetic field synthesis and analysis, micro electromagnetic system, and optimization technology. He has published more than 50 conference and journal papers in these areas.

Contents

Preface

Magnetic information storage systems plays a very important role in this era of digital information. All computer literate people of the world use these system to save billions of bytes of digital information and to access them as and when required; both of these services are available at the touch of our fingertips. Magnetic hard disk drives constitute the lions share of these storage systems.

The hard disk drive (HDD) industry is slightly more than half a century old. The five decades of this industry have experienced many excellent and triumphant technological innovations. It took creativity and hard work of many scientists and engineers to transfer from the earliest HDD of 1956 to the current status. In comparison to the disk drives we see these days, the earliest disk drive was a monstrous device occupying a large floor. It was used in the system called *Random Access Method of Accounting and Control* (RAMAC) produced by IBM. Today's drives are tiny compared to that, however, the storage capacity of the monstrous RAMAC can not even be described by the word *tiny*. The journey of the HDD industry that began in 1956 is yet to reach its end. The demand for more and more online storage created by the new and developing digital technologies is growing and will continue to grow in the near future. The ability to meet such demand at relatively low cost makes HDD the undisputed candidate for online, direct access, non-volatile storage of a computing system. The HDD industry has so far met successfully the challenges of high density storage at low cost, and it continues to progress at lighting speed. The future demand for online storage of huge digital information brings in promises for this industry as well as new challenges.

The HDD industry owes to many disciplines of knowledge for the growth it has experienced so far. Advances in magnetics, material science, analog and digital electronics, motor and actuator design, mechanics, tribology, signal processing, servomechanism and control, manufacturing technology etc have propelled the growth of this industry. All of these branches of knowledge remain vital for sustaining the growth and facing the future challenges. Three direct links of contribution exist between the HDD industry and the subjects of precision mechatronics and control - (1) precision control of read-write head, (2) control of spindle motor, and (3) creation of precision magnetic pattern on disks during manufacturing. This book is focused on these aspects of HDD technology. It presents the fundamental aspects in each of these areas, explains the problems associated with them, discusses the solutions currently in use, and highlights the future challenges.

Many excellent books on the exciting technology of HDD are available in the market, most of which are listed in the reference of this book. These books provide very good account of the primary technology involved in HDD, i.e., the digital magnetic recording. *Magnetic Recording* [142], edited by C.D. Mee and E.D. Daniel is an excellent reference for engineers and scientists involved in the HDD industry and for students specializing in magnetic data storage technology. These books also provide essential information about the mechatronics components of HDD. *Magnetic Information Storage Technology* [202], a book jointly authored by S.X. Wang and A.M. Taratorin, is a self-contained reference for magnetic storage technology and provides great deal of information on developments that took place in the 1990s. A chapter in the book titled *Digital control of dynamic systems* [54] authored by G.F. Franklin et al and the book *Hard Disk Drive Servo Systems* [29] by B. M. Chen et al provide useful insight of the control problems encountered in HDD servomechanism. Though these books highlight the control aspects of HDD, we humbly felt that there is a need for a textbook focusing on the applications of mechatronics and control in the HDD industry, which would cover not only the actuator control problems but also many other aspects of HDD using mechatronics. This book is aimed to provide both theoretical and practical accounts of mechatronics and control in hard disk drives as well as in the manufacturing of HDD. The contents of this book are planned to address the needs of the following readers:

- Graduate and undergraduate students studying mechatronics and control,

- Researchers working in the areas of precision mechatronics systems and control,

- Engineers involved in the research and development in HDD industry,

- Entry-level engineers at the HDD industry who wish to acquire quickly the background knowledge of the HDD servomechanism and spindle motors,

- Researchers in the area of control who are interested to take up challenging practical problems.

We tried our best to make the book self-contained so that the readers can follow the contents without referring to other technical papers. However, many challenging issues considered by us as beyond the scope of this book are just mentioned in simple words. Technical references required to have an in-depth understanding of such issues are provided for interested readers.

This book is organized as follows. The first chapter provides an introduction to the fascinating world of HDD technology. It includes (1) a brief account of the history of the industry, (2) definitions of several jargon used in the community related to this industry, (3) description of different components that make an HDD, (4) introduction to the access mechanism of HDD, and (5) the trends in the industry. Chapter 2 is dedicated to the servomechanism used to access data in a hard disk drive. The head positioning servomechanism controls the position of the read-write head which are used to record binary information on the magnetic media coated on rotating circular disks and to retrieve information from the disks. This chapter provides (1) a description of the mechanism used in the head positioning servo system, (2) a dynamic model of the actuator, (3) an explanation of the methods used in HDD to generate feedback signal, (4) an account of sources of noise and disturbance affecting the performance of the servomechanism, (5) objectives of the closed loop servo, and (6) the near time-optimal control used in HDD. Details of the design issues related to the head positioning servomechanism are provided in Chapter 3. Starting from a simple model of the actuator to design a controller, this chapter explains the difficulties faced by the servo engineer, provides solution to each of them, thus reaches finally to a more complex design of the controller. Important issues related to the spindle motor that facilitates spinning of the disks are explained in Chapter 4. Keeping with the objective of making the book self contained, this chapter begins with the fundamentals of electromechanical energy conversion. Principle of operation is explained for brushless DC motors, key concerns related to the spindle motor and spindle motor drives are highlighted, and solutions are provided for these issues. The performance of the spindle motor affects directly the precision obtained by the head positioning servomechanism. The feedback signal for the head positioning control loop is derived from the pre-written spatial patterns on the disk surface. Any lateral or vertical movement of the disks induced by the motion of the spindle motor has consequences on the performance of the head positioning servo. Acoustic noise that spindle motor generates and several ways to minimize it are also discussed in this chapter. Problem of acoustic noise gets more and more attention in the industry as the HDD finds its way in new applications such as MP3 player and video camera. Chapter 5 of the book discusses the issues related to another mechatronic system, the *Servo Track Writer*, which is not a part of disk drive but plays a vital role in the process of manufacturing HDD. The reference patterns that define the tracks and sectors of an HDD are created at this stage. These patterns are used later by the HDD servomechanism. Creating these patterns reliably with the required precision is a pre-requisite for successful realization of ultra-high precision control of the read-write head in the HDD. The precision and accuracy demanded of the servo-track writing process are higher than those achieved by the HDD servomechanism, and this process involves mechatronics and control with very stringent specifications.

Great efforts have been put in editing the book but, in our opinion, the book is far from being perfect. There are inevitably some errors undetected by the authors. If you notice any such error, or have a comment, please contact the authors at eleaam@nus.edu.sg. We shall be very grateful.

The writing of this book would not be possible without the help and support of many individuals. We extend our sincere thanks and gratitude to Prof. T.H. Lee and Dr. K.K. Tan of the National University of Singapore (NUS) for the encouragement they provided at the beginning of this endeavor. We wish to thank Prof. B.M. Chen (NUS), Prof. S.Z. Ge (NUS), and coworkers in DSI Dr. Chunling Du, Mr. Jingliang Zhang, Dr. Branislav Hredzak, Dr. Eng Hong Ong, Dr. Zhimin He, Dr. Qide Zhang, Dr. Lin Song, Dr. Jiang Quan, Mr. Soh Cheng Su and Mr. Tan Choon Keng. We appreciate the collaboration of Dr. J.Q. Mou, Mr. Yi Lu and Dr. J.P. Wang (all from DSI) in the works on MEMS. Our gratitude is extended to Ms. Nora Kanopka of Taylor and Francis Publications, who expedited the process of reviewing the book proposal and prepared the publications contract in a very short time, and to Prof. Frank Lewis (University of Texas, Arlington) for his consent to include this book in the series edited by him. We wish to thank Dr. J.J. Ritsko and Dr. S.M. Sri-Jayantha, both from IBM, for their help in securing the permission to publish copyrighted materials from IBM publications. We are also grateful to Mr. Shashi Kumar of International Typesetting and Composition (ITC) for his cordial help and prompt support to solve the problems encountered in the course of preparing the manuscript. We extend our sincere appreciation to members of the CRC press who painstakingly read through the final manuscript to make it free of any error.

Abdullah Al-Mamun expresses his sincere thanks and gratitude to Prof. T.S. Low and Dr. M.A. Jabbar for introducing him to the fascinating world of disk drive technology, to Prof. T.H. Lee for introducing him to the world of control engineering, and to Dr. Siri Weerasooriya (Western Digital), Dr. K.T. Chang (DSI), and many other researchers for creating a conducive environment of learning and research in the early days of Magnetics Technology Center (later transformed into DSI). He expresses his sincere thanks to his colleagues in the ECE department of NUS, especially those in the Drives, Power and Control Systems division. Al-Mamun takes this opportunity to thank his wife and two children for the loving support they rendered and sacrifices they made during the course of writing the book.

Guoxiao Guo wishes to thank Prof. Youyi Wang (Nanyang Technological University, Singapore) for guidance over the years, Prof. Long Gao (Tsinghua University, China) for introducing automatic control theory to him, Prof. T.C. Chong, Executive Director of DSI for giving him the opportunity to lead the DSI mechatronics and control group that helped him to see the hard disk drive mechanics as a system instead of viewing it as an isolated control problem, and Dr. You Huan Yeo (DSI) for his support. Guoxiao would express his deep gratitude to his mother and late father for their love; they have always been his role models for joyous life and work.

Chao Bi would like to express his gratitude to Prof. T.S. Low (NUS) for providing him with opportunities in high performance motor research which thereafter extended into disk drive motor research and technologies. Chao Bi would also like to express his appreciation and gratefulness to Prof. T.C. Chong (DSI) for his guidance and trust in Chao Bi's research on spindle motor and electromagnetic technologies; Dr. Y.F. Liew, Dr. K.T. Chang and Dr. Y.Y. Huan for their extensive support and advice in his research work in DSI. Chao Bi would like to use this chance to express his indebtedness to his father, Prof. Bi Houjie and late mother, Prof. Zhu Lisheng, for their love, care and concern in his life and career.

List of Acronyms

ABS	Air-Bearing Surface
ADB	Aerodynamic Bearing
ADC	Analog-to-Digital Converter
AFC	Adaptive Feedforward Control
AGC	Automatic Gain Control
BLDC	Brushless DC
BPI	Bits per Inch
CD-ROM	Compact Disk - Read Only Memory
DAC	Digital-to-Analog Converter
DASD	Direct Access Storage Device
DFT	Discrete Fourier Transform
DISO	Dual Input Single Output
DMS	Decoupled Master Slave
DSP	Digital Signal Processor
DTR	Discrete Track Recording
DVD	Digital Versatile Disk
EMF	Electromotive Force
EMI	Electro-magnetic Interference
FDB	Fluid Dynamic Bearing
FIR	Finite Impulse Response
FOH	First-Order Hold
GA	Genetic Algorithm
GB	Gigabytes
GM	Gain Margin
GMR	Giant Magneto-Resistive
HAMR	Heat Assisted Magnetic Recording
HDD	Hard Disk Drive
ID	Inner Diameter
IDE	Integrated Device Electronics
ITAE	Integral of Time multiplied by Absolute value of Error
IVC	Initial Value Compensation
LBA	Logical Block Address
LDV	Laser Doppler Vibrometer

LMI	Linear Matrix Inequality
LQG	Linear Quadratic Gaussian
LQR	Linear Quadratic Regulator
LTI	Linear Time Invariant
LTR	Loop Transfer Recovery
MASSC	Maximum Allowable Stable Step Change
MEMS	Micro Electro-Mechanical Systems
MIMO	Multi Input Multi Output
MMF	Magnetomotive Force
MR	Magneto-resistive
MSC	Mode Switching Control
NIL	Nanoimprint Lithography
NRRO	Non-repeatable Runout
OD	Outer Diameter
PCB	Printed Circuit Board
PD	Proportional plus Derivative
PES	Position Error Sensing
PI	Proportional plus Integral
PID	Proportional-Integral-Derivative
PLL	Phase Locked Loop
PMACM	Permanent Magnet AC Motor
PMSM	Permanent Magnet Synchronous Motor
PTOS	Proximate Time Optimal Servomechanism
PZT	Lead (Pb) Zirconium Titanate
RAM	Random Access Memory
RAMAC	Random Access Method of Accounting and Control
RNS	Random Neighbourhood Search
ROM	Read Only Memory
RPM	Revolutions per Minute
RRO	Repeatable Runout
SISO	Single Input Single Output
SNR	Signal-to-Noise Ratio
SQP	Sequential Quadratic Programming
SSTW	Self-Servo Track Writing
STM	Servo Timing Mark
STW	Servo Track Writer
TFI	Thin Film Inductive
TMR	Track Mis-Registration
TPI	Tracks per Inch
UMP	Unbalanced Magnetic Pull
VCM	Voice Coil Motor
ZBR	Zoned-Bit Recording
ZCP	Zero-crossing Point
ZOH	Zero-Order Hold

List of Figures

List of Tables

Chapter 1

Introduction

The *hard disk drive* or HDD plays an important role in the modern era of digital technology. The HDD industry began its journey in 1956, and since then, it has traveled through a history of extra-ordinary achievements which is rivaled only by the semiconductor revolution. Storage capacity of the HDD has grown from mere 5 MB (Mega Bytes) in 1956 on fifty 24-inch disks to more than 100 GB (Giga Bytes) stored on one disk of $3\frac{1}{2}$ inch diameter. During this relatively short period, the HDD industry has fostered excellent innovations in various scientific and technological disciplines related to the design and manufacturing of HDD. Mechatronics and control played a vital role in this path of achieving rapid growth in the capacity of HDD and continuously decreasing cost.

The term *Mechatronics*, originated in Japan in late 1970s, describes a branch of engineering that is firmly established now. According to the Mechatronics Forum (UK), *Mechatronics is the synergistic integration of mechanical engineering with electronics and control in the design and manufacturing of product process* [138]. A mechatronic system is neither just a marriage of electrical and mechanical systems nor just a control system; it is a complete integration of all of them. Everyday we find systems and devices that involve mechatronics, e.g., a camera with auto-focus and auto-exposure, an automatic cash machine, a printer, a robot, and an automatic production line etc.

Many mechatronic systems demand for ultra-high precision in controlling the output of the system. The HDD is one such system where the tolerance limit for position error is only few nanometers. The HDD includes several subsystems some of which are mechatronic systems, and the integration of all these subsystems to realize a practical product is a challenging task. The mechatronic parts of HDD include the servomechanism that controls the position of the read-write heads of the HDD and the spindle motor system that spins the disks at precisely regulated speed. The challenging task of the HDD servo engineers can be visualized using the following analogy, which helps one to comprehend the difficulties faced in making an HDD and thus to appreciate

1

the achievements of researchers and engineers from a variety of disciplines [81].

Imagine an airplane flying at 5M miles per hour but only $\frac{1}{16}$ inch above the ground on a highway with 100,000 lanes where the width of each lane is only fraction of an inch. The challenge of the problem is further intensified by the fact that the airplane is expected to switch lanes frequently and then follow the new lane with the same precision. A scaled down version of this scenario is what one finds in the head positioning servomechanism of an HDD.

We live in an era of information technology where every aspect of our life is affected by some kind of information processing or information storage. Modern computing systems use different technologies to store information, either temporarily or permanently. These are semiconductor memories such as ROM (*Read Only Memory*), RAM (*Random Access Memory*) etc, magnetic storage such as hard disk, floppy disk, tape etc, and optical storage such as CD-ROM (*Compact Disk - ROM*), DVD (*Digital Versatile Disk*) etc. Important attributes of a storage device considered by users include cost, rate of data transfer, access time, and reliability. If low cost is the main consideration while selecting the storage device for a specific application then one must accept less desirable features such as slower response, lower transfer rate and poorer reliability.

It is more efficient for the processor to access and store information in semiconductor RAM. The average access time (*time taken by the process of recording or retrieval of data*) is the shortest for this type of memory. However they are the most expensive, constitute the least of the storage volume associated with an information processing system, and form the highest level of the storage pyramid shown in Figure 1.1. The lowest level of this pyramid consists of removable storage devices such as magnetic tapes, zip-disks and floppy disks using magnetic recording, and CD-ROM, DVD etc employing optical recording technology. Removability is the main advantage offered by this class of storage devices. Magnetic tapes and floppy disks are cheap but very slow and, therefore, not suitable for on-line direct access of data or programs. Optical disks are widely used for applications like program distribution, library and archive, entertainment systems etc, but they are not suitable for on-line storage due to their slow performance and high cost per read/write element. This segment of storage market was dominated by magnetic tapes in the early days of computing, but the emergence of optical recording technology caused the tapes to be replaced gradually by more cost-effective CD-ROM, DVD etc.

Hard disk drives sit in the middle of the storage pyramid, between the semiconductor memories and removable drives, and occupy the non-removable on-line data storage niche. They provide direct access to large amounts of non-volatile storage (*no power is required to preserve the data*). Speed of data access in HDD is much higher than the removable, non-volatile storage, and its cost per gigabyte is only a fraction of that of non-volatile, direct access semiconductor memory such as "flash". Hard disk drives are also known as

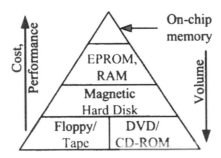

Figure 1.1: Storage hierarchy pyramid in a computing system

Direct Access Storage Devices (DASD, pronounced as *daz-dee*).

In hard disk drives, a binary bit is stored in a tiny segment of the surface of a circular disk by magnetizing the medium coated on the surface with the help of an inductive head. In a majority of hard disk drives, the disk is spun at constant angular velocity by a spindle motor when the bits are written, and the head traces a circular path (*Track*) on the spinning disk. Saturated magnetization of the media is used and it is magnetized in one of the two possible polarizations. The transitions between two opposite polarizations in the magnetic medium can be sensed by a sensor held over the track of a spinning disk. The disks are spun at the same speed during both writing and reading. The read head and write head are fabricated on a single slider whose surface facing the disk is profiled such that an *air bearing surface* (ABS) is produced between the spinning disk and the slider. As a result, the slider is lifted and is not in contact with the disk. The separation between the slider and the spinning disk, known as the *flying height*, is maintained as constant as possible. Characteristics of the flying height depend on many factors such as profile of the slider surface, smoothness of the disk surface, rotating speed of the disk etc. Flying height has direct effect on the achievable *areal density* - a key parameter defining the storage capacity and is equal to the number of bits recorded in unit area of the disk surface. Demand for higher areal density has always been and still is the driving force behind the dramatic growth of the magnetic storage technology. Areal density in magnetic recording has grown by a factor of $5,000,000$ over last four decades.

Some Commonly used Terminologies:

- **Data Track**- Concentric circular (not perfectly) tracks on the disk where binary bits are stored sequentially.

- **Track Pitch** - Distance between two adjacent tracks.

- **Track Density** - Inverse of track pitch, i.e., the number of tracks in unit length of radius of the disk. It is usually defined in units of *Tracks per*

Inch* (TPI).

- **Bit Density** - Number of bits recorded per unit length of a track, defined in units of *Bits per Inch* (BPI).

- **Areal Density** - Number of bits recorded per unit area of the disk surface. It is equivalent to the product of track density and bit density, and is defined in units of *bits per square inch*.

- **Access Time** - This is the time required to retrieve a block of data from the disk and is equal to sum of seek time and average latency, both defined below.

- **Seek Time** - Time taken by the head positioning servomechanism to move the read/write head from one track to another.

- **Single Track Seek Time** - *Seek Time* for moving the head from one track to the adjacent track.

- **Average Seek Time** - Exact *Seek Time* depends on the seek length, i.e., the difference between the initial track and destination track. Average Seek Time is an average of seek times for all possible seek lengths.

- **One-third Stroke Seek Time** - *Seek Time* for moving head over a distance equal to one third of the maximum stroke.

- **Latency** - The process of reading or writing can not be initiated immediately after positioning the head over the destination track as the exact location of the track may not be under the head at that moment. The read-write process must wait for some time before the desired sector of data is available. This waiting time is the latency, and it contributes to the access time.

- **Average Latency** - Each data retrieval process has a different latency. *Average latency* is the time equal to half the time required for one revolution of the disk.

During the process of seek from one track to another, the error between the position of the head and the destination track gradually becomes smaller. However, it is practically impossible to bring the error to zero and maintain it there from that time onward. Even though the head positioning servomechanism tries to make the head follow the center of a track while reading or writing data, it is practically not possible to make the error zero. So the end of a seek process does not imply zero position error. In fact, the seek is assumed to come to an end if the position error remains less than some pre-specified limit for few consecutive samples. The limit is typically 15% of the track pitch before a reading operation is allowed, and 10% of track pitch for a seek prior to

---

*Imperial units are widely used in the HDD industry.

writing data. This is a major difference from the typical definition of settling time in control system step responses. In HDD servo mechanism, the error must be less than 10%(writing) or 15%(reading) of a single track irrespective of the number of tracks traversed by the seek operation. Let us consider a seek command asking for movement of the head from track N to track $N + 100$, which is equivalent to a step response with $y_{ref} = 100$. According to the definition of 5% settling time in linear control system, the *settling time* is equal to the time it takes to bring the position error within ± 5 tracks. And it is equal to ± 10 tracks for a seek command of 200 tracks. However, in HDD servomechanism, the limit of position error is 10% (writing) or 15% (reading) of one track for all seek lengths.

1.1   History of HDD Technology

The HDD industry has a relatively short but fascinating history. In four decades it evolved from a monstrosity with fifty 24-inch diameter disks storing only 5 MB of data to today's drives storing close to 100 GB (100,000,000,000) of data on one surface of a disk in a $3\frac{1}{2}$ inch drive. This enormous growth was made possible by developments in diverse fields of knowledge including materials, tribology, mechanics, servo control, signal processing and electronics. Drives of first generation were significantly different from the drives we see now in aspects like size, capacity, and data transfer rate as well as in the technologies used. For a comprehensive reading on the history of magnetic recording in general and hard disk drive in specific, interested readers may refer to many published articles such as [5], [74], [186].

1.1.1   The Early Days

The fascinating journey of this marvelous device began with a huge, monstrous equipment called RAMAC (*Random Access Method of Accounting and Control*), which used the very first non-volatile DASD introduced by IBM in 1956 [151]. The disk drive of RAMAC contained fifty disks, each 24 inch in diameter, and could store 5 Megabytes of data at a recording density of 2K *bits/in²*. The track density and linear density were 20 TPI and 100 BPI, respectively. The disks used to be spun at a speed of 1200 RPM (*revolutions per minute*) and the rate of data transfer was 8.8 kilobytes per second (KB/s) only [74]. The RAMAC did not use any closed loop control for the head positioning mechanism. IBM, the only company designing and building hard disk drives in the early years, was the sole contributor to the growth of this industry in those days. It designed in 1961 the first drive using air bearing heads, and in 1963 the first removable disk pack drive. All these drives used either motor-clutch mechanism or hydraulic actuators for the head positioning system operated under open loop control. The first HDD with a closed loop servo control was produced in 1971 (IBM 330 Merlin Drive), and it used the position of the head relative to the track sensed from the disk. In 1973, IBM introduced another model of HDD (IBM 3340) that used head and slider fabricated on ferrite. It contained 2 or 4 disks of 14-inch diameter. The storage capacity was 35 MB for a drive with 2 disks; data transfer rate was 0.8 MB/s. The areal density on the recording medium reached 1.69M *bits/in*^{2}. These drives were known as *Winchester drives*. This was the first HDD to use the servo control loop found in disk drives nowadays [157].

1.1.2 Emergence of Desktop Computers

Neither computers nor the storage devices were mass-produced items in the early days of computing. Information processing systems were available at specialized facilities only. Computers started emerging as a system used by a wide range of people in the beginning of 1980s. Development of desktop computers was the beginning of the new era of information technology. This form of computers initiated a new demand in the data storage industry. Need for DASD with physical dimensions suitable for desktop computers became evident and hard disk drives with $5\frac{1}{4}$ disks emerged. This was also the first non-IBM HDD product; the ST506 series (storage capacity 5 MB) was introduced by *Seagate Technology*, a company founded by an ex-member of the IBM family, Allan Shugart. For desktop market, ST506 drives were shipped with 4 disks, but drives for large storage requirement used to have up to 16 disks. An entire drive bay of the original IBM PC was required to house one ST506. Comparing to today's standard, these drives were primitive. But they opened a new world of volume storage on PCs. The ST506 series used stepper motors to move the read/write head positioning actuator. It was early 1980's when the first $5\frac{1}{4}$ inch drive with *voice coil motor* (VCM) actuator used in today's HDD was produced in volume, but stepper motor drives continued to be present for few more years.

The first HDD with $3\frac{1}{2}$ inch disks came from a company named Rodime in 1983. These drives started as a device on a plug-in expansion board. The drive was on the controller card which eventually evolved into IDE (*Integrated Device Electronics*) HDD, where the controller was incorporated into the printed circuit board (PCB) on the back of HDD. Quantum Corporation made the first IDE drive in 1985. The first $3\frac{1}{2}$ inch HDD with VCM actuator was volume produced by Conner Peripherals in 1986. In the same year, PrairieTek shipped the first $2\frac{1}{2}$ inch drive. However, the $5\frac{1}{4}$ inch drives continued to dominate the market until the late 1980's or early 1990's. Large number of competitors emerged to produce HDDs for the growing PC market. The $5\frac{1}{4}$ inch drives started to be phased out giving space to $3\frac{1}{2}$ inch and $2\frac{1}{2}$ inch drives, the later targeting the growing laptop market.

1.1.3   Small Form Factor Drives

Hard disk drives are designed to be installed inside of a PC, and are produced in one of few standard sizes and shapes. These standards are called hard disk *form factors* and refer primarily to the external dimensions. Compatibility is the main reason for such standardization of size as well as of the interface used in the drive. This allows the producers of computing systems to buy HDD from any manufacturer. There were attempts to make drives smaller than $2\frac{1}{2}$ inch form factor in mid 1990s. It was not economically viable at that time as the market was not ready with appropriate application for such storage devices. Integral Peripheral's 1.8-inch drive and Hewlett Packard's Kittyhawk (1.3-inch) are two examples of these attempts. The Kittyhawk supported a capacity of 20 MB on two 1.3-inch disk platters. Recently, IBM introduced a new product (*micro drive*) that uses 0.85 inch disks. This drive can store 1 GB of data on a disk of the size of an American quarter. The world's first gigabyte-capacity disk drive, the IBM 3380, introduced in 1980, was the size of a refrigerator, weighed approximately 250 kg, and had a price tag of $40,000. A comparison between the *micro drive* and IBM 3380 is a proof of the magnanimity of the achievements that the HDD industry can be proud of. Now the $2\frac{1}{2}$ inch format shows promising future with the growing demand of new applications in the consumer electronics and PDA market. The $3\frac{1}{2}$ inch drives are still the dominant secondary storage in the server and desktop applications, where as laptop market is the niche for $2\frac{1}{2}$ inch drives. Small form factor drives, such as the micro drive, will provide the storage solutions for the growing market of PDA, digital entertainment, and other similar products.

1.2   Components of a Hard Disk Drive

Components used in HDD can be broadly classified into 4 categories - magnetic components, mechanical components, electro-mechanical components, and electronics. The magnetic components, i.e., the media and the head are the principal components that enables storage and retrieval of binary information. In HDD, information bits are stored in concentric data tracks on a rotating disk coated with magnetic media; the information is recorded as well as retrieved using the read/write head. Practical realization of such non-volatile storage and retrieval of binary bits, however, involves many other essential components, e.g., a motor to spin the disks, an actuator to make read/write head access the desired data etc. Figure 1.2 shows the important components found in a typical HDD. Functions and special features of some of these components are briefly explained here.

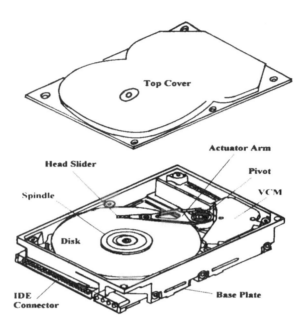

Figure 1.2: Typical components found in HDD.

1.2.1   Head and Disk

Data is recorded on a continuously spinning disk made of aluminum or glass and coated on both sides with a thin layer of magnetic material. The disk is mounted through a hole at the center on the shaft (spindle) of a motor that spins the disks. In desktop application, the disks are spun at 6,000 or 7,200 RPM. The spinning speed can be 10,000 RPM or beyond in high performance HDD. Disk is coated with several layers of other materials. Details of these layers can be found in any textbook on magnetic recording such as [36] and [202]. Two separate elements, the *write* and *read* heads, are used for writing data to or reading data from the disks. These two heads are fabricated together on a larger structure called the *slider* that serves several important purposes. The slider provides electrical connectivity to both heads, and helps to place the read and write heads in close proximity to the magnetized bits by *flying* over the surface of the spinning disk. Well defined aerodynamic surface is created on the surface of the slider facing the disk to achieve the desired flying characteristic. The air moving along with the spinning disk and entrained between the disk and the slider's aerodynamic surface produces an air bearing that makes the slider float.

The surface of the disk must be very smooth to produce uniform readback signal from the heads flying few nanometers above the disk. However, smooth disk gives rise to a different kind of problem. As no air bearing is formed

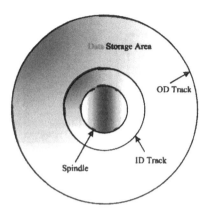

Figure 1.3: Spindle with disks showing data storage area.

when the disk is stationary, the slider touches the disk surface and a stiction force is produced. Smoother the disk surface more is the stiction between the slider and the stationary disk. The stiction between sliders and disks opposes the applied torque during the spinning up of the spindle, and large starting current is required to overcome this torque. This problem was solved in the earlier drives by creating appropriate texture on a small annular ring, known as *landing zone*, on the disk near the center hole. The sliders are pushed to the landing zone before the spindle is spun down so that they rest on the textured surface when the drive is spun up next time. The area of the landing zone can not be used for data storage. An alternative method, the *Dynamic Load/Unload*, was later adopted to solve the problem of stiction between head and disk [7]. This method avoids contact between sliders and stationary disks by bringing the sliders out of the disk surface prior to spinning down. A lift tab extending from the arm engages a ramp structure as the actuator moves beyond the outer radius of the disk. The ramps lift ('unload') the heads from the disk surfaces as the actuator moves to the parking position. Starting and stopping of the spindle motor occurs only with the heads in this unloaded state. During spin up, the actuator arm is pushed over the ramp after the disk attains the specified speed so that the sliders fly.

There is no need to reserve an area for landing zone when dynamic load/unload is used. However, a small ring near the spindle shaft is not used for storage of data due to many other factors such as limit of accessibility by the read/write head, EMI (electro-magnetic interference) generated by the motor coils etc. The innermost track on the disk surface used for storing data is known as the ID (*inner diameter*) track (Figure 1.3). The OD (*outer diameter*) track is created as close as possible to the edge of the disk. However, the magnetic coating in the region near the edge is often not as uniform as in the inner region. The quality of the magnetic layer must be taken into consideration while deciding the radius of the OD track.

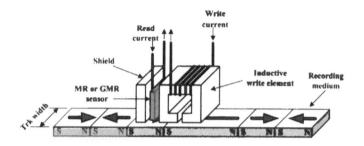

Figure 1.4: Thin film inductive head and MR sensor.

The write head is a thin film coil structure that puts out a magnetic field when current is passed through the coil. This head is commonly known as *thin film inductive head* or TFI head. The core of this head has a tiny gap which flies above the disk in close proximity (Figure 1.4). The magnetic field created in this gap polarizes the medium in the area of the disk that passes under the write head when current is allowed to flow through the coil. Polarity of the magnetic field, and therefore the magnetization of the media, can be reversed by changing the polarity of the current flowing through the coil. Whenever such reversal is made, a transition of magnetization is created on the medium. Since the disk is spinning, these polarized areas or bits are arranged along concentric, circular tracks. In older generation drives, inductive heads were also used for reading the recorded data bits. Once a sequence of transitions is created on the track (using inductive write head), it becomes a spatial distribution of series of tiny magnets with alternating polarity. Changes in the polarity reverse the flux emanating from these tiny magnets. When the disk spins, this spatial distribution becomes a temporal distribution sensed by the read head placed above the track. If an inductive head is used as the read sensor, it experiences the flux reversal as function of time. According to the principles of magnetic induction, a voltage is produced between the two terminals of a coil when it is placed in a time-varying magnetic field. The voltage induced in the coil of the inductive read sensor can be expressed mathematically as $V_{ind} = -N\frac{d\Phi}{dt}$, where N is the number of turns in the coil and Φ is the magnetic flux. Since the flux has a spatial distribution, we can say that $\frac{d\Phi}{dt} = \frac{d\Phi}{dx}\frac{dx}{dt} = v\frac{d\Phi}{dx}$, where '$v$' is the relative velocity between head and medium. So the output of the inductive read sensor is proportional to the velocity of data track with respect to the head.

The read head is usually made narrower than the width of the data track to avoid interference from adjacent tracks while reading. The dimension is about 60%-80% of the track width. As the track pitch continues to get smaller, so does the width of the read head. Unfortunately, sensitivity of an inductive read head decreases with reduction in its dimension as narrower head allows fewer turns of the coil and, therefore, smaller amplitude of the induced voltage. The *Magneto-resistive* (MR) sensing provides the solution for designing narrow read head. In modern HDDs, the read head is a thin-film, metalized structure that exhibits magneto-resistive effect, that is, its resistivity changes when it is brought under the influence of a magnetic field. A current (I_{MR}) is sent through the MR head. Voltage across the MR sensor is $V_{MR} = I_{MR}R_{MR}$, where the resistance R_{MR} of the MR element varies as a function of magnetic field in the vicinity of the sensor. The voltage V_{MR} is a direct measurement of the magnetic field produced by the binary bits written on the disk surface. Sensitivity of the MR head is much higher than that of an inductive head. Moreover, unlike in the case of inductive head, the amplitude of the voltage signal produced by the MR sensor does not depend on the media velocity. In hard disk drives, the disks are spun at a constant angular velocity (ω rad/s) and the media velocity, i.e., relative velocity between head and data track ($v = \omega R$) depends on the radius (R) at which the track is located. This velocity (v) is smaller when the head is over an inner track than that for an outer track. As a result, the amplitude of the read back signal in an inductive head becomes smaller and smaller as the head moves towards ID. Drives currently available in the market use the giant magneto-resistance (GMR) heads, dual spin valve heads, and tunnel junction heads. These advanced heads utilize new technologies to enhance the magneto-resistive effect and thus the read head sensitivity.

1.2.2 Electromechanical Components

Two different electromechanical components are used in a hard disk drive - (1) a spindle motor to spin the disk or disks and (2) an actuator to re-position the read-write heads on the desired data track and to maintain its position precisely over the track while data is being read or written.

Spindle Motor

Brushless DC motor is used to spin the stack of disks in an HDD. Both ends of the spindle are fitted with pre-loaded ball bearing. High end, high performance drives (spindle spinning at 10,000 RPM or more) use fluid dynamic bearing or aerodynamic bearing spindles. The spindle speed was 1200 RPM for RAMAC, and 3600 RPM for the 14-inch drives. Earlier generations of drives with form factors $5\frac{1}{4}$ inch and $3\frac{1}{2}$ inch, designed for desktop applications, had their disks spinning at 3,600 or 4,200 RPM. Many models of drives for the desktop market and mobile market still use spindle speeds of 5400 RPM or 6000 RPM. Higher spinning rates such as 7,200 RPM, 10,000 RPM or even 15,000 RPM are being used these days for high performance drives.

Speed of the motors must be controlled precisely to ensure conformity of the rate of read back bits. Variation of speed affects the performance in many ways. Firstly, the *flying height* of the slider varies affecting the bit density. Secondly, if the spindle speed during reading is different from its speed during writing of information bits, then the rate of bits while reading differs from the actual rate at which they were written. Thirdly, the variation in spindle speed is a major source of disturbance in the tracking servo loop as it alters the spectrum of disturbance related to the imperfections in the shape of the track [4]. The variations in the track shape enters the servo loop as a disturbance, and the spectrum of such disturbance depends on the spinning speed.

Ball-bearing spindle motors comprise a significant number of shipments in HDD today. Demand for higher areal density and faster spindle speed are the reasons for adopting *fluid dynamic bearing* (FDB) spindle motors in HDD. Majority of drives today use FDB motors. In ball bearing motors, mechanical contact exists between the ball and race of the bearing. It is impossible to make the ball and race of the bearings perfect and free from defects. Any contact with these inherent defects found in the geometry of the race ball interface and the layer of the lubricant film produces lateral movements of the spindle shaft and, therefore, of the disks. Since such movements are random in nature and not synchronized with the rotation, they can not be modeled and compensated exactly. These are known as *non-repeatable runout* or NRRO. The NRRO is the main contributor to *Track Mis-Registration* (TMR), the error between the position of the read head and the center of track when the head positioning servomechanism tries to follow a track. The NRRO is a bottleneck in achieving higher track density and there is an upper limit at which the ball bearing design can no longer overcome the NRRO problem. Currently with ball bearings, NRRO is in 0.1 μ-inch range.

Besides, the FDB spindle motors prove to be more advantageous than the ball bearing motors in term of non-operational shock resistance. When the motor is stationary, the FDB spindles have a larger area of surface to surface contact compared to the ball bearing motors. So the FDB spindle can withstand larger non-operational shock. The dynamic motions of the spindle motor components and disks are the main contributors to the acoustic noise in HDD. Sound energy generated by the spinning disks, motor magnets, stator components, and bearings is transmitted through the spindle assembly to the HDD base casting and the top cover. In an FDB spindle, one of these sources (the bearing) is eliminated. There exists no contact between the rotating surface and the stationary surface of the spindle motor. Moreover, the viscosity of the lubricant between the two surfaces is much higher than that for the lubricant used in ball bearings. As a result of these features, the FDB spindle motors generates less NRRO. Furthermore the damping effect of the lubricant of relatively higher viscosity between the stator and the rotor of the FDB spindle attenuates noise. As a result, the FDB spindle shows better acoustic performance with approximately 12-15 dBA reduction of noise level compared to ball bearing spindles. Other inherent properties of the FDB spindle include higher damping, reduced resonances, and greater speed control.

Clamps are used to hold the disks firmly on the spindle shaft. Annular rings known as spacers are placed between two disks to ensure desired gap between them. Spindle of a typical HDD used for desktop application supports 4 disks. However, actual number of disks depends on the desired capacity of the drive and the capacity of individual disks. Several years ago, a single disk could record approximately 10 GB of data, and 4 such disks were used in a 40 GB drive. Technology involved in producing head and media has improved tremendously over last decade. Servomechanism and channel electronics were also improved. Together they made it possible to record more than 60 GB on a single surface of a 3.5 inch disk. An HDD of 60 GB capacity doesn't even need to use both surfaces of the disk.

Read-Write Head Positioning Actuator

Movement of the read-write head between different radii of the disk surface is effectuated by an actuator. The actuator used in RAMAC supported a single pair of head, and moved radially to access different tracks on a surface and vertically to reach different surfaces. It used cables and pulleys. Such mechanism was replaced by hydraulic actuators in the series of drives (IBM 2314) introduced in the late sixties and early seventies. Some of the early generations of HDD used stepper motors as actuator. These actuators, controlled open loop, performs well within the specifications when track density is low, i.e., the space between two adjacent tracks is quite large. With increasing density of data tracks, open loop control failed to perform and closed loop control using VCM replaced the stepper motors. The first linear VCM actuator was developed by IBM in 1965, followed by the rotary VCM actuator [5]. The VCM is a moving coil type actuator, in which a coil is held suspended in the magnetic field produced by pairs of permanent magnets fixed to the casing of the HDD. The suspended coil is free to move within a restricted area. When a current is passed through the coil, it moves (the motion is governed by Faraday's Law). The actuator arm is glued to this coil; a movement of the coil makes the actuator move. In a linear VCM, the actuator arm moves in and out of the yoke holding the permanent magnets. On the contrary, the coil of the rotary VCM moves sidewise and, with the arm pivoted, the tip of the actuator arm moves on an arc (Figure 1.5).

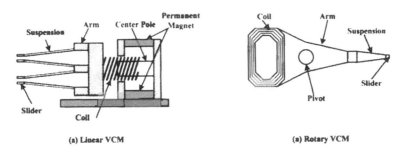

(a) Linear VCM   (a) Rotary VCM

Figure 1.5: Linear (left) and Rotary (right) voice coil motor actuator

The actuator arms are made of solid steel or aluminum of significant thickness not suitable for holding the read-write heads over the disk surface. An extended arm, known as the *suspension*, carries the head slider. The suspension is made of thin sheet of stainless steel and is attached to the actuator arm. In linear actuator, the suspension arm and the movement of the actuator arm are collinear. The first generation rotary actuators had their suspension arms turned sideways to mimic the motion of linear actuator. Though the actuator arm moves within a restricted angle (about 30°), the motion of the suspension arm follows a nearly straight line along the radius of the disk. In modern days, the actuator arm and suspension of the rotary actuator are collinear making the movement of the slider follow an arc and not a straight line. Each head slider is attached to the tip of a suspension. Usually, there are as many sliders in an HDD as the number of disk surface. All suspension arms are attached to a single piece of actuator. Even if it is intended to move only one head from an initial position at radius r_i to a final position at radius r_f, all heads are moved together.

The slider floats over the spinning disk which has a certain degree of roughness relative to the fly height. The suspension provides a force on the slider in the direction into the disk to counteract the upward aerodynamic forces of the air bearing surface. This force must act precisely in the proper location; otherwise a twisting force acting on the slider will cause one of its corners to be too close to the disk and the other too far from the disk surface. In addition to providing this downward force, the suspension must allow the slider to rotate in the pitch and roll directions so that it can stay close to the surface despite the presence of asperities on the disk surface. Any torque applied by the suspension on the slider negatively affects the flying characteristics of the slider. A good suspension gimbal design has very low rotational stiffness so that the magnitude of the torque caused by slight deviation of static attitude from the nominal is minimized. This asks for compliance in roll and pitch directions of the suspension, but the servomechanism asks for high stiffness in the other direction so that it can be swang back and forth rapidly during tracking and seeking without producing excessive vibration. It should be pointed out here that the HDD servomechanism uses error signal sensed by the read head attached to one end of the actuator/suspension arm and the input torque for positioning the head is generated by the VCM on the other end of the actuator. The servo controller cannot accurately control the position of the head if the structure connecting the sensing point and the actuating point is excessively flexible. The stiffer this intervening structure is, the more rapidly and precisely the servo controller can position the head to compensate for off track disturbances, i.e., more servo bandwidth.

In order to address this issue of competing requirements, low stiffness for the gimbal and high stiffness for tracking, suspension arms are usually made of two separate components - the *flexure* and the *load beam*. The slider is adhesively bonded to the flexure made of thin material and designed to give

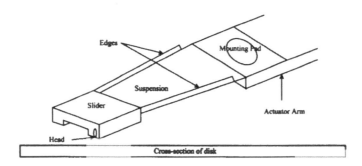

Figure 1.6: Read/Write head slider on disk

low pitch and roll stiffness. The load beam is made relatively thicker compared to the flexure to provide high stiffness in the other directions that are required for the servo. The flexure is laser welded to the load beam. The edges of the load beam are often bent to form a rail and, therefore a half-I beam, to increase its stiffness (Figure 1.6). The load beam ends at the *base* of the suspension, which is a relatively thick rectangular piece. The base provides a mechanism to swage it to the end of the actuator arm. The edge of the load beam is removed at the end near the base plate so that there is room for up-and-down motion of the suspension arm. This vertical movement of the suspension is necessary to accommodate unevenness of the disk surface and wobbling of disk. Whenever the slider encounters a slowly increasing bump on the disk surface, the aerodynamic force pushes the slider away from the disk. The narrow scope of vertical movement of the suspension accommodates these motions.

More recent tracking issues come from the increase in disk rotational speeds. The turbulent nature of the air between the disk and actuator causes additional off track motion at the head. Once again, a suspension with higher stiffness works better to reduce the response to this windage energy. In addition, low profile designs extract less energy from the turbulent air.

Like the other components of the HDD, the sliders have gone through evolution. The dimension of the slider of the Winchester drive introduced in the 1960's was 5.6mm × 4mm × 1.93mm. This is also known as the *Full-size slider*. The next generation sliders, known as *100% minislider*, were introduced in 1975 and had the dimensions of 4mm × 3mm × 0.8mm. Both of these sliders were made of ferrite. This was followed by the *70% micro slider* (2.84mm × 2.23mm × 0.61mm) in 1987, the *50% nano slider* (2.05mm × 1.6mm × 0.43mm) in 1990, and the *30% pico slider* (1.25mm × 1.0mm × 0.3mm) in 1995. These three classes of sliders are made of $Al_2O_3 - TiC$. The air-bearing features of the sliders were machine railed for full-size, mini slider and micro sliders, and etched railed for nano slider and pico slider. The industry is currently preparing for a transition to the *20% femto slider* (0.85mm × 0.7mm × 0.23mm).

1.2.3   Mechanical and Electronic Components

The drive enclosure is the external casing of the hard disk drive. It provides features for mounting the drive in the drive bay of the host system, and supports other components of the HDD such as spindle motor, actuator, PCB etc. There are two parts of the enclosure - *top cover* and *base plate*. All components are assembled on the base plate. Enclosure is then covered using the top cover. Gasket is used to seal the contact between base plate and top cover. The environment inside the enclosure must be maintained clean. Any particle at the head-disk interface can cause abrasion of the disk resulting in loss of data and increase in number of particles. Therefore, assembly of the drive is done in a clean room to ensure particle-free enclosure. Particles created during the operation of drive by sudden contacts between disk and slider are thrown out of the spinning disk by the centrifugal force and eventually trapped in the filter, placed in the empty space inside the enclosure.

A special feature of both the base casting and top cover is the *crash stop*. These are small mechanical protrusions from the base plate and top cover used to restrict the movement of the actuator beyond the desired space.

Electronic components of an HDD can be categorized according to the following functions:

1. Electronics for reading/writing also known as the *channel electronics*

2. Electronics for spinning the disks and positioning of the read/write head also known as the *servo channel*

3. Electronics for controlling various operations (such as read data, write data, transfer data between HDD and host etc) of the disk or the *disk controller*

4. Electronics for interface with the host system, and

5. RAM, ROM etc.

Several of these functional components are often combined in a single chip. As a result, we do not see many ICs on the PCB (printed circuit board) of an HDD. One IC that is not placed on the PCB but is kept inside the drive enclosure is the pre-amplifier. It is put as close to the read/write heads as possible to avoid amplification of noise, and is mounted on the flex-cable that carries signal between heads and PCB. The output of the preamplifier is sufficiently large ensuring good signal-to-noise ratio at the input of the PCB.

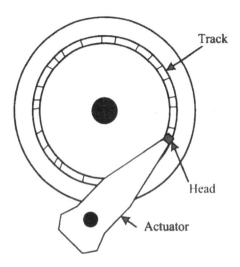

Figure 1.7: A track is created by recording binary bits on a spinning disk.

1.3 Accessing Data in HDD

1.3.1 Arrangement of Data on the Disks

Bit is the smallest unit of recorded information on magnetic media. It is a tiny piece of the disk surface and contains binary information. During writing, the read/write head is positioned at radial distance from the center of the disk while the disk is spinning. The read/write head is moved to the desired location and is positioned there with as low variance is possible with the help of the actuator operated under closed loop servo control. Recording/writing of information is achieved by alternating the polarity of the current in the write-head's coil. Since the disk is spinning and the head is held at a point, the write current magnetizes a circular path on the disk with alternating polarity of magnetization. Type of magnetization on the medium depends on the polarity of write current. A transition in the write current waveform creates a transition of magnetization on the disk. The circular pattern of magnetization created on the disk is called a *track* (Figure 1.7). A new track can be created by repositioning the write head to a new point on the disk radius. In a typical $3\frac{1}{2}$ inch HDD, 70,000 to 100,000 tracks exist on each surface of a disk.

The polarity of write current is altered according to the binary bits to be recorded. A 1 in the binary data causes the polarity to be reversed, otherwise it is unchanged. If the data is recorded at a rate of b bits per second then a clock signal of frequency $f_W = b$ Hz is used to change the polarity of the write current. Minimum separation between two consecutive transitions in the write current waveform (one from +ve to -ve and the other from -ve to +ve) is $T_W = 1/f_W$. The minimum distance between two magnetic transitions on the track (in the direction along the track or *down-track* direction) is $L_{bit} = vT_W$, where v is the relative speed between the write head and medium. Dimension of the write head is one of several factors that affect the dimension of a track in the *cross-track* direction, i.e., along the radius of the disk. The head positioning error in the HDD servo loop is another factor that affects the decision on track-to-track spacing. We should not create a new track that may erase significant part of an adjacent track. Similarly, while data is read from a track, the interference from the magnetic transitions recorded on an adjacent track should be as low as possible. Maximum error in positioning the read/write head by the closed loop servomechanism sets the limit on the allowable proximity of two adjacent tracks. Let σ_{pes} be the standard deviation of the tracking error in the head positioning servomechanism, then $3\sigma_{pes}$ is widely accepted in the industry as a measure of the minimum track pitch. It should be noted that actual track pitch is larger than the $3\sigma_{pes}$ achieved by the head positioning servo. Different factors such as spindle eccentricity, mechanical vibration, environmental disturbances etc contribute to the tracking error. If the track width is W_{trk} and minimum separation between two transitions is L_{bit}, then $A_{bit} = L_{bit}W_{trk}$ is the area occupied by a single bit of information. The storage density or *bit density* of a magnetic recording system is the inverse of A_{bit}. Inverse of L_{bit} and inverse of W_{trk} are known as the *linear density* and *track density*, respectively. Imperial units are widely used in the HDD industry and *bits per inch*^{2}, *bits per inch* and *tracks per inch* are the units of areal density, linear density, and track density, respectively.

1.3.2   Locating Data

When the host system sends data for recording, it is recorded in chunks of 512 bytes. Each of these chunks is called a *data block*. Bits of a data block are recorded sequentially along the track. In order to locate a data block on the surface, they are tagged with an identification number. From system level point of view, each data block is assigned with a *Logical Block Address* or LBA, starting at 0 and ending at a number appropriate for the capacity of the entire drive. These block addresses, however, are not suitable for low level access to the data. Access at the low level uses *head number*, *cylinder number*, and *sector number* assigned to each LBA. An HDD may contain one or more disks with data recorded on both surfaces of a disk. Data on each surface is accessed (for reading as well as writing) using a separate head for that surface. If the number of disk surfaces used is S in an HDD, there will be S head sliders in it. Each used surface of the disk stack is identified by the corresponding head number; for an HDD with 8 usable surfaces, heads are numbered 0 to 7. There are tens of thousands of tracks on each surface, numbered 0 on the outermost track and increasing inward. If we consider a stack of disks, then track 0 of all disks form a cylinder and is identified as *cyl 0*. One of the cylinders is shown in Figure 1.8. Each cylinder is assigned with a unique identification number.

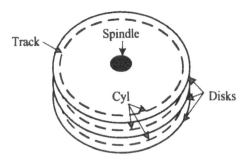

Figure 1.8: Tracks and Cylinders

Each track is further divided into sectors, using some special magnetic patterns written on the disks at the time of manufacturing. These special patterns are known as *servo sectors* (Figure 1.9) which divide a track into equal segments, and data is written in these segments. Number of servo sectors per track is the same on all surfaces in an HDD. There are typically 100-200 servo sectors in any HDD produced these days.

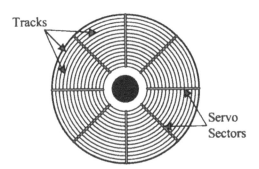

Figure 1.9: Tracks and servo sectors on a disk.

Tracks and sectors are identified using special magnetic patterns written on the disks during the production of HDD. In the earlier generation drives with four or more disks, entire surface of one disk used to be dedicated for recording these special patterns known as the *servo pattern*. Since all heads are moved simultaneously by a single actuator, it can be assumed that when the head on the servo surface (surface containing the servo patterns) is positioned on the Nth track, all other heads are also positioned on the Nth track of their respective surfaces. This assumption started to fall apart with increasing track density. Slight difference between the thermal properties of arms on the servo surface and a data surface gives rise to different expansion of two arms causing an offset between the positions of those two heads. This offset can no longer be neglected in high track-density drives. Moreover, the scheme of dedicated servo surface is not suitable for drives with few disks. Earlier

generations of HDD used to contain 4 or more disks. Current state of the
art in recording technology allows storage of approximately 100 Gbytes on
a single disk; for many applications a drive with one disk meets the storage
requirement. For a drive with 4 disks (8 surfaces), *servo overhead* is 12.5%, i.e.,
12.5% of the storage area is occupied by the servo patterns. Servo overhead is
increased if fewer disks are used. Both of these issues, thermal expansion and
increasing servo overhead, associated with HDDs with a dedicated servo surface
can be resolved using an alternative servo scheme where the servo patterns are
written on every track interleaved with the data blocks. With this scheme in
place, the servomechanism can control the position of any head using servo
information written on the corresponding surface as the feedback. Different
thermal expansion of different arms is no longer a problem, and servo overhead
is independent of the number of disks used. However, unlike in the scheme with
dedicated surface, feedback signal is available only at discrete sampling points.

The method used in the earlier generations of drive with position information encoded on a dedicated surface is called the *Dedicated Servo* scheme,
whereas the other scheme having position information encoded on all surfaces
is called the *Embedded Servo* or *Sectored Servo*. The segment of the track
containing the servo information in an embedded servo drive is known as *servo
sectors*, and the section between two servo sectors is allocated for storing data
bits. Servo patterns, both in dedicated and embedded case, are created during
manufacturing of the drive and the firmware of the HDD takes care not to
overwrite them in any situation.

1.3.3   Track Seek and Track Following

The position of the read/write head is controlled by a closed loop servomechanism that uses the feedback signal generated by decoding the information
written on the disks. There are two modes of operation for this control loop -
(i) moving the head from one track to another in shortest possible time, and (ii)
regulate the position of the head such that the relative offset between the head
and the track-center is as small as possible. The first of these modes is known
as *Track Seek* while the second mode is called the *Track Following*. Design
objectives of these two modes of operation are significantly different. Besides,
there must be a smooth transfer between the two modes. It is impossible to
meet the specifications of both modes using a single control law. Two controllers can be made to produce desired performances if each is designed and
tuned independent of the other. However, while designing and implementing
such controller, special attention must be paid to ensure that sudden change in
the amplitude of control signal does not occur at the time of switching between
modes. Sharp discontinuity in the control signal excites the lightly damped
resonances of the actuator. Occurrence of such jerk increases the time it takes
to settle and, therefore, must be avoided.

1.3.4 Zoned Bit Recording

If the rate of data transfer between the electronics and the media is kept constant irrespective of the radial position of the track then the read/write electronics can be optimally designed to cater for this single data rate. Such a scheme is called *constant data rate* recording. Frequency of the clock signal used during reading or writing of data remains the same on all tracks, i.e., T_W and therefore $f_W = \frac{1}{T_W}$ is constant. If the angular speed of the disk is kept constant, the linear velocity $v = \omega r$ of the medium with respect to the head increases with increasing radius. Then the dimension ($L_{bit} = vT_W$) of a single bit depends on the radius of the track where the bit is written; inner is the track smaller is this dimension. If the linear density is optimized on the outermost track then the transitions on an inner track are too close to each other and can not produce significant read back voltage. On the other hand, if the clock frequency is chosen to achieve optimum linear density on the innermost track, the transitions are sparsely created on outer tracks. In the constant data rate recording, same number of data blocks are stored on all tracks. However, the circumference of the outer track is larger than that of an inner track and, therefore, it makes better sense to store more number of data blocks on outer tracks.

It is possible to achieve the ideal solution to this problem if either radius-dependent clock frequency or radius-dependent spindle speed is used. In CD-ROM, the speed of the spindle motor is continuously adjusted as the head moves from one track to another. This ensures constant linear density of recording and hence constant areal density. Radius-dependent clock frequency is not used in any storage device. Hard disk drive employs a scheme, *Zoned Bit Recording* (ZBR) or *Zoned Density Recording*, that groups the tracks into several annular zones. Each zone has its own recording frequency which optimizes the linear density on the innermost track of that zone. Frequency of recording is increased from inner zone to outer zone. All tracks within a zone use constant data rate recording and contain equal number of data blocks. A schematic illustration of zoned-bit recording in Figure 1.10 shows the surface of a disk divided into 8 zones. There are more data blocks per track in an outer zone than an inner zone, which is clearly shown in this diagram.

More is the number of zones, better is the utilization of storage space. The extreme end is to assign one track per zone and the clock frequency is optimized for each track individually to achieve optimum linear density on all tracks. This also results in constant areal density. The marginal improvement due to increase in number of zones is significant when few zones are used. There is approximately 13.63% improvement in storage when we increase from 2 zones to 3 zones, but increasing number of zones from 7 to 8 gives only 1.59% improvement. Commercially available drives use 16-32 zones.

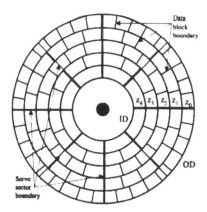

Figure 1.10: Schematic illustration of a disk surface with zoned-bit recording

1.4   Trend in HDD Industry

The disk drive industry has evolved through dramatic changes in the five
decades of its existence. Demand for larger capacity, need to have smaller
dimensions for specific applications, requirement of data transfer rate compatible for fast on-line applications etc are some of the driving forces behind
the extraordinary growth of this industry. Desire to have large capacity in a
smaller dimension is directly linked to the demand for ever increasing storage
density. Areal density and dimensions of the drive are two factors affected by
the mechatronics and control of actuator servomechanism, where as the design
and operation of the spindle motor has direct effects on the data transfer rate.

1.4.1   Areal Density Growth

The most obvious change that took place in the hard disk drive industry over
last four decades is the phenomenal increase in the storage capacity of HDD. As
recent as in early 1990s, a typical PC used to be shipped with an HDD capable
of storing approximately 100 megabytes of data. Today, even a computer
for home or personal use comes with HDD storage of 80 gigabytes or more.
Demand of storage capacity caused by larger size of programs and multimedia
data has driven the manufacturers to increase the capacity of their products.
While the capacity continued to increase, the price of HDDs experienced a
continuous fall. This was made possible by increasing the amount of data
stored on each surface of the disk, i.e., by increasing the areal density.

The first hard disk drive (RAMAC) supported an areal density of 2000
$bits/in^2$ only. Today the density has escalated to a high 100 $Gbits/in^2$, which
is an increase by a factor of 5 million (Figure 1.11). This phenomenal increase
in areal density have been achieved by a coordinated efforts in improving all
aspects of the hard drive. Head technology has been continuously improved

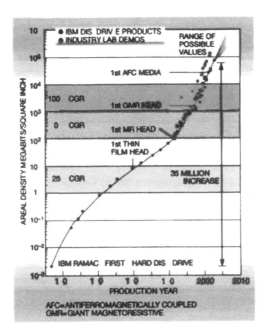

Figure 1.11: Trend in HDD areal density (From Grochowski, E., IBM Systems Journal, 42(2), 2003. With permission.).

to be able to read and write smaller bits. Improved technologies have enabled new generation media to reliably hold magnetic domains of smaller size. Track density has been increased by reducing component disturbances that move the head off track and by improving the servo controller's ability to regulate the head on narrower tracks. Smoother disk surface, better quality of lubricant and better air bearing technologies allow the slider to fly in closer proximity of the disk so that the bit size is reduced. Typical fly heights in 1997 were 25 nanometers (nm). Today they are about 5 nm. The read-write electronics and data encoding schemes have played their part in improving bit density by enabling detection of information reliably from ever smaller data signals contaminated by the surrounding noise. Excellence in design and production maintained the steady growth in the areal density which was accelerated time to time by availability of new technologies. Introduction of new technologies has always made an impact on the strive for improving areal density. Cumulative average growth rate of the areal density was 25% in the seventies and eighties. Drives in those days used inductive heads for both reading and writing. MR heads, introduced in early 1990s, boosted the growth rate to 60%. The giant MR heads (GMR) appeared in the drives in later half of 1990s, and the industry experienced a growth rate of more than 100%.

The ever-increasing areal density is sustained by increasing both track density and linear density. The industry has been experiencing 100% per year cumulative growth rate for areal density for the last several years and 30% in BPI over last one decade or so. Starting in the later part of 1990s, the track density (TPI) has been growing at a rate faster than that of the growth in linear density (BPI). Table 1.1 shows TPI and BPI of several products in the first few years of this decade. It is evident that both the densities have been increasing. However, the *Bit Aspect Ratio* (BAR) or the ratio between BPI and TPI has been showing steady decrease during this period. This trend in BAR suggests that the increase in track density is taking place at a rate higher than that of linear density. Capability of sustaining the rate of increase in TPI will play a more important role in pushing continuously the recording density higher. Compared to the data shown here, the track density (20 TPI) and linear density (100 BPI) of RAMAC were very small. It is generally believed in the HDD industry that areal density still has a lot of room to grow, and it is expected to reach 50 Terra bytes per squared inch in the future. The growth in track density must be sustained at an appropriate rate to make this projection a reality.

Table 1.1: Trend in track density, linear density, and bit aspect ratio

Year	Company	Capacity (GB)	kTPI	kBPI	BAR
2000	Maxtor	61.4	27.3	412	15.09
2000	IBM	40.0	35	415	11.86
2000	Maxtor	81.0	34	402	11.82
2000	Quantum	80.0	35.4	417	11.77
2000	Quantum	73.4	40	448	11.2
2001	Seagate	40.0	50	540	10.8
2001	Maxtor	15.0	46	489	10.63
2001	Maxtor	40.0	54	524	9.72
	Hitachi	180.0	72	632	8.77
2003	Seagate	200.0	98	671	6.84
2004	Seagate	300.0	105	658	6.26

1.4.2 Trend in Drive Form Factor

Introducing a new form factor standard requires coordination between manufacturers of computers, producers of HDD, and other support industries that produce components for HDD as well as for computers. There is a natural resistance in the industry to changes in form factor unless there is a compelling reason to do so. At the emergence of laptop computers, the HDD industry created new, smaller drives to save space as well as power, a very important consideration in the world of mobile computing. This was a necessity that the HDD industry and other support industries eagerly acted to meet. There have been only a few different form factors in the entire history of HDD. The most common form factors prevailing today are $3\frac{1}{2}$ inch and $2\frac{1}{2}$ inch.

The form factor usually refer to the width of the drive enclosure or the diameter of the disks used. However, in some cases, form factor represents neither of the two. For example, the width of a $3\frac{1}{2}$ inch HDD enclosure is 4 inch and the disks used in these drives have diameter larger than 3.5 inch. This particular form factor got its name from the fact that the size fits well in the space originally allocated for 3.5 inch floppy disk drive.

Phenomenal increase in areal density achieved over last few decades allows the manufacturers to increase storage capacity with simultaneous decrease in the size of hard disk drive suitable for applications such as laptop computers, cameras, and other small devices. The trend in form factors is downward: to smaller and smaller drives. The first form factor used in a PC ($5\frac{1}{4}$ inch) have now all but disappeared from the mainstream PC market, and the $3\frac{1}{2}$ inch form factor dominates the desktop and server segment. For laptop market, the dominant form factor is $2\frac{1}{2}$ inch. HDDs of smaller form factor is the most desirable choice for the emerging market of digital entertainment with devices such as digital camera, MP3 etc. The *micro drive* of IBM is less than 0.25-inch thick and uses disk one inch in diameter. Continuous growth of areal density will initiate soon a transition to the $2\frac{1}{2}$ inch form factor for the desktop and server drives. The reasons for this shrinking trend include the enhanced rigidity of smaller platters, reduction of mass to enable faster spin speeds, and improved reliability due to enhanced ease of manufacturing. Faster spin speed increases the rate of data transfer between the media and read/write electronics, reduces latency, and therefore, improves data access performance. IBM's micro-drive uses disks of only 0.85-inch diameter and can store sufficient data useful for hand held and entertainment devices such camcorder, PDA, and portable MP3 players. The capacity of this drive was 2 Gbyte in 2004 and 4 Gbytes in 2005. Toshiba announced their plan to introduce 8 Gbyte micro drives in 2006 which will use perpendicular recording technology. Dimensions for drives of different form factors are tabulated in Table 1.2.

Smaller form factor drives usually come with lower performance than a larger drive. Spindle speed is usually lower in these drives, for example, IBM micro drive uses 3600 RPM spindle motor. However, because of the small size of disk(s), it can be spun up very fast. A small drive can spin up to full speed in less than half a second. This makes it possible to spin down the drive frequently, which is an essential feature for portable computers.

1.4.3 Trend in Data Transfer Rate

Another significant change that the HDD industry experienced in the past is the trend in the access time and data transfer rate. For applications that require faster data rates, speeding up the disk rotational speed has reduced the latency component of access time and increased the speed of data flow from the heads. There has been a steady progression over the years from

Table 1.2: Dimensions in drives of different form factors

Form factor		Length (in)	Width (in)	Height (in)	Comments
$5\frac{1}{4}$ inch	Full Height	8.0	5.75	3.25	Early 1980
	Half Height	8.0	5.75	1.63	Early 1980-90
	Low Profile	8.0	5.75	1.00	mid to late 1990
	Ultra Low Profile	8.0	5.75	0.75-0.8	mid to late 1990
$3\frac{1}{2}$ inch	Half Height	5.75	4.0	1.63	
	Low Profile	5.75	4.0	1.0	most common form factor for PC
$2\frac{1}{2}$ inch	19 mm height	3.94	2.75	0.75	for Laptop
	17 mm height	3.94	2.75	0.67	for Laptop
	12.5 mm height	3.94	2.75	0.49	for small Laptop
	9.5 mm height	3.94	2.75	0.37	for very small Laptop
1.3 inch	Kittyhawk	3.37	2.13	5mm or 10.5mm	1992, didn't survive
1.0 inch	Micro drive	1.6	1.43	5mm	1998

4,500, 5,400, 7,200, 10,000, to now 15,000 RPM hard drives. This has posed challenges to keeping the windage off-track disturbances to an acceptable level. Fluid dynamic bearings have replaced ball bearing spindles to reduce runout (off-track movement of the head or disk) at high RPM. The higher data rates coming into the head due to the higher rotational speed and bit density have introduced challenges in drive electronics to be able to reliably process the data.

The time required to move the head to a new track position and get it ready for reading or writing is called access time. It is the sum of the time required to find the new track (seek time), time required to settle on it (settling time), and latency. Latency is defined as half of the time the disk takes to make one rotation as, on the average, the desired data is located 180° from the position where the head settles onto the track. One-third stroke seek times are around

3 milliseconds on high performance drives making spindle latency the most significant contributor to the access time. Low access time is very important in computer applications because the number of data transfers is so high that a small increase in the time required for each transfer causes considerable overall delays in processing data or running programs.

1.5 Alternative Recording Technologies

Hard disk drives are expected to maintain their position as the primary on-line, non-volatile storage device for computing systems in the foreseeable future because of their advantages of large capacity with fast access but at low cost. Though access time is faster in semiconductor memory, its higher cost per stored bit makes it less attractive for mass storage. This is expected to be continued in the future. However, the semiconductor devices will have their usage for low capacity functions such as on-board RAM and flash memory for small and portable applications such as digital camera and MP3 player. The optical storage devices cannot compete with hard drive technology in either storage capacity or data access speed and will continue to fill the niche functions of high capacity data portability and program distribution. The burgeoning market segment of off-line storage will continue to grow providing better market opportunities for high performance, quick access hard drive arrays. In addition, with the proliferation of consumer-oriented devices for which data storage is a critical capability, non-computer applications are expected to contribute significantly to future growth in overall disk drive demand.

There still exists wide opportunity for the amazing technological development of hard drives to continue. Today's leading edge areal densities are close to 100 gigabits per square inch. A consortium of industry, academic, and government participants (*Information Storage Industry Consortium*, NSIC) has recently targeted 1000 gigabits per square inch for their new magnetic storage demonstration development project. However, this effort to make the bit size smaller and smaller is leading towards a situation which is constrained by the super paramagnetic effect; the grains of the media becomes so small that they interfere with one another and thus loose the ability to retain their magnetic orientations. As a result, magnetic north and south poles of a grain suddenly and spontaneously reverse corrupting the stored data and therefore, making the storage device unreliable. Alternative technologies to overcome this problem include perpendicular recording, heat assisted magnetic recording and recording on patterned media.

Perpendicular Magnetic Recording:

At present, the HDDs employ *Longitudinal recording* which, as its name indicates, aligns the data bits horizontally, parallel to the surface of the disk. In contrast, in *perpendicular recording*, the bits are aligned vertically, perpendicular to the disk, which allows additional room on a disk to pack more data, thus, enabling higher recording densities. In March 2005, Hitachi Global Storage Technologies demonstrated an areal density of 230 gigabits per square inch (Gb/in^2) on perpendicular recording technology. This accomplishment represents a doubling of todays highest data densities on longitudinal recording technology. Such products with perpendicular recording is expected to greet the market as early as in 2007. Projection made by Hitachi suggests the availability of 1-inch micro drive with 20 gigabytes capacity and $3\frac{1}{2}$ inch products with terabyte capacity.

Heat Assisted Magnetic Recording (HAMR):

HAMR shows the promises to be the key enabling technology that will increase the areal density to a level breaking through the so-called super paramagnetic limit of magnetic recording. This technology is expected to deliver storage densities as high as 50 terabits per square inch. If disk drives are produced to have such a great areal density, one can store the entire printed contents of the Library of Congress on a single disk drive.

If the phenomenal growth rate of bit density continues, the size of an individual bit will soon reach such a small dimension that the bits become magnetically unstable. At that stage, though the bits can be written very tiny, they may not be suitable for information storage as some of them may flip into different polarization. This phenomenon is known as *super paramagnetism*. This problem can be overcome by heating the medium with a laser beam at the precise spot where a data bit is being recorded and subsequently cooling the spot rapidly to stabilize the written bit. Heating makes it easier to write on the medium. This heat assisted recording can increase the recorded density dramatically.

Patterned Media:

Another promising approach to circumvent the density limitations imposed by the super paramagnetic effect is the use of patterned media. Conventionally, the disk is coated with a thin layer of magnetic alloy. If the disk surface is examined at high magnification, it becomes apparent that within each bit cell there are many tiny magnetic grains. These grains are randomly created during the deposition of the magnetic film. Each grain behaves like an independent magnet whose magnetization can be flipped by the write head during the data writing process. In patterned media, the magnetic alloy is not coated on the entire disk surface. The layer is created as an ordered array of highly uniform tiny islands, each island capable of storing an individual bit. Each bit is stored in a single deliberately formed magnetic switching volume. This may be one grain, or several exchange coupled grains, rather than a collection of random decoupled grains. Single switching volume magnetic islands are formed along circular tracks with regular spacing. Magnetic transitions no longer meander between random grains, but form perfectly distinct boundaries between precisely located islands. Since each island is a single magnetic domain, patterned media is thermally stable, even at densities far higher than can be achieved with conventional media. Though the concept of patterned media looks simple, realization of this to achieve high recording density is immensely challenging. For an areal density of 100 Gbits/square inch, the center to center spacing between two islands need to be 86 nanometers. For 10 terrabits/square inch density, this spacing is only 9 nm. Creating islands of such dimension is beyond the capabilities of optical lithography. E-beam lithography and nano imprint replication are considered to be two approaches that can be used to realize patterned media commercially.

The HDD industry will soon embrace these and other technologies to manufacture commercially hard disk drives with extremely high areal density. This makes the design of the head positioning servomechanism more challenging. Shrinking bit size also means narrower track pitch. Many disturbances ignored today will ask for special attention at such high track density. Ultra high areal density will also require the head to fly very low such that occasional contact between head and disk will become inevitable. The servomechanism must be robust enough to withstand these unpredictable disturbances.

Chapter 2

Head Positioning Servomechanism

When an HDD is powered up, the disks are spun to a precisely regulated speed and the heads are allowed to move radially over the disk surfaces. Limited vertical movement within a very small range, self-regulating by the formation of an *air bearing surface* (ABS) between the head and slider is also allowed. Accurate and precise control of radial position of the head slider is done by the *head positioning servomechanism*. This servomechanism is a feedback system consisting of a sensing element that measures the displacement of the head, a servo motor and actuator, an amplifier, and a controller controlling the movement of the actuator. In the early generations of HDD, the controller used to be implemented using analog electronics but all modern drives come with digital controller. Nowadays almost all practical servomechanisms, HDD head positioning servo being one of them, use μ-controller or digital signal processor (DSP) to implement the controller. In this chapter, functions and principle of operations of different components of the HDD servomechanism are discussed, sources of noise and disturbances are explained, and the basic guidelines for design of controller are presented.

2.1 The Servo Loop

Binary bits are stored in an HDD by setting a small area of magnetic material coated on the disk to one of two possible polarities. This tiny area, called a *bit cell*, consists of several grains of the magnetic material alloy. If all grains in a bit cell are magnetized in the same polarity, it is said to be storing a binary '0'. On the other hand, a bit cell where a transition of magnetization takes place is considered as storing a binary '1'. The bit cells are created by the write head while the disk spins causing the bits to be arranged in concentric

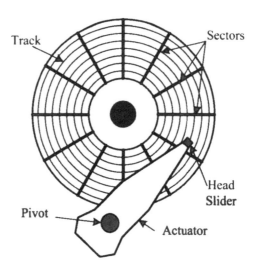

Figure 2.1: Tracks, Sectors, and Actuator.

circles, known as the *tracks* [Figure 2.1]. There can be as many as 100,000 tracks on each surface of a disk used in a $3\frac{1}{2}$ inch HDD. Recording (or writing) of the bits and playback (reading) is performed with a write head and a read head, respectively. The disks are spun at a precisely controlled speed when the operation of writing or reading is performed. The two heads are fabricated on a single slider, which is epoxy-bonded to a stainless-steel or aluminium gimbal at the end of a long and thin structure known as the *suspension* arm.

Each surface of an HDD is accessed by a dedicated head slider mounted at the tip of a suspension arm. Suspension arms carrying sliders for different surfaces are attached to a single actuator, driven by a motor popularly known as *Voice Coil Motor* or VCM. The movement of the head sliders between any two tracks is effectuated by the VCM actuator. It also regulates the position of the head over the center of a track while data is being written on or read from that track. As the disk spins while the head is regulated over the center of a track, the read/write head scans the entire track in one revolution of the spindle. It is desired to have the head positioned above the track-center with minimum variance before reading recorded data off the disk or writing new data on the disk can be performed. The tracking error during regulation of head position must be less than 10% of the track pitch (distance between two adjacent tracks) for data writing.

When the disk spins at high speed, an ABS is formed between the slider and the spinning disk that makes the slider float above the disk surface. The suspension arm is designed such that it produces precise load force and damping required by the slider to interact with the ABS formed. The movement of the slider perpendicular to the disk surface is self-regulated by the interaction between ABS, load force and damping. The movement of the slider in direction parallel to disk surface is effectuated by the torque generated by VCM. This is the motion a sliders go through during repositioning of the head over a new track, as well as during the track following, i.e., when the head is regulated over a track. Both of these operations, track seek and track following, uses the same actuator to create the motion parallel to disk surface. Error tolerance during the track-following is in the scales of nanometers, and it must be achieved in presence of various disturbances acting on the slider, suspension and actuator arm. On the other hand, the transfer of head from one track to another is expected to be performed in few milliseconds.

The HDD servomechanism is a unique example of practical applications that demonstrate the degree of precision achieved in a mechatronics system. Current state of the art in HDD industry enables laying out of data tracks on disk surfaces at density greater than $100,000$ tracks per inch (TPI), that is, the centers of two adjacent data tracks are separated by 10 μ-inch or 0.25 μ-m. The read/write head is expected to fly above the center of data track as precisely as possible while writing binary information on the data track or retrieving it from the track. Deviation of the head from this desired position increases the probability of occurrence of erroneous bits by either accidental overwriting on adjacent track or unwanted interference from the adjacent track, and hampers the reliability of the disk drive.* Head-positioning error tolerated in an HDD is typically less than 10% of track pitch which is equivalent to 1 μ-inch or 25 nm in a $100,000$ TPI drive. Projections suggest that track density will reach $400,000$ TPI in laboratory demonstration by the year 2009 and in production by 2013, particularly for small form factor drives, i.e., the drives using disks of diameter 2.5-inch or smaller. Desired error tolerance of the head positioning servomechanism for such drives will be 0.25 μ-inch or 6.3 nm.

The HDD market is now dominated by $3\frac{1}{2}$ inch form factor drives; but the smaller form factors have shown a growth potential comparable to those of the $3\frac{1}{2}$ inch drives in the early years of 1990s. Starting in 2003, the growths of $2\frac{1}{2}$ inch, 1.8 inch, and 1 inch drives are 36%, 380% and 55%, respectively. Global shipment of small form factor drives ($2\frac{1}{2}$ inch and below) was 50 million units in 2003 and is expected to reach 100 million units in 2006. Insatiable demand of notebook PC and application in consumer electronics, e.g., MP3, video camera etc are the main driving force behind this growth in the small form factor drives. As HDDs are being used in new applications, they are expected to meet more stringent performance specifications. For example, drives to be used in PDAs, camcorders, or automobiles must be able to withstand much larger vibration than those experienced by drives used in PC.

A comprehensive illustration of the closed loop head positioning servomechanism of HDD is shown in Figure 2.2. The VCM actuator moves the read-write head between tracks (track seek mode) and regulates the position of the head (track following mode). During track following, the head must follow the ref-

---

*Error probability greater than 10^{-10} is not acceptable in a properly functioning HDD; this means only one bit in error is permitted out of $10,000,000,000$ bits read from the disk.

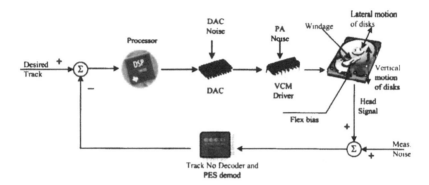

Figure 2.2: Closed Loop Head Positioning Servomechanism.

erence marks that define the center of the track; these reference marks are created on the disk at the time of servo track writing.[†] These references are special magnetic patterns written in designated areas on the disk surface known as *servo sectors*. The servo sectors are created at the time of manufacturing and are never overwritten or erased. The closed loop servomechanism uses the feedback signal generated by decoding the information written in these sectors. The servo sectors and the demodulation of the written information are explained later in section 2.3. The servo information and user data are multiplexed in space around the track. When the disks spin, this spatial multiplexing becomes temporal multiplexing. The feedback signal constitutes of both digitally coded track number and *Position Error Sensing* (**PES**) signal. The PES signal is proportional to the radial distance between the track-center created during servo track writing and the actual position of the read head. The composite feedback signal is used by the control algorithm which is implemented on a digital processor. There are separate heads for reading and writing, but the two heads are fabricated on a single slider. Since the suspension arms accessing different surfaces are mounted on the same actuator, all heads are moved simultaneously even though the tracks and sectors are accessed one surface at a time. The suspension provides a preload to press the slider down towards the surface of the disk. Care must be taken while designing the suspension so that the effects of different torsion and sway modes of the drive mechanics on the servo loop are minimized.

The head positioning servomechanism moves the read/write head as fast as possible from one track to another when asked by the host system (*Track Seek*). Once the head reaches the target track, it is regulated precisely over the track so that the PES is minimized (*Track Following*). Smooth *settling*, i.e., transition between the track seek and the track following modes without any jerk is another important feature expected in HDD servomechanism.

[†]This process is carried out by another high precision mechatronics system, *Servo Track Writer*, which is discussed in chapter 5.

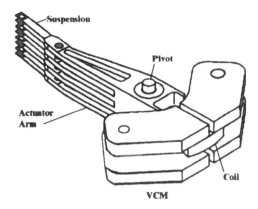

Figure 2.3: Rotary VCM actuator.

2.2 The Actuator

The VCM is the torque producing component of the head positioning servo-mechanism. When current is passed through the coil of VCM suspended in the magnetic field produced by permanent magnets, a force (torque) is generated. The force (torque), proportional to coil current, can be controlled by changing the amplitude and polarity of the current. There are two types of VCM actuator - (i) Linear VCM and (ii) Rotary VCM. These are shown in Figure 2.3.

In the first of these types, the coil is wound around a central yoke placed between two permanent magnets. The coil, when energized, is free to move forward and backward. As a result, the actuator arm attached to the coil structure moves in and out of the yoke. The VCM is fixed rigidly to the base plate outside the area of the disk, and the movement of the arm takes place along a radius of the disk (Figure 2.4). With this arrangement, the orientation of the slider with respect to track remains the same at all radial position of the slider. In a rotary VCM, the actuator arm is pivoted at a point between the coil structure and the suspension arm. The coil is attached using epoxy glue to one end of the arm and the suspensions carrying sliders to the other end. The pivot point is nearer to the coil which is suspended in the magnetic field of permanent magnets. Force is generated whenever the coil is energized by allowing current to pass through it. This force makes the coil move in a way that generates torque around the pivot point and causes the sliders to move on an arc. The angle between the slider's orientation and the track underneath varies as a function of the radial position of the track (Figure 2.5). The orientation of the head gap with respect to the track affects the amplitude of the readback signal. However, this effect is insignificant and all HDDs available in the market today use rotary actuators. Another effect of this variation in orientation of the slider is the changes in *micro-jog* distance

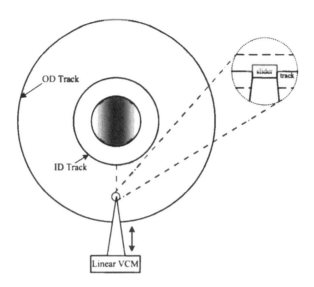

Figure 2.4: Movement of suspension arm for linear VCM.

as a function of track, which is explained next.

We can use the diagram in Figure 2.6 for a better visualization of the problem of micro-jog distance. It was mentioned earlier that two different heads fabricated on a single slider are used for reading and writing of data - bits are written with the help of a *thin film inductive* (TFI) head while the read head is a *magnetoresistive* (MR) sensor. As a result, there always exists a physical gap between the read sensor and the write head. The read head is also used to sense the servo patterns from the servo sectors which is used to derive the position feedback signal. The read head is used as the position sensor. During the operation of data reading, it is the read head whose position is regulated by the servomechanism, making the sensor and point of control collocated. On the contrary, during write operation, the point of control is the write head but position feedback comes from the read head. If the gap between the read head and the write head is known then that information can be taken into consideration as offset while regulating the position of the write head. This offset is known as the *micro-jog* distance. When a rotary actuator is used, the micro-jog distance is different for different radial position of the slider, which is illustrated in Figure 2.6. This figure shows the head slider positioned by a rotary actuator over two different tracks, Trk_m and Trk_n. Since the slider moves on an arc, the micro-jog distances for these two locations (d_m and d_n) are different; $d_m < d_n$ for the illustration shown in Figure 2.6. In this case, the head positioning servomechanism must use track-dependent micro-jog distance to compensate for the offset. Many factors such as head geometry, distance between actuator pivot and spindle-center affect the micro-jog distance and its variation as a function of radius. The micro-

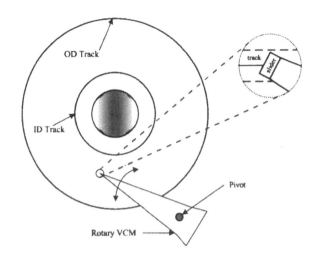

Figure 2.5: Movement of suspension arm for rotary VCM.

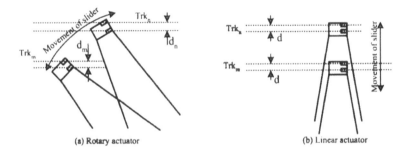

Figure 2.6: Micro-jog in HDD servomechanism.

jog distance must be calibrated as a function of disk radius for each drive using a built-in function in the initialization firmware of the drive. Linear VCM actuators were already extinct when the HDD industry adopted MR head technology. However, had there been any drive with MR head and linear VCM actuator, the micro-jog distance would be constant for all radial position and would require simpler calibration algorithm.

If we let a current flow through the coil of the VCM, it experiences an electromagnetic force as shown in Figure 2.7 and the actuator arm which is attached to the suspended coil and pivoted is subject to a rotating torque. The electromagnetic force acting on the coil is produced by the interaction between the magnetic field of the permanent magnet and the field produced by the coil current. If we assume the field from the permanent magnet constant for the entire range of operation, then the magnitude of the torque depends on the field produced by the coil current and the actuator geometry. Since the

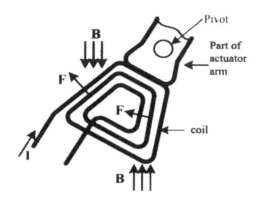

Figure 2.7: Generation of torque in rotary VCM actuator.

geometry is fixed for a given actuator, the torque it generates is function of coil current only and we can assume this to be $K_t I$, where K_t and I are the torque constant of the VCM and the coil current, respectively. In practice, the torque constant may vary with the position of the coil, i.e., $K_t(\theta)$ is a function of actuator position θ. However, the change in the magnitude of torque constant is usually very small and insignificant. We shall assume throughout this book a constant value for this parameter of VCM.

The motion of the actuator arm, defined according to the Newton's second law of motion, can be modeled as

$$\ddot{\theta}(t) = \frac{K_t}{J} I(t), \qquad (2.1)$$

where J is the moment of inertia of the rotating arm, and $\ddot{\theta}$ is the angular acceleration of the actuator's motion. If the distance between the pivot center and the read head is L inches,[‡] then the linear displacement of the read head corresponding to an angular displacement (θ) is $x = L\theta$. It is very common in the HDD industry to express the displacement of read head in units of track, i.e., $y = D_{trk} L\theta$, where D_{trk} is the track density in units of *Tracks per Inch* (TPI). Taking all these factors into consideration, the rigid body dynamics of the VCM actuator is given by

$$\ddot{y}(t) = \frac{D_{trk} L K_t}{J} I(t) = K I(t). \qquad (2.2)$$

The corresponding transfer function model is $G_v(s) = \frac{K}{s^2}$.

If the VCM is driven by a voltage amplifier, shown on the left of Figure 2.8, the output (V_O) of the amplifier is proportional to the input, i.e., $V_O = K_{VA} u$. The current in the VCM coil and the applied voltage are related to each other

---

[‡]Imperial units of measurement are widely used in the HDD industry.

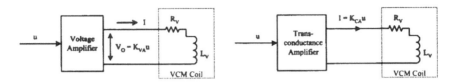

Figure 2.8: Amplifiers for VCM driver: Voltage source (left) and Current source (right).

by the differential equation

$$V_O(t) = R_v I(t) + L_v \frac{dI(t)}{dt}. \tag{2.3}$$

In this case, the transfer function between the input (u) and the coil current (I) is

$$\frac{I(s)}{U(s)} = \frac{K_{VA}}{L_v s + R_v}, \tag{2.4}$$

and the overall transfer function is

$$G_{v,v} = \frac{Y(s)}{U(s)} = \frac{K K_{VA}}{s^2 (L_v s + R_v)}. \tag{2.5}$$

On the other hand, if a current amplifier(shown on the right of Figure 2.8) is used, i.e., $I = K_{CA} u$, then the overall transfer function is

$$G_{v,c} = \frac{Y(s)}{U(s)} = \frac{K K_{CA}}{s^2}. \tag{2.6}$$

In HDD, the VCM driver is usually implemented as a voltage controlled current amplifier. A sensing resistor of very low ohm is connected in series with the VCM coil. Voltage across the sensing resistor is proportional to the coil current, which is then used as feedback to control the coil current. Circuit representation of a typical VCM driver is shown in Figure 2.9.

The current in the VCM driver is proportional to the input voltage as long as the amplifier operates in the unsaturated mode. If the amplifier is saturated, the output current can not be increased anymore. So one can model the VCM driver as a current amplifier with an upper limit bounding the amplitude of the current. The fact that the amplitude of current is upper bounded must be taken into consideration while designing the closed loop feedback controller. Assuming that the output current of the amplifier operating in the linear region is proportional to the input, setting an upper bound on the input u is equivalent to setting an upper bound on the current I. An working model of the VCM actuator plus the driver is

$$G_{v,rigid}(s) = \frac{Y(s)}{U(s)} = \frac{a}{s^2}; \quad given \quad |u| \le U_m. \tag{2.7}$$

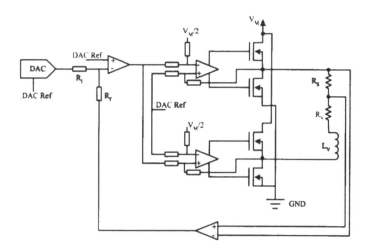

Figure 2.9: Circuit representation of a typical VCM driver.

This model represents only the rigid body dynamics of the actuator. The actual dynamics, however, is more complex and includes higher order dynamics representing different torsional and sway modes of the arm, lightly damped flexible modes of the gimbal-slider assembly, and modes of coil structure etc. It may need a transfer function of order as high as 40 to accurately model the dynamic behavior of the head positioning actuator [54]. Frequency response of a typical HDD actuator is shown in Figure 2.10. The response of an identified model that includes a double integrator plus 10 poles and 10 zeros is also drawn on the same figure [6].

2.2.1   Measurement of Frequency Response

In the head positioning servomechanism of HDD, position feedback is obtained from the readback signal produced by sensing special magnetic patterns written on the disks. These patterns, which are explained later in section 2.3, are created by a process known as servo-writing[§]. This position signal is available only in an assembled and servo-written HDD. However, to obtain a dynamic model of the head positioning actuator, one may use other means to measure the displacement. The use of interferometry to measure changes in position is well known [181],[18]. The interferometer optics split the laser light into a reference path and a measurement path. These lights are reflected using two retroreflectors - the reference beam from a stationary reflector but the measurement beam from a retroreflector attached to the object whose change in position is to be measured. Recombination of the two reflected beams creates an interference signal. A difference of one wavelength between the lengths of

---

[§]Servo-writing and related issues are discussed in chapter 5

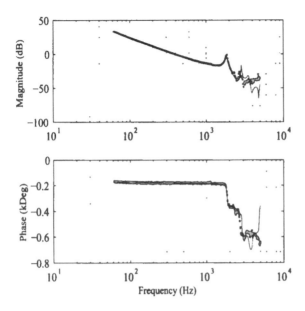

Figure 2.10: Frequency response of a typical VCM actuator.

the two paths results in a phase change of 360° in the interference signal. The measuring electronics measures and accumulates the phase and provides a position output. This method has some drawbacks for application in the measurement of displacement of HDD actuator. When a retroreflector, which is usually quite heavy, is attached to the actuator, dynamics of the actuator arm is significantly modified. Besides, our aim is to measure the displacement of the head slider, which is too small to carry the load of the retroreflector. One possibility is to attach the retroreflector on the E-beam of the actuator, but then the measured displacement does not reflect the dynamics of either the suspension or the slider-gimbal assembly.

Laser Doppler Vibrometers (LDV) are optical instruments for accurately measuring velocity and displacement of vibrating structures completely without contact. A rugged laser head is mounted on a large vibration-free platform with the laser beam pointing to the object whose displacement or velocity is to be measured. Christian Doppler was the first to describe the frequency shift that occurs when sound or light is emitted from a moving source. If a laser beam of precise frequency is incident on a moving reflector then, according to Doppler effect, the frequency of the reflected beam is different from that of the incident beam. The velocity of the moving reflector can be measured by measuring the change in frequency between the incident and reflected beams. The LDV measures the velocity according to this principle and then integrates it to provide displacement measurement. Since the measurement with LDV requires only a reflecting surface on the moving object, we can attach a tiny

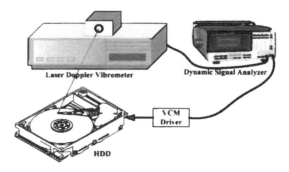

Figure 2.11: Experimental setup for frequency response measurement.

reflecting sheet on the head slider without causing any significant change in the dynamics of the actuator mechanism.

Frequency response shown in Figure 2.10 is measured using an experimental setup involving an LDV, a Dynamic Signal Analyzer, a VCM driver and an HDD actuator. The schematic diagram of the setup is shown in Figure 2.11. The Signal Analyzer generates a swept-sine signal, which is used to excite the VCM actuator. The excitation signal, which is the input to the VCM driver, and the displacement measurement from the LDV are fed to two channels of the signal analyzer. The signal analyzer computes the gain and phase at each of the frequencies of swept-sine signal. It should be noted that the measured gain is in the unit of V/V as the output of LDV is a voltage signal. However, one can easily change the unit of gain by taking into consideration the resolution of LDV ($V/\mu m$).

Care must be taken while setting the amplitude of the swept sine signal. Small amplitude of input signal results in small displacement of the slider and, therefore, low signal-to-noise ratio (SNR) in the output of LDV. Too large an amplitude, on the other hand, may cause the actuator to move beyond the range of LDV. Since the head slider moves on an arc, the reflected beam is not in line with the incident beam. If the angle between the two beams is large, the reflected beam is not received well by the measurement electronics. The gain of the actuator is higher in the lower frequencies and is expected to decrease with increasing frequency. That means the amplitude of the input excitation should be lower for in the low frequency range and should be increased as the frequency of the input goes higher. For the result shown above, input amplitude is kept reasonably small (below 100 mV) for the lowest range of frequency, and was increased to almost 1 V for high frequency. Dynamic signal analyzers available in the market these days come with the capability of automatic adjustment of the input amplitude.

2.2.2 Identification of Transfer Function Model

Frequency response in the low frequency range (Figure 2.10) show approximately -40 dB/decade slope in magnitude and $-180°$ phase suggesting a rigid body, double integer model $\frac{k}{s^2}$. The zero cross-over of the magnitude response occurs at about 400 Hz, i.e., $\left|\frac{k}{(j\omega_c)^2}\right| = 1$ for $\omega_c = 2\pi \times 400$ or $k = 6.3 \times 10^6$. The measurement was carried out with the LDV resolution set to $0.1V/\mu m$. So if we define the transfer function in units of $\mu m/V$ then $k = 6.3 \times 10^7$.

This double integrator model or, as used in chapter 3, a second order model with poles on the left hand side of the complex plane is often used as the nominal model for the sake of controller design. However, knowledge of the flexible mode dynamics is also crucial. There are well established methods for identification of a transfer function from frequency response data [163], [172]. These methods finds the coefficients of the transfer function $G(s) = \frac{B(s)}{A(s)}$ such that the frequency response of the identified transfer function matches as close as possible to the frequency response obtained experimentally. The frequency response data include two vectors:

1. the vector $[\omega_k]$ for $k = 1, \cdots, N$ contains all frequencies for which magnitudes and phases are measured and

2. the vector $[G_{fr}(\omega_k)]$ contains the frequency response measured at each of the N frequencies.

The transfer function model $G(s)$ is the ratio of two polynomials of Laplace Transform parameter s,

$$G(s) = \frac{B(s)}{A(s)} = \frac{b_0 s^m + b_1 s^{m-1} + \cdots + n_m}{s^n + a_1 s^{n-1} + \cdots + a_n}, \tag{2.8}$$

with $m \leq n$ for proper transfer function. The frequency response of a system is equal to its transfer function evaluated at the points along the positive imaginary axis of the complex plane, i.e., the response at any frequency ω_k is

$$\frac{b_0(j\omega_k)^m + b_1(j\omega_k)^{m-1} + \cdots + b_m}{(j\omega_k)^n + a_1(j\omega_k)^{n-1} + \cdots + a_n} = G_{fr}(\omega_k). \tag{2.9}$$

The parameters of the transfer function G can be obtained by solving the least squares estimation problem that minimizes the error criterion

$$\sum_{k=1}^{N} \left|\frac{B(j\omega_k)}{A(j\omega_k)} - G_f r(\omega_k)\right|^2. \tag{2.10}$$

This is a nonlinear least squares problem and can be solved iteratively. Commercial softwares are available for solving such problems, e.g., *Frequency Domain Identification Toolbox* from $MATLAB^{TM}$ [115].

One can rewrite equation 2.9 as,

$$b_0(j\omega_k)^m + b_1(j\omega_k)^{m-1} + \cdots + b_m = G_{fr}(\omega_k)\left((j\omega_k)^n + a_1(j\omega_k)^{n-1} + \cdots + a_n\right), \tag{2.11}$$

or, equivalently

$$b_0(j\omega_k)^m + \cdots + b_m - G_{fr}(\omega_k)\left(a_1(j\omega_k)^{n-1} + \cdots + a_n\right) = (j\omega_k)^n G_{fr}(\omega_k). \tag{2.12}$$

Using vector notation,

$$\phi^T(j\omega_k)\begin{bmatrix} b_0 & b_1 & \cdots & b_m & a_1 & a_2 & \cdots & a_n \end{bmatrix}^T = x(\omega_k) + jy(\omega_k), \tag{2.13}$$

where
$x(\omega_k) + jy(\omega_k) = (j\omega_k)^n G_{fr}(\omega_k)$ and
$\phi(j\omega_k) =$
$\begin{bmatrix} (j\omega_k)^m & (j\omega_k)^{m-1} & \cdots & 1 & -G_{fr}(\omega_k)d_1(j\omega_k)^{n-1} & \cdots & -G_{fr}(\omega_k)d_n \end{bmatrix}^T$. For
each frequency, ϕ is a vector of complex numbers. Equating real part of ϕ with
x and imaginary part of ϕ with y, we can rewrite equation 2.13 as

$$\begin{bmatrix} \phi_R^T(\omega_k) \\ \phi_I^T(\omega_k) \end{bmatrix} \cdot \theta = \begin{bmatrix} x(\omega_k) \\ y(\omega_k) \end{bmatrix}, \tag{2.14}$$

where, $\theta = \begin{bmatrix} b_0 & b_1 & \cdots & b_m & a_1 & a_2 & \cdots & a_n \end{bmatrix}^T$ represents the parameter
vector.

The frequency response is measured for N different frequencies, and we get
N sets of the above equation. That is,

$$\Phi_{2N \times np}\Theta_{np \times 1} = Y_{2N \times 1}, \tag{2.15}$$

where np is the number of parameters to be identified. This is a *linear in the
parameters* (LIP) model and can be solved using linear least-squares method,
i.e., to find the estimate $\hat{\Theta}$ of the parameter vector by minimizing the cost
function,

$$J_{LS} = (Y - \Phi\hat{\Theta})^T(Y - \Phi\hat{\Theta}). \tag{2.16}$$

Solution of this least squares problem is,

$$\hat{\Theta} = (\Phi^T\Phi)^{-1}\Phi^T Y. \tag{2.17}$$

It should be noted that the measurement vector $Y_{2N \times 1}$ in this LIP model
is obtained by multiplying the experimental frequency response by $(j\omega_k)^n$. In
other words, true measurement of the frequency response is weighted by a
frequency dependent factor which increases as ω_k increases. The estimation
algorithm puts higher weights on the measurements in the higher frequency.
This becomes problematic particularly in cases where measurement data span
several decades of frequencies. Because of the multiplication by $(j\omega_k)^n$, both
Φ and Y differ widely in magnitude. This may lead to failure to achieve
good fit between the measured response and model's response. A remedy is
to filter Y and Φ by some approximation to the filter $\frac{1}{A(j\omega)}$ and iterate this
procedure [103].

2.3 Feedback of Position Signal

The feedback signal for the head positioning servo loop is obtained by decoding spatially coded magnetic patterns written on the disks. These patterns, known as *servo pattern*, are created at the time of manufacturing HDD after the spindle, disks, actuator and heads have been assembled inside the drive enclosure. The process of writing servo patterns is known as *Servo Track Writing* (STW) and is carried out using a very high precision equipment that controls the position of the actuator of HDD and writes the servo patterns on the disks. For all HDDs manufactured these days, the enclosure is covered and sealed after servowriting and, as a result, the same head-suspension-actuator assembly and disk-spindle assembly are used for normal operation of HDD. In the past, however, a different scheme was in use where only one surface of one disk was pre-written with servo patterns. Writing of servo patterns were performed in bulk, several tens of disks at a time. One servo-written disk and several other blank disks were then assembled inside the drive enclosure. In an HDD assembled in such fashion, there is one surface of a disk containing servo pattern. This scheme of creating servo information is known as *dedicated servo*, where only one side of one disk in a multi-disk HDD contain the spatial servo patterns. In such drive, the signal necessary for position feedback came from the head accessing the servo surface. Since a single VCM actuator controls the motion of sliders on all surfaces simultaneously, moving the slider on the servo surface to a desired track is equivalent to moving any other head to the same track. The main assumption for proper functioning of this scheme is that sliders on all surfaces are displaced by precisely the same amount which, in reality, is impossible. The suspension arms get elongated due to thermal expansion if the drive is kept ON for long time and the amount of elongation may be different for different arms. For older generation drives, the acceptable tracking error (20 μ-inches or more) was much larger compared to the differences between the thermal expansions of arms. The dedicated servo scheme worked well for drives of the past, but its limitations started to surface with the trend of increasing track density when the disparity between thermal expansions became comparable to error tolerance. The need to overcome the thermal expansion related problems to pave the way for higher track density gave birth to a new scheme of servo encoding - *embedded servo* or *sectored servo*, in which servo patterns are created on all surfaces and same head is used for accessing both servo and data. Instead of one dedicated servo surface, the embedded scheme puts servo patterns on all surfaces interleaved with the data blocks. The servo sector, a small segment of the track containing servo patterns, are created at regular intervals, and the space between two servo sectors is designated for storing data. The embedded servo scheme with interleaved data blocks and *servo sectors* is illustrated in Figure 2.12.

Besides the problem of thermal expansion, large *servo overhead* is another drawback of dedicated servo scheme when only a few disks are used in an HDD. Servo overhead is the percentage of available area that is consumed by

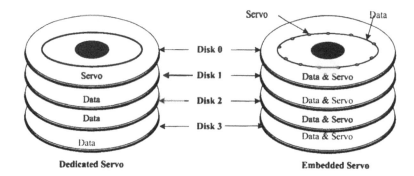

Figure 2.12: Dedicated Servo and Embedded Servo.

servo information. For an HDD with only 2 disks, which is quite common nowadays because of large areal density, one out of four surfaces are used for servo information if dedicated servo is employed, i.e., 25% of the available storage space is consumed by servo. On the contrary, servo sectors in each track of an embedded servo HDD occupies less than 10% of the track. So the servo overhead is less than 10% for any surface, and does not depend on the number of disks used in the HDD.

2.3.1   Servo Bursts

In an embedded servo HDD, the servo sectors are placed on all tracks interleaved with data blocks. Small section of a track is illustrated in Figure 2.13. This figure also shows different fields in a servo sector; each field contains a specific pattern of magnetization. These fields are,

- DC-gap field,

- Automatic gain control (AGC) field,

- Servo timing mark (STM) field,

- Grey coded track number field, and

- PES burst pattern field.

Each of these fields in the servo sector has specific function. Only two of them are directly related to the generation of position feedback signal and are explained here. These are the track number field and burst pattern field. The grey coded track number is the identification of a track; the outermost track is tagged as $Trk0$ and it increases inward. So each track has a unique grey coded track number field, and the same pattern is repeated in the track number fields of all sectors of a track. It remains unchanged within the width of the track and, therefore, does not provide any scope to measure the off-track

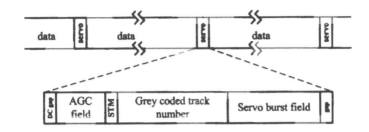

Figure 2.13: Different fields in a servo sector.

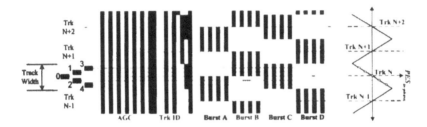

Figure 2.14: Illustration of magnetic pattern for servo burst.

displacement of the read head from the center of the track, i.e., the fraction of track-pitch. The readback waveform generated by the head scanning the PES burst pattern is decoded to measure the fractional off-track error.

The schematic layout of the magnetic patterns in the servo burst field is illustrated in Figure 2.14. This represents only a tiny segment of the disk surface. Shaded and clear segments are used in this figure to differentiate between areas of the disk magnetized in opposite polarity. Moving from left to right in this figure or vice versa is equivalent to moving in the direction along the track, i.e., the *down-track direction*, and moving up or down is the *cross-track direction*. This illustration shows the track-centers of 4 consecutive tracks with track numbers increasing upward. Definition of the track-centers will become clear after the following analysis of the signals obtained from these patterns.

When the read sensor scans a magnetic transition, a voltage pulse is produced. The polarity of the pulse depends on the type of transition. An example is shown in Figure 2.15 for two consecutive magnetic transitions. The readback waveform shows two similar pulses of opposite polarity. The amplitudes of the pulses depend on the magnetic flux linking the read head and hence on the distance of the head with respect to the transitions, both in the vertical plane as well as on the plane parallel to the disk surface. With the help of this simple explanation on the amplitude of readback pulse, it is easy to deduce the readback waveform produced when the read head scans the servo burst patterns at different cross-track positions.

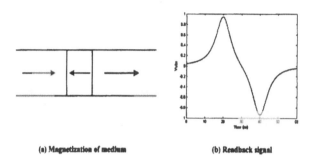

(a) Magnetization of medium   (b) Readback signal

Figure 2.15: Two consecutive transitions (top) and corresponding readback signal (bottom).

When the disk spins, the entire track containing all data block and servo sectors is scanned by the head. The read head positioned at the point marked 0 in Figure 2.14 senses the burst patterns scanning them along the dashed line and produces a series of voltage pulses with alternating polarities. For this case, the read head senses maximum flux from burst C and no flux from burst D. For bursts A and B, the flux linkage is smaller than that for burst C. As a result, the amplitude will be maximum for the burst C, zero for burst D, and non-zero but less than maximum for bursts A and B. Corresponding waveform is shown in Figure 2.16.

Following the same method, one can easily find the waveforms generated by read head scanning the servo patterns along different off-track positions, marked 1, 2, 3 and 4 in Figure 2.17[¶].

It is revealed from the observation of waveforms shown in Figure 2.16 and Figure 2.17 that the amplitudes of different burst waveforms vary with the cross-track location of the read head. The information contained in the amplitudes of these bursts is used to measure the displacement of the read head with respect to the burst patterns. A signal proportional to the off-track error, called the *Position Error Sensing* (PES) signal, is obtained by demodulating the burst waveforms. A down-track line (that is line along the track) is designated as the center of a track if the amplitudes of burst A and burst B are equal when the head scans the burst along this line. The difference between the amplitudes of these two bursts is called the *in-phase PES* signal,

$$PES_{in-phase} = A_A - A_B. \qquad (2.18)$$

A_A and A_B are the amplitudes of the waveforms of burst A and burst B, respectively. When the read head moves away from the center of the track, one of these amplitudes become larger than the other and the $PES_{in-phase}$ becomes non-zero. Further the head is from the track-center larger is the

---

[¶]The lines shown in this figure are fictitious; in reality there is no such physical line along the track.

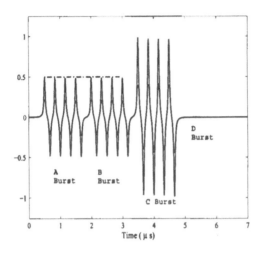

Figure 2.16: Burst signal for zero off-track.

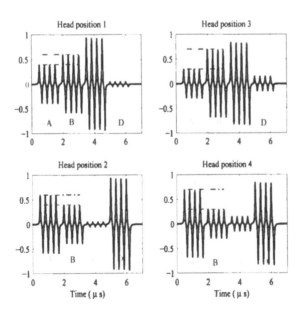

Figure 2.17: Burst signal for nonzero off-track.

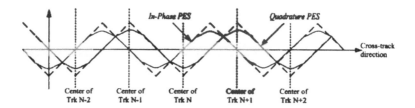

Figure 2.18: PES signal Vs off-track distance.

magnitude of $PES_{in-phase}$. The $PES_{in-phase}$ at different positions of the head are shown using circles at the extreme right end of Figure 2.14. It also shows an interpolated plot of $PES_{in-phase}$ versus cross-track displacement or off-track position. It is easily understood that the $PES_{in-phase}$ signal has alternating positive and negative slopes in adjacent tracks.

The plot shown in this illustration is obtained assuming ideal conditions and it does not represent the $PES_{in-phase}$ in a drive. The dimension of the read head is usually smaller than the width of a track, and if the head moves far from the center of the track it senses flux emanating from only one of the two bursts (either burst A or burst B) but not both. As a result, small movement of the head around such position has no or little effect on the amplitudes of two bursts. So the $PES_{in-phase}$ signal tends to saturate at large off-track distances, as shown in Figure 2.18. This issue of nonlinearity in the measurement of off-track displacement is resolved by creating a second pair of burst patterns, burst C and burst D, placed in spatial quadrature with respect to the pair of burst A and burst B, shown in Figure 2.14. The difference in amplitudes of these two bursts (C and D) is called the *quadrature PES* signal,

$$PES_{quad} = A_C - A_D. \qquad (2.19)$$

A_C and A_D are the amplitudes of the waveforms of burst C and burst D, respectively. The PES_{quad} signal as a function of off-track error is also shown in Figure 2.18. Its dependence on the off-track displacement is similar to that of $PES_{in-phase}$, but its zero-crossings coincide with the boundaries between two adjacent tracks. Appropriate manipulation of *in-phase* and *quadrature* PES signals produces an error signal proportional to the distance between read head and the center of a track. The feedback signal used by the servo loop is the combination of PES signal and track number obtained from the grey code field.

In disk drives using embedded servo, the position feedback is available only at discrete points in time and the servo control is also implemented in discrete-time. However, the designer of servo controller is not at the liberty of selecting the sampling frequency arbitrarily. The position signal is available at a frequency $S\frac{N_{disk}}{60}$, where S and N_{disk} are the number of servo sectors per track and rotational speed of disks in units of *revolutions per minute* or RPM, respectively. The sampling frequency can be increased either by spinning the disk faster or by including more servo sectors. Each of these options comes with its disadvantages. Larger is the number of servo sector, more is the *servo overhead.* Increase in spindle speed, on the other hand, generates more heat which in turn requires better cooling mechanism. Increased speed also translates into larger rate of data transfer between media and electronics that demands for expensive electronics in the read-write channel. Besides, higher spindle speed pushes the spectrum of disturbances related to disk rotation to higher frequency and, as a consequence, higher bandwidth is required for actuator servo. As a result, the disk drive servomechanism remains as an example of control system which demands for as large a bandwidth as possible but comes with severe restrictions on the sampling frequency.

2.3.2   Servo Demodulation

A good estimate of the burst amplitude is the most important consideration for reliable generation of the PES signal. Until very recently, the servo demodulation used non-coherent analog method. In this method, the burst waveform is first processed through a full-wave rectifier. In a method called the *peak-detected servo demodulation*, a circuit that can detect and hold the peak amplitude of the rectified burst waveform is used. Another method, known as the *area demodulation*, finds the area under the rectified waveform. The analog area detection method uses several precautionary measures to minimize detection error. The zero-crossings of the burst waveform are first detected. Once a zero crossing is found in a burst, a demodulation window is opened. The window is closed after a pre-defined number of cycles of the burst waveform have elapsed. The rectified waveform that falls inside the window controls a charge pump that, in turn, charges a capacitor. The capacitor voltage is proportional to the area of the burst falling inside the demodulation window. The process is illustrated in Figure 2.19. Once the area of one burst is obtained, the amplitude of the capacitor voltage is converted into a digital number and latched to a register. The capacitor is then discharged to zero voltage before the demodulation window for the next burst is opened. The same capacitor is used for all four burst waveforms to eliminate the effect of component variances. The first few cycles in the burst are not used for charging the capacitor. During this period, the zero-crossing detector synchronizes the charging process. Each burst is, therefore, made longer than the number of cycles actually used for meaningful area detection. Moreover, inter-burst space is essential to provide sufficient time for analog-to-digital conversion of the capacitor voltage and for discharging of the capacitor. The peak-detected servo demodulator uses simpler circuit, but it is prone to error in presence of noise. Area detection, on the other hand, is an averaging process and provide better immunity to broadband noise.

Availability of higher processing power in modern disk drives encouraged the use of digital algorithms to find the burst amplitude. The first approach

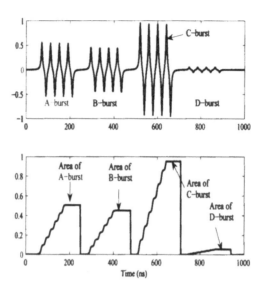

Figure 2.19: Area detector for servo burst demodulation.

in this direction is the *digital area detection* which samples the burst signals and adds the absolute values of pre-defined number of samples [164]. The result of the summation is proportional to the burst area. The burst waveform is filtered before it is sampled to reduce aliasing effects. Let $y_A(kT_S)$ be the samples from the A-burst waveform, sampled at intervals of T_S. Then the area of this burst is,

$$A_{area} = \sum_{k=0}^{N-1} |y_A(kT_S)|. \tag{2.20}$$

The number of samples (N) is selected such that the summing window is equal to an integer number of periods of the burst waveform. If the sampling of the burst is not synchronized with the zero-crossings of the waveform, the estimated area differs from its true value, and the reliability of the PES becomes questionable.

Two other methods proposed for estimating amplitude (or area) of a burst waveform from its samples are *digital maximum likelihood detection* [216], and *coherent detection with selective harmonics* [3]. The estimated burst amplitude using these two methods are shown below for the samples $y_A(kT_S)$.

Digital Maximum Likelihood Detection:

$$A_{ml} = \frac{y_A^T d_0}{d_0^T d_0}. \tag{2.21}$$

The vector y_A contains the samples of the burst waveform and the vector d_0 consists of the samples from a ideal model signal of the burst.

Coherent Detection using Selective Harmonics:

$$A_{coh} = \frac{y^T m}{m^T m}.$$ (2.22)

This is a special case of the maximum likelihood (ML) detection. The model signal (d_0) in the ML detector contains all harmonics of the nominal noise-free signal. In this scheme, the model signal ($m(k)$) is created with weighted sum of selected harmonics,

$$m(k) = r_0 + \sum_p (r_p \sin(pwkT_S)),$$ (2.23)

where, ω is the fundamental frequency of the burst signal, and p and k are the indices for harmonics and samples, respectively.

2.3.3 Recent Developments

For an HDD operating at 10,000 rpm with 120 servo sectors per track, the sampling frequency is 20 kHz. This sampling frequency is good enough to meet the specifications for the head positioning servomechanism during track-following. However, the relentless effort in reducing track pitch demands for ever improving precision and accuracy for the head positioning servomechanism. Since there is no end to the demand for higher bandwidth, the servo engineers always remain in the pursuit of higher sampling frequency as it affects the achievable bandwidth in a discrete-time control system.

Sampling frequency can be increased in the HDD head-positioning servomechanism in two possible ways:

1. spinning the disks at higher RPM, and

2. adding more servo sectors per track.

Spinning the disk at higher RPM also reduces average latency. However, the speed can not be increased arbitrarily. Increased RPM comes at the cost of higher power consumption and generation of excessive heat inside drive enclosure. Better design of the spindle motor and other mechanical features inside the enclosure is an ongoing effort to solve these problems. There is yet another problem with higher RPM drives that directly affects the head-positioning servo. Many of the disturbances discussed later in this chapter are synchronous to the spindle motor. Spinning the disk at higher RPM shifts the spectrum of these disturbances to higher frequency. As a result, higher bandwidth is required for satisfactory rejection of those disturbances. Increasing sampling frequency with higher spindle speed is therefore not a solution that

Figure 2.20: Dual frequency burst pattern.

is always possible. The second method of increasing sampling frequency by adding more servo sectors per track does not affect the spectrum of the noise and disturbances. Neither it generates more heat. However, it takes up more space of the disk that can otherwise be used for storing data. If the length of the servo sectors can be reduced without compromising the quality of de-modulation of bursts then more servo sectors can be added with no additional servo overhead. One scheme with patterns of different frequencies for different servo bursts reduces the length of servo sector. Only difference between this scheme and the conventional embedded servo is in the burst pattern; all four bursts in the conventional scheme uses patterns of same frequency.

Dual Frequency Burst Pattern

In the conventional method, the four servo bursts use the same pattern of magnetization, i.e., a series of transitions created at regular spacing along the track (down-track direction). Since these bursts have identical patterns, they can not be distinguished from one another if two of them are aligned in the radial direction (cross-track direction). This makes it absolutely essential to place the bursts shifted circumferentially from one another, leaving many voids, i.e., area with no pattern, in the servo sector (see Figure 2.14). If the pattern of transitions in one burst is made different from that of another burst then these bursts can be distinguished easily. Two such bursts can be placed along the same radius of the disk with no circumferential offset, and yet without any loss of their identities. Then the voids otherwise found between the bursts of conventional patterns are eliminated. Schematic representation of the dual frequency burst pattern is shown in Figure 2.20. The diagram shows both in-phase and quadrature bursts, which are essentially the same pattern but are placed with radial offset by half a track.

In the conventional method, waveforms of all four bursts are identical. But for dual-frequency burst pattern, the two bursts of the in-phase (or quadrature) burst-pair produce different waveforms. Both of them contain series of pulses with alternating polarities, but the intervals between two adjacent peaks in one burst waveform differ from those of the other. If the read head senses both patterns simultaneously, that is when the head is at or around the junction of these patterns, the readback waveform $y_{dual}(t)$ is the superposition of the burst waveforms $y_1(t)$ and $y_2(t)$ of two different frequencies:

$$y_{dual}(t) = m_1 y_1(t) + m_2 y_2(t). \tag{2.24}$$

The amplitudes $m_1(x)$ and $m_2(x)$ of the individual burst waveforms vary linearly as a function of the off-track displacement x of the read head from the center of the track.

The area detection method can not be applied directly for estimating the amplitude from the dual-frequency burst waveform. One possible solution is to use two band-pass filters to separate the two frequencies, and then applying area detection method to the samples at the outputs of two filters. Maximum likelihood detection and coherent detection using selective harmonics can be used to estimate individual amplitude from the samples of the dual-frequency burst waveform. Both of these methods are sensitive to jitter in sampling clock. If the clock is not synchronized, the phase error between the sampled signal and model signal contributes to error in the estimate of amplitude. Such error can be eliminated if both sine and cosine are used in the model signal. The *Coherent Detection using DFT* of estimated amplitude of a periodic signal with fundamental frequency ω from its samples is given by

$$\begin{aligned}
m_s(k) &= \sin(\omega k T_S); \\
m_c(k) &= \cos(\omega k T_S); \\
A_s &= y^T m_s; \quad A_c = y^T m_c; \\
A_{dft} &= \sqrt{A_s^2 + A_c^2}.
\end{aligned} \tag{2.25}$$

This method, which extracts the amplitude of the fundamental frequency of the burst signal, is equivalent to discrete Fourier Transform (DFT) with interest in one frequency only. The burst signal is superposition of two periodic signals of two different fundamental frequencies, ω_1 and ω_1. The coherent detection using DFT can be employed to estimate the amplitude of each of these frequencies present in the burst signal. However, in reality, the servo burst signals y_1 and y_2 contain not only the fundamental frequencies but also other odd harmonics and noise [202]. It should be noted that the burst signals are odd signals and therefore contain sine waves only. Applying the above mentioned method for estimating amplitude of the fundamental frequencies, we get

$$\begin{aligned}
A_s &= \sum_{k=0}^{N-1} [a_1 \sin(\omega k T_S) + a_3 \sin(\omega 3 k T_S) + a_5 \sin(\omega 5 k T_S) \\
&\quad + ... + n(k)] \sin(\omega k T_S + \theta); \\
A_c &= \sum_{k=0}^{N-1} [a_1 \sin(\omega k T_S) + a_3 \sin(\omega 3 k T_S) + a_5 \sin(\omega 5 k T_S) \\
&\quad + ... + n(k)] \cos(\omega k T_S + \theta);
\end{aligned} \tag{2.26}$$

where, θ represents the synchronization error due to clock jitter, and $n(k)$ is additive white Gaussian noise (AWGN). The number N is chosen such that integer number of full cycles of the fundamental frequency are sampled. Then it can be easily proven that

$$\sum_{k=0}^{N-1} [\sin(\rho\omega k T_S)\sin(\omega k T_S + \theta)] = 0;$$

$$\sum_{k=0}^{N-1} [\sin(\rho\omega k T_S)\cos(\omega k T_S + \theta)] = 0; \quad \forall \rho = 3, 5, 7, ... \qquad (2.27)$$

Moreover, if the noise $n(k)$ is zero-mean AWGN then the following ensemble averages are also equal to zero,

$$E\sum_{k=0}^{N-1} [n(k)\sin(\omega k T_S)] = 0,$$

$$E\sum_{k=0}^{N-1} [n(k)\cos(\omega k T_S)] = 0. \qquad (2.28)$$

So,

$$A_s = \sum_{k=0}^{N-1} [a_1\sin(\omega k T_S)\sin(\omega k T_S + \theta)] = a_1\cos(\theta),$$

$$A_c = \sum_{k=0}^{N-1} [a_1\sin(\omega k T_S)\cos(\omega k T_S + \theta)] = a_1\sin(\theta). \qquad (2.29)$$

The estimated amplitude of the burst waveform of fundamental frequency ω rad/s is $A_{dft} = \sqrt{A_s^2 + A_c^2}$. This method can be applied to estimate the amplitudes of both frequency components of the dual-frequency burst.

Estimating Head Position from Data Block Signal

When the head reads information from the data blocks, the readback signal is converted into a sequence of '0's and '1's by the *partial response maximum likelihood* or PRML read channel. The operation of the PRML channel generates mean squared error (MSE) between the readback signal and the most likely candidate signal. This MSE is representative of absolute value of the position offset of the read head from the track-center. An inverse mapping of the MSE gives rise to two possible values of PES. An algorithm, referred to as ACORN estimator in [95] and [96], uses the MSE from the PRML channel together with the PES reading from the servo channel to provide accurate estimates of position error at high sampling frequency.

Another method of obtaining information on the position of the read head scanning the data block was proposed in the patent [208]. The method proposed in this patent modifies the process of writing by assigning different bit intervals on adjacent tracks. As a result, the readback waveform from the data block of one track differs in frequency or phase from the readback waveform of the adjacent track. Samples from the read waveform are processed by a discrete Fourier transform type algorithm to determine the magnitude of the frequency component associated with the track being scanned. The output of this process provides an indicator to the position of the head. This result is further smoothened using a simple first order filter.

One drawback of these methods of estimating head position from data block is that it can be used only during read operation. The readback waveform is available when a data block is being read and, therefore, can be further processed to estimate PES at high sampling rate. However, during a write operation, the write head is enabled and read sensor is disabled. The readback signal is not available for any kind of processing and the data track is continuously being modified with the new data overwriting the old pattern of magnetization. Realization of either of the two methods would require major change in the head-slider configuration by inserting an additional read sensor for servo only.

2.4   High Frequency Dynamics

The rigid body model of equation 2.7 represents the dynamics in the mid-range frequencies, from about 50-60 Hz to approximately 1.5 kHz, for most actuators used in modern hard disk drives. However, the frequency response of a practical VCM actuator shown in Figure 2.10 suggests that the structure of actuator is anything but rigid. The frequency response measured in frequencies above 1.5 kHz shows large gain and phase changes in a narrow range of frequencies indicating presence of dynamic modes with low damping coefficients. These flexible modes of the actuator are contributions from various bending modes, torsional modes and sway modes of the suspension, the VCM coil, and the gimbal with which the slider is attached to the suspension. The torsional mode of the suspension twists it along the center line of the load beam causing a small amount of in-plane head motion. The first torsional mode of commercially available suspensions lies typically around 3 kHz. There is a second torsional mode in the frequency range between 5 kHz and 8 kHz. The sway mode is caused by the in-plane deformation of the load beam; it is the result of in-plane bending of the suspension. It was explained earlier that part of the load beam is left without edge so that the suspension arm has necessary compliance to accommodate vertical disk runout. This section of the load beam is the weakest part of the suspension. The sway mode produces large amount of radial motion and contribute to the off-track error. Typical frequency of the sway mode lies in the range of 8 kHz - 12 kHz. The sway mode is a greater problem in the rotary actuator as the suspension in this case is moved side-to-side unlike the linear actuator whose movement is along the length of the load beam. Modes other than the torsion and sway modes have small amplitude or lie in frequencies very far away.

It is hard to control precisely the position of the head using an actuator with lightly damped resonant modes, which are subject to variation from drive to drive and in a single drive over time. As a result, these resonances limit achievable bandwidth of the servomechanism. Servo engineers try to eliminate the effects of these resonances by using notch filters in series with the amplifier driving the VCM. But it is not very effective to use analog notch filters when the resonance frequencies are not fixed. Application of digital filters is suitable for adaptation to such variations, but implementation of digital notch filter is restricted by relatively low sampling frequency in hard disk drives. This led to the design of multi-rate notch filter for HDD servomechanism. Design of notch filter as well as multi-rate compensator are explained in details in chapter 3. An alternative solution to the problems of resonance is to use other materials that make stiff and yet light weight suspension arms.

The relatively heavier part of the actuator to which all the suspension arms are attached is known as the E-block. If it is rigid then the dynamics of one suspension is not coupled to that of another. In such case, each suspension resonates by itself and does not interfere with other arms. Frequency and damping of the resonant mode of one suspension arm are slightly different from the resonant frequency and damping of another arm due to manufacturing tolerances. If the E-block is flexible, resonant modes of different suspensions interact with one another degenerating into two new modes - an in-phase mode at lower frequency and an out-of-phase mode at higher frequency. The in-phase mode further limits the bandwidth of the servo system.

2.5   Noise and Disturbances

The disturbances affecting the performance of HDD head positioning servomechanism are contributed by some internal components of the drive as well as sources external to the HDD. The external disturbances are typically in the form of shock and vibration that come from the environment. For example, a moving vehicle or a machine running in a factory or an accidental hit by the user of desktop computer or carrying a laptop computer causes the servo loop to be subjected under external forces. Besides these external disturbances, drive components and their interconnections also give rise to several disturbing forces acting on the control loop. Even the interaction of the aerodynamic forces with different structures inside the drive enclosure affects the performance of the servomechanism.

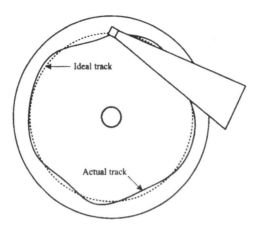

Figure 2.21: Effect of runout: deviation of true track from the ideal circular track.

2.5.1 Repeatable and Non-repeatable Runout

Some internal sources give rise to disturbing forces that are repetitive in nature. Most of the lateral and vertical movements of the fast-spinning disk platters appear as repetitive in the sense that they have definite temporal pattern and they remain the same every time the disks are spun. These disturbances are contributed mostly by mechanical factors. For example, disturbances caused by misalignment between center of the spindle shaft and the disk center, wobbling of the disk platters, vibration modes of the disk platters, defects in the inner and outer races of the ball-bearing etc, repeat with revolution. The misalignment between the disk-center and shaft center makes the shape of a track elliptical and not circular. Wobbling and vibration of the disk cause the track move away or come closer to the head slider affecting the PES. A defect in the inner or the outer race of bearing produces a lateral force whenever a ball hits that defect. All these disturbing forces occur in every revolution making them repetitive. However, if there is a defect in one ball of the bearing, the lateral force is created only when that defect comes in contact with the inner or the outer race. Occurrence of such event is rather random. Similarly flutter caused by the aerodynamic interaction of the fast flowing air with the mechanical components is also random in nature. These internal disturbances, whether repeatable or not, are always present in the HDD servomechanism. As a result, if a track is created by holding the write head steady while the disk spins, the point below the head forms a trajectory that is not perfectly circular but a wavy circle as shown in Figure 2.21.

The mechanical disturbances mentioned above, such as bearing defects or spindle-disk misalignment, are also present during the process of *Servo Track Writing* (STW) when the servo sectors, i.e., the reference marks for the tracks, are written on the disk. (More in-depth explanation of the process of servo track writing is given and various mechatronic challenges associated with it are discussed later in chapter 5.) The burst patterns created in the servo sectors form the reference of the track center, and this reference is used by the servo control loop of HDD during its normal operation. Because of the disturbances present during servo track writing, the created reference for the track-center deviates from being perfectly circular in shape. However, same actuator and disk-spindle assembly are used in both STW and HDD servo controller for embedded servo scheme. Mechanical imperfections manifested in the form of repeatable and non-repeatable disturbances are present at the time of STW, and therefore, are written-in on the tracks in the form of non-circular shape of the track reference. The repeatable disturbances are the same on all tracks. Non-repeatable disturbances are however, written-in differently on different tracks. So a track created by the STW records the non-repeatable disturbances present while that particular track was written.

If the head is positioned on any servo-written track, it follows naturally the deviations from perfectly circular track caused by repeatable mechanical disturbances present during STW. In other words, repeatable disturbances present at the time of STW disappear during the operation of HDD servomechanism. Deviations caused by non-repeatable mechanical distortions at the time of STW are already written-in and appear in the HDD servomechanism as disturbances that are synchronized with the spindle speed. These deviations in the track reference constitute the *repeatable runout* (RRO) for the head positioning servomechanism. The non-repeatable mechanical disturbances are still present during the operation of HDD and they cause deviations of track-center reference with respect to the position of the head slider in a non-repetitive manner. These factors constitute the *non-repeatable runout* (NRRO) affecting the performance of the head positioning servomechanism. The deviations in two servo-written tracks are illustrated in Figure 2.22. The references of the servo-written tracks are shown using solid lines in this figure. The head is expected to follow these references. It should be noted that these mechanical distortions of the track are to be followed, not rejected, by the head positioning servomechanism. However, due to limited bandwidth and presence of random noise and disturbances, the read head can not follow the track-center perfectly. The envelope around the path followed by the head is shown by dashed lines in Figure 2.22. Since the track-references created by the STW is not perfectly circular and since the non-repeatable variations of the references of one track may differ from that of the adjacent tracks, it is possible to have one track encroaching another at some points. Track misregistration due to this factor of track-to-track "squeeze" is known as *write-to-write track Mis-Registration* or WWTMR. When a head tries to follow the center of a track defined by the servo patterns written by STW, it wanders around the center of the track because of various disturbances affecting the servo loop. This misregistration between the reference and the read-write head is known as *write-to-read track Mis-Registration* or WRTMR.

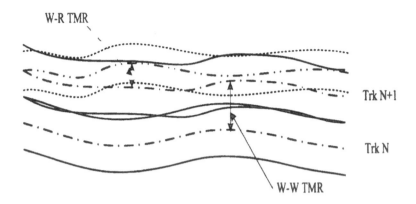

Figure 2.22: Write-to-Write and Write-to-Read Track Misregistration.

2.5.2 Pivot Friction

The friction of the actuator pivot has detrimental effect on the performance of the HDD servo control loop. This problem, which is particularly serious in small form factor drives, demands attention of mechanical engineers as well as control engineers. For the drives in early days, the inertia of the actuator arm was larger than that of today's actuators and the specifications for position error during track following was less stringent; such specifications could be met without paying much attention to the problem of friction. Inertia plays the dominant role in the dynamics of large actuators, and therefore effect of friction is negligible. With the continuous development of HDD, the actuator technology experienced an appropriate evolution. To meet the demand for smaller size of drives and faster response, the inertia of the actuators has to be reduced and the issue of friction becomes more noticeable. The friction of the pivot produces a torque that opposes the torque generated by VCM. The friction torque is a nonlinear function of actuator's velocity and position. In the frequency domain, the effect of friction is manifested as reduction in gain at low frequency, below 100 Hz in most of the actuators used. Reduction in gain depends on the amplitude of the excitation signal used to measure the frequency response.

Performance of the track-following servo controller can be significantly enhanced by proper compensation of the pivot friction. The disturbance torque arising from pivot friction can be recreated from the knowledge of inertial torque and actual acceleration of the actuator. Inertial torque is proportional to the VCM coil current. By including two additional sensors, one for coil current and the other for acceleration, friction torque can be estimated. In [99], the difference between the coil current and acceleration of the actuator is filtered through a disturbance filter. The output of this filter is an approximation of the friction torque and can be used to cancel the effect of pivot friction.

An alternative approach is to get a model of friction off-line and then use the model on-line to estimate the friction force. Classical models of friction include preload model and Dahl model [40] as defined by the equations below.

Preload model: In this model, the friction force is expressed as,

$$F = F_c sgn(v) + F_v v \qquad (2.30)$$

F_c and F_v are the Coulomb friction and viscous friction, respectively. Coulomb friction is constant irrespective of the velocity (v) of motion, but the magnitude of viscous friction increases with increasing velocity.

Dahl Model: This model defines the derivative of friction force with respect to position as,

$$\frac{dF}{dy} = \sigma |1 - \frac{F}{F_c} sgn(\dot{y})|^i \qquad (2.31)$$

where,
σ = rest slope,
F_c = rolling torque, and
i = exponential factor.

The preload model of equation 2.30 is independent of the position variable (y) whereas the Dahl model (equation 2.31) defines the friction as a function of position only. Experimental results suggest that the frictional behavior of the actuator pivot is not determined solely by velocity or position, but by a combination of the two [1], [201]. This conclusion is reached after examining the experimentally obtained frequency response of VCM actuator. The frequency response of the double integrator model is expected to show -40 db/decade slope and $-180°$ phase at the lower end of the frequency axis. But the experimentally obtained frequency response show 0 dB/decade slope and 0° phase at low frequency suggesting a transfer function with no integrator (Figure 2.23). In other words, the friction force, that opposes the applied force, is a function of both position and velocity. The experiments also show variations in the low frequency gain with varying amplitude of the input signal used while measuring frequency response, which requires a nonlinear model to describe the actuator.

The method for finding a nonlinear model of the pivot friction described in [1] exploits the relationship between the frequency response measurement and describing function. Several models have been suggested that take feedback of both velocity and position into consideration. Some of these models are given below.

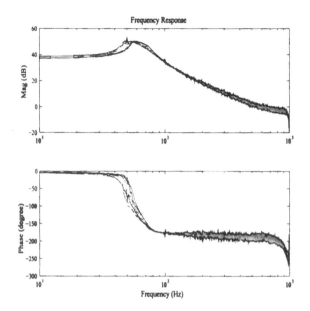

Figure 2.23: Frequency response of actuator with friction.

Preload plus Two-slope Spring Model:

$$F = F_c sgn(v) + F_v v + K_a y; \quad |y| \leq S_a$$
$$F = F_c sgn(v) + F_v v + K_b y + (K_a - K_b)S_a; \quad |y| > S_a \quad (2.32)$$

F_c and F_v are Coulomb friction and viscous friction, respectively. This model uses two different spring constants, one for small displacement and the other for large displacement. The model is linear in all parameters except S_a, the break point in the spring stiffness. The model is found to produce very good match between observed response and model response in frequency domain, but not in the time domain hysteresis curve of input voltage versus displacement.

Classical Dahl Model plus Viscosity: The classical Dahl model does not give a good match between the experimental observations and actuator model response in both time and frequency domains. Adding a viscosity term improves the results with the exception at the point where velocity reversal occurs.

Hysteretic Two-Slope Model: This model combines two-slope spring behavior and the hysteretic type behavior of the Dahl model. This model has only 4 parameters, but is not suitable for on-line identification.

These models show better result than the classical model of friction, but fail to match the response of the system in both time and frequency domain.

One possible reason for the discrepancy is the assumption that the velocity and position feedback are decoupled.

2.5.3 Flex Cable Bias

A flat sheet of plastic, known as *flex cable*, is used to support thin conducting wires carrying signals between the PCB and the heads. The flex cable is attached to the side of the actuator arm. This flat sheet gets deformed when the actuator moves in one direction or the other; it is bent for one direction of movement and is relaxed when the actuator moves in the opposite direction. The deformation in the structure of the flex cable exerts a reaction force on the actuator. This reaction depends on the amount of deformation of the flat sheet and, therefore, is dependent on the actuator's position. However, during the track following mode, the variation in flex cable reaction force is very negligible as the movement of the actuator arm is in the scales of nanometers, and hence, the reaction force can be taken constant during track following.

2.5.4 External Shock and Vibration

These are the disturbances contributed by factors external to HDD. Operating heavy machinery in the proximity of the computer, accidental hit by user, carrying the laptop etc are factors that cause this kind of disturbances. As drives become smaller with lighter actuators and suspensions, these disturbances turn to be prominent issues. The HDD servomechanism must have the capability to reject such disturbances or, at least, to detect the occurrence and suspend reading/writing operation. Typical solution includes the use of an accelerometer to sense the disturbance and move the actuator accordingly. The use of accelerometer feedforward was found to improve the disturbance rejection capability of HDD. But these methods have not been widely practiced because of cost factor and reliability of the accelerometers.

2.5.5 Other Sources of Noise

Besides the noise and disturbances discussed so far, many other sources of noise affect the performance of the servo loop of HDD. Airflow at the interface of disk and head, interaction of airflow with the suspension and actuator arm, noise associated with the PES generation process, quantization noise of the ADC, noise introduced by truncation of words in the processor, DAC, and power amplifier - all of these have detrimental effect on the performance of the servomechanism. Airflow induced disturbances are hard to model but several attempts has been made to obtain them. These models, mostly based on finite element analysis, are good for understanding the process better and to identify the factors contributing to the process. Disk diameter, mechanical properties of the disk, aerodynamics of the slider-disk interface, aerodynamics of actuator, actuator position, properties of the enclosure, air pressure inside the enclosure, mechanical features built inside the enclosure are the factors that determine the nature of the flow and therefore the characteristics of the airflow induced disturbances [91]. The spectrum of these disturbances is also related to the spindle speed. With high spindle speed $(7,200 - 15,000$ RPM) in high performance disk drives, these disturbances have become critical. However, the models available for these disturbances are mostly empirical and their properties are very sensitive to small changes. As a result, solving the problem of flow-induced disturbances remains to be a challenge.

Media noise at the domains of servo bursts introduces a significant amount of noise in the servo loop. The servo burst patterns are nothing but a sequence of magnetic transitions on the media. These transitions, though depicted earlier using straight lines, contain magnetization fluctuations near the recorded transitions. These fluctuations make the readback waveform noisy; this noise is known as *transition noise*. Media also introduces a noise, called *particulate or granularity noise*, caused by random dispersion of the grains in magnetic media. When a read head scans the servo burst patterns, the media noise affects the burst signals. The burst signal is also affected by the MR head noise and the noise from the electronics. It is further aggravated by demodulation, the process to extract PES from the burst signals.

2.6 Track Seek Controller

There are two distinct and often competing objectives to be achieved by the head positioning servo controller in HDD.

- *Track Seek:* track-to-track maneuver of the heads must be performed in as short a time as possible, and

- *Track Following:* the position of the head must be regulated above the center of the track with minimum variance during reading or writing of data.

It may be appropriate to point out here that, on the contrary to the popular use of the term *tracking control* in optimal control literature to represent controllers that track a signal, the track-following in HDD is essentially a regulator problem. Seek control, on the other hand, is a point-to-point control.

It is impossible to meet both the requirements of minimum-time seek and minimum-variance track following using a linear controller when the magnitude of control signal is upper bounded. A linear controller can be optimized to meet the specifications of disturbance rejection and good tracking performance. Gains of the controller required to meet these specifications when the position error is low, typically below 10% of the track pitch, cause actuator saturation when the reference command is several tracks. Performance of the linear control deteriorates under such condition of control saturation. The *time optimal control* or *bang-bang control* is a well known solution for point-to-point maneuver with limited control authority. Practical realization of the bang-bang control has several drawbacks such as *control chatter* and sensitivity to parameter variations. In practice, the head positioning servomechanism of HDD uses variants of bang-bang control that achieve near-time-optimal performance but avoid problems associated with time-optimal control. These solutions often combine time-optimal control for large errors with linear control for small errors and ensure smooth transfer between the two.

2.6.1   Time Optimal Control

The nominal dynamics of the VCM actuator can be modeled as a double integrator

$$\ddot{y} = au, \tag{2.33}$$

with an upper bound on the amplitude of the control input, i.e., $|u| \leq U_m$. Since the magnitude of the input is upper bounded, the magnitude of the acceleration is also limited by a maximum value ($|\ddot{y}| \leq aU_m$). The objective of the seek control is to move the states of the head positioning servomechanism from its initial values to some final values in the shortest possible time. The state vector (x) includes the position (y) of the read-write head and the velocity (v). If we assume the actuator to be at rest before and after such minimum-time maneuver then $v_i = v_f = 0$. So the seek control transfers the state vector from $[y_i \quad 0]^T$ to $[y_r \quad 0]^T$ in minimum time. Defining the position error $e = y_r - y$, the double integrator model is re-written as

$$\ddot{e} = -au. \tag{2.34}$$

Then the time-optimal seek control brings the error from $e(0) = y_r - y_i$ at $t = 0$ to $e(T_f) = 0$ so that T_f is minimized. Simple analysis of Newton's 2nd law of motion reveals that for a double integrator, time-optimal maneuver is achieved by applying maximum acceleration to the actuator for exactly half of its travel and then decelerating with maximum effort for the remaining half of the travel. This is equivalent to driving the actuator with $+U_m$ (or $-U_m$) for half of the travel followed by $-U_m$ (or $+U_m$) for the remaining half. The sign of the accelerating or decelerating input depends on the direction of displacement the head is expected to go through. Many text books such as [23] and [11] provide in-depth analysis of the time-optimal control law, not only for a double integrator model but also for a general plant model.

If the actuator of equation 2.34 is subject to maximum acceleration for T second followed by maximum deceleration for another T second then the total displacement of the read/write head is aU_mT^2 (according to Newton's 2nd law of motion). For a given initial error $e(0)$, the time required to complete a time-optimal maneuver is

$$T = \sqrt{\frac{e(0)}{aU_m}}. \tag{2.35}$$

Open loop realization of the bang-bang control involves (1) pre calculation of T of equation 2.35, (2) applying maximum acceleration for $T/2$ seconds and (3) switching to deceleration and applying maximum deceleration for the remaining $T/2$ seconds. Closed loop realization, on the other hand, controls the velocity to follow a pre-defined deceleration velocity profile which is a function of error

$$f_{toc} = -sgn(e)\sqrt{2aU_m|e|}. \tag{2.36}$$

When the actuator is subject to maximum deceleration from any initial state, the state trajectory follows a path parallel to this profile shown in the phase plane of Figure 2.24. The solid line in this figure represents the deceleration velocity profile. The trajectories x_1 and x_2 (dashed lines) represent vectors of error and error velocity for 2 different initial conditions but both with $u = +U_m$, while the trajectories x_3 and x_4 (dotted lines) are traced when $u = -U_m$.

For $u = +U_m$, the trajectories from any initial condition follow a path parallel to that of x_1 or x_2. The upper part of the deceleration error profile is also a trajectory with $u = +U_m$ and it passes through the origin of the phase plane. Similarly, $u = -U_m$ from any initial condition traces a trajectory parallel to that of x_3 or x_4, and the lower part of the deceleration trajectory is one of them. From any initial condition not lying on the deceleration velocity profile, it is possible with appropriate polarity of input u to make the trajectory eventually intersect the deceleration profile. If $u = -U_m$ is applied when the initial condition lies anywhere in the region below the deceleration profile, the trajectory will intersect in finite time with the positive segment of the deceleration profile. Similarly for any initial condition above the solid line, $u = +U_m$ ensures intersecting the negative half of the velocity profile. If switching from acceleration to deceleration is performed when the state vector $[e \ \dot{e}]^T$ is exactly on the deceleration profile, then the trajectory moves along this curve and reaches the origin $[0 \ 0]^T$ in finite time.

The actuator can be assumed to be at rest at the beginning of a track-to-track seek maneuver, that is, the initial state lies on the horizontal axis. If maximum acceleration is applied, the state trajectory follows a parabolic path until it meets the deceleration profile when control is switched to the decelerating phase so that the trajectory follows the deceleration velocity profile and reaches $[0 \ 0]^T$ in finite time. The control input is set to zero as soon as the origin is reached. The control signal is generated using the rule

$$u = U_m sgn(f_{toc} - \dot{e}), \tag{2.37}$$

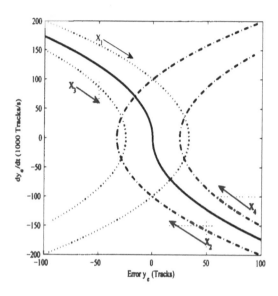

Figure 2.24: Phase plane diagram for time optimal control of double integrator.

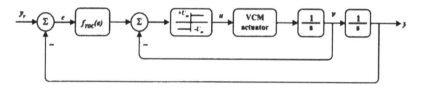

Figure 2.25: Time optimal control: Block diagram.

where

$$sgn(s) = \begin{cases} 1, & x > 0 \\ -1, & x < 0,. \\ 0, & x = 0 \end{cases} \qquad (2.38)$$

Time optimal bang-bang control is very sensitive to noise and variation in actuator parameters. Any small change in the acceleration constant a or any noise in the measurement of output y makes the control signal switch between two extreme levels, a phenomenon commonly known as *control chatter*. Moreover, the above mentioned control law is derived for a double integrator model. A model with a real-axis pole and an integrator is a more appropriate choice for majority of the electromechanical actuators used in practice, including the VCM actuator. These actuators also manifest high frequency torsion and bending modes. Though it is not impossible to derive theoretically the time optimal law for such models, these formulae become increasingly more complex and more difficult for practical realization for higher order plants. In all

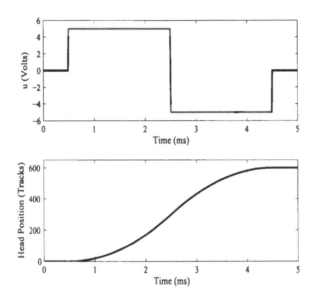

Figure 2.26: Time optimal control: Simulation result.

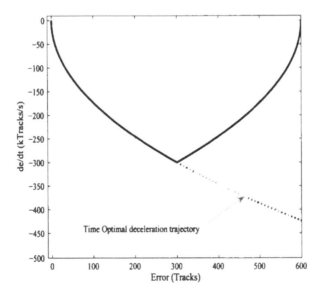

Figure 2.27: Time optimal control: Phase Plane (Simulation result).

practical situations, a near time-optimal solution is implemented by combining time-optimal control for large errors, where it make sense, with linear control for small errors, and ensuring smooth transition between the two modes. One such solution, *Proximate Time Optimal Servomechanism*, was first proposed in 1987 [209]. Realization of the PTOS for continuous-time and discrete-time systems were presented in [210] and [211], respectively.

2.6.2 Proximate Time Optimal Servomechanism

The proximate time-optimal servomechanism (PTOS) provides a smooth, bump-free merger of linear state feedback control for small errors with nonlinear time-optimal control for large errors. Linear state feedback controller is used for position errors bounded by an upper limit, i.e., $|e| \leq e_l$,

$$u_l = Kx = [k_1 \quad k_2] \begin{bmatrix} x_1 \\ x_2 \end{bmatrix} = k_1 x_1 + k_2 x_2. \tag{2.39}$$

The state variables are the position error, $x_1 = e = y_r - y$, and the error velocity $x_2 = \dot{e} = -\dot{y}$. For error larger than e_l, an approximation of time optimal solution is applied. To implement this, the velocity profile and control law are modified according to

$$f_{PTOS} = \begin{cases} sgn(e)(\sqrt{2a\alpha U_m |e|} - \frac{U_m}{k_2}), & |e| > e_l, \\ \frac{k_1}{k_2} e, & |e| \leq e_l \end{cases}, \tag{2.40}$$

and

$$u = U_m sat \left\{ k_2 \frac{f_{PTOS} - x_2}{U_m} \right\}. \tag{2.41}$$

The saturation function $sat(w)$ is defined as

$$sat(w) = \begin{cases} sgn(w), & |w| \geq 1, \\ w, & |w| < 1. \end{cases} \tag{2.42}$$

The state feedback gains are selected such that the conditions given in equation 2.43 are satisfied. These conditions ensure continuity of the velocity profile as well as its first derivative with respect to e at $|e| = e_l$.

$$k_2 = \sqrt{\frac{2k_1}{a\alpha}} \tag{2.43}$$

$$e_l = \frac{U_m}{k_1}.$$

The acceleration discount factor α ($\alpha \leq 1$) in equation 2.40 allows us to accommodate uncertainty in the acceleration factor a of the plant. Using $\alpha < 1$ however makes the response slightly slower. With $\alpha = 1$, the nonlinear part of

f_{PTOS} reduces to the velocity profile of time-optimal control with magnitude shifted by U_m/k_2.

The velocity profile of equation 2.40 and the control law of equation 2.41 divides the phase plane into 5 regions. These are labeled in Figure 2.28 as,

S^+ : area above the upper dash-dot line,
S^- : area below the lower dash-dot line,
U^+ : area bounded by two dash-dot lines and on the right of e_l,
U^- : area bounded by two dash-dot lines and on the left of $-e_l$,
L : area bounded by two dash-dot lines and two vertical dotted lines.

When the state trajectory is inside the linear region L, the velocity profile is a linear function of the error ($f_{PTOS} = \frac{k_1}{k_2}e$), and the control signal is $u = k_1e+k_2\dot{e}$ which is same as the linear state feedback control of equation 2.39 $u_l = k_1x_1 + k_2x_2$. The areas labeled S^+ and S^- are the areas of saturated control. If the state trajectory is in one of these areas, then maximum control effort is applied (U_m in S^+ and $-U_m$ in S^-). Such an input forces the state trajectory move in a parabolic path to reach eventually the band around the velocity profile f_{PTOS}, marked U^+ or U^- or L. The control signal is unsaturated when the trajectory is inside the areas marked U^+ or U^-, but it is computed using nonlinear feedback rule. A trajectory, once inside U^+ or U^-, remains inside this band and finally enters the linear region L in finite time. Nature of the trajectory inside the region L depends on the state feedback gains k_1 and k_2.

Simulated results for $PTOS$ applied to a VCM actuator are shown in Figure 2.29. The double integrator model identified in section 2.2.2 is used for the simulation,

$$G(s) = \frac{k}{s^2} = \frac{3.6 \times 10^7}{s^2} \quad \left(\frac{\mu m}{Volts}\right). \tag{2.44}$$

It is very common in the HDD industry to use *Tracks* as the unit of displacement instead of *metres* or μm; the simulation results shown here also use this unit.

Figure 2.29 shows displacement, control input, velocity, and phase-plane for a 100-track seek. It is evident from the plot of the control input (u) that saturated control is used only during the acceleration phase. Once the phase-plane trajectory enters the band of unsaturated control, the control signal is less than the maximum value and the control loop tries to make the trajectory follow the deceleration profile of PTOS. Trajectory, deceleration profile and the upper and the lower bounds are illustrated in Figure 2.30. Note that the phase plane is plotted as velocity of the head-slider versus position error, not $\frac{de}{dt}$ versus e shown in Figure 2.28.

If *tracks per second* is used as the unit of the velocity of head-slider then the magnitude of velocity is extremely large. It is a common practice to express

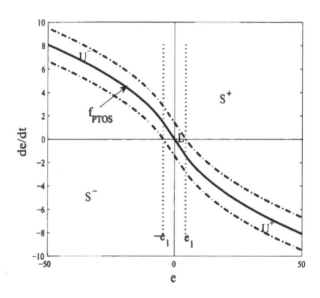

Figure 2.28: Phase plane diagram for proximate time optimal control.

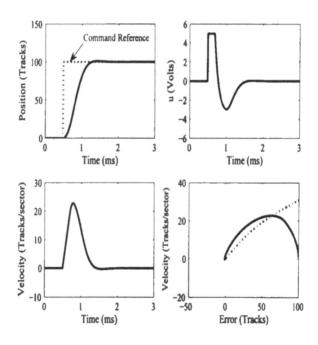

Figure 2.29: Proximate time optimal control: Simulation results.

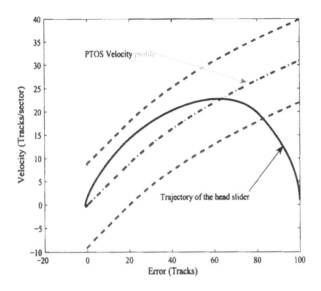

Figure 2.30: Phase plane for PTOS: Simulation result.

head velocity in *tracks per sector*, which is followed in the simulation results shown. Position error for the 100-track seek is shown in Figure 2.31 with a magnified scale so that settling behaviour of the seek is clearly seen. As mentioned earlier, position error less than 10% or 15% of a track is considered as the end of seek. For the simulation of 100-track seek, the error enters into the band of ±10% at approximately 1.835 ms. Since the seek was initiated at 0.5 ms, time taken to complete a 100-track seek is approximately 1.335 ms. Parameters of the PTOS controller are not designed optimally for the example shown here. These simulation results are intended for an explanation of the PTOS design, and the parameters are selected such that the linear controller is applied for error less that 20 tracks, i.e., $e_l = 20$.

The state feedback controller, which is used for the linear part of PTOS, does not perform well in HDD during track following as it fails to reject the effect of input disturbance. The input disturbance in VCM actuator originates mainly from the flex cable. Rejection of such input disturbance requires high gain at low frequency which can be realized using integral control. But the linear part of the PTOS is PD (*Proportional plus Derivative*) control and does not increase the low frequency gain substantially. But the input bias can be estimated using an observer. It should be pointed out here that use of observer is very much essential in HDD servomechanism as position is the only measured state variable. The velocity of the read-write head slider, a variable required by the seek control algorithm, must be estimated. Besides estimating the input bias, its effect can also be cancelled by implementing a state feedback controller

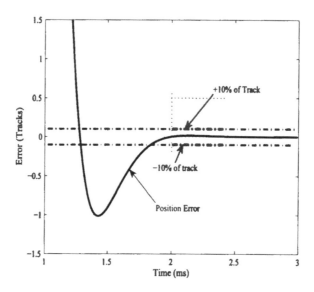

Figure 2.31: Position error for 100-track seek: Simulation result.

that includes integral control. These two methods are explained later in this section.

The State Estimator

Considering the rigid body model ($\frac{a}{s^2}$) of the VCM actuator, corresponding continuous-time state space model is

$$\frac{dp}{dt} = \begin{bmatrix} 0 & 1 \\ 0 & 0 \end{bmatrix} p + \begin{bmatrix} 0 \\ a \end{bmatrix} u, \quad y = [1 \ \ 0] p, \quad (2.45)$$

where $u(t)$ and $y(t)$ are the input and output of the VCM, respectively. The state vector $p = [p_1 \ \ p_2]^T$ includes the position (p_1) and velocity (p_2) of the head-slider effectuated by VCM's motion. In this state-space model, the unit of velocity is tracks per second. Considering the speed of motion in HDD servomechanism and the magnitude of displacement (maximum velocity can be hundreds of thousand tracks per second for corresponding displacement of few tens of tracks), this representation results in large magnitude disparity between the two states. As a result, one of the feedback gains becomes very large compared to the other. Such case requires extra care while implementing the controller using fixed point processor. The issue can be resolved by suitable transformation of variables. If we redefine the velocity in units of tracks per sector (time between consecutive sectors is few μs only), its magnitude is scaled down by a factor of $\frac{1}{T_S}$, where T_S is sampling interval. This is equivalent to transforming state vector p into a new state vector x as,

$$x = \begin{bmatrix} x_1 \\ x_2 \end{bmatrix} = \begin{bmatrix} 1 & 0 \\ 0 & T_S \end{bmatrix} \begin{bmatrix} p_1 \\ p_2 \end{bmatrix} \qquad (2.46)$$

where, x_1 and x_2 are state variables in units of tracks and tracks/sector, respectively. The sampling interval T_S is the same as the time between two servo sectors in HDD. Then the transformed state equation is,

$$\frac{dx}{dt} = \begin{bmatrix} 0 & \frac{1}{T_S} \\ 0 & 0 \end{bmatrix} x + \begin{bmatrix} 0 \\ aT_S \end{bmatrix} u, \quad y = [1 \quad 0]x, \qquad (2.47)$$

Corresponding discrete-time state space model is

$$\begin{aligned} x(k+1) &= \Phi x(k) + \Gamma u(k), \\ y(k) &= Hx(k). \end{aligned} \qquad (2.48)$$

This nominal model is good enough for designing an observer that can estimate the two states, position and velocity, required by the PTOS algorithm. But we also want to estimate the input disturbance. Augmenting the nominal state space model to include the disturbance as an additional state is a common practice. In the case of HDD actuator, the input disturbance is assumed constant during the linear range of operation. If w is the input disturbance, then $w(k+1) = w(k)$. Combining this with the nominal discrete state space model of equation 2.48 we get,

$$\begin{bmatrix} x(k+1) \\ w(k+1) \end{bmatrix} = \begin{bmatrix} \Phi & \Gamma \\ 0 & 0 \end{bmatrix} \begin{bmatrix} x(k) \\ w(k) \end{bmatrix} + \begin{bmatrix} \Gamma \\ 0 \end{bmatrix} u(k) \qquad (2.49)$$

$$y(k) = [H \quad 0] \begin{bmatrix} x(k) \\ w(k) \end{bmatrix}$$

or

$$z(k+1) = \Phi_a z(k) + \Gamma_a u(k); \quad y(k) = H_a z(k). \qquad (2.50)$$

Here $z = [x^T \quad w]^T$ is the augmented state vector, $\Phi_a = \begin{bmatrix} \Phi & \Gamma \\ 0 & 0 \end{bmatrix}$, $\Gamma_a = \begin{bmatrix} \Gamma \\ 0 \end{bmatrix}$, and $H_a = [H \quad 0]$. The prediction observer using this model is

$$\bar{z}(k+1) = \Phi_a \bar{z}(k) + \Gamma_a u(k) + L_p[y(k) - H_a \bar{z}(k)], \qquad (2.51)$$

where $\bar{z}$ is the estimate of the augmented state vector z. The observer gain L_p must be chosen to satisfy the stability condition, i.e., all eigenvalues of $\Phi_a - L_p H_a$ are inside the unit circle. There are many standard methods to select the observer gains such as Ackerman's formula.

2.6.3 Rejection of Input Disturbance

Integral control is the most widely acknowledged solution for eliminating steady state error. The state-space design presented earlier in conjunction with PTOS does not automatically include integral action. Two possible augmentations of the state-space design can be used to nullify the effect of input bias:

- Add an integrator and augment the plant model by considering the output of integrator as a new state, and

- Augment the state space model of the open loop plant by taking input bias as a new state.

These methods are briefly explained below and realizations of PTOS with these augmentations are shown.

Realization of PTOS with Integral Control

The first of the two methods for rejection of input disturbance includes an integrator in the feedback path to generate a new state, the integral of error. For the discrete-time state space model of equation 2.48, the error is

$$e(k) = y_r(k) - y(k), \qquad (2.52)$$

where, $y_r(k)$ is the reference input. Let us define the integral of this error as a new state $x_I(k)$. Then according to the discrete realization of integrator

$$x_I(k+1) = x_I(k) + y_r(k) - y(k). \qquad (2.53)$$

This is equivalent to,

$$x_I(k+1) = x_I(k) - Hx(k) + y_r(k). \qquad (2.54)$$

Combining equations 2.48 and 2.54, we get the augmented state space equation,

$$\begin{bmatrix} x(k+1) \\ x_I(k+1) \end{bmatrix} = \begin{bmatrix} \Phi & 0 \\ -H & 1 \end{bmatrix} \begin{bmatrix} x(k) \\ x_I(k) \end{bmatrix} + \begin{bmatrix} \Gamma \\ 0 \end{bmatrix} u(k) + \begin{bmatrix} 0 \\ 1 \end{bmatrix} y_r(k),$$

$$y(k) = [(H \quad 0)] \begin{bmatrix} x(k) \\ x_I(k) \end{bmatrix}. \qquad (2.55)$$

The state feedback gain K can be selected such that the eigenvalues of the following matrix satisfy the desired design specifications,

$$\begin{bmatrix} \Phi & 0 \\ -H & 1 \end{bmatrix} - \begin{bmatrix} \Gamma \\ 0 \end{bmatrix} K. \qquad (2.56)$$

Realization of PTOS with integral control included is shown in Figure 2.32.

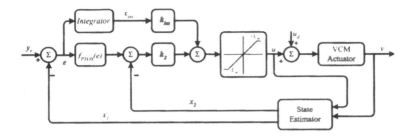

Figure 2.32: Schematic diagram of PTOS with integral control.

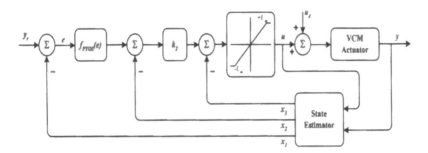

Figure 2.33: Schematic diagram of PTOS with bias estimator.

Realization of PTOS with Bias Estimator

It is explained earlier that an augmented state observer can estimate the states of the actuator as well as the input disturbance. The estimates of the states x_1 and x_2 are used to generate the control signal using the feedback law of the PTOS algorithm. However, it should be noted that although the augmented system is observable, it is not controllable. The input bias can be estimated by the augmented observer but can not be controlled through feedback. Even though the augmented observer gives estimates of three states, the state feedback control is still designed using a second order model of the actuator with the states x_1 and x_2 only. The third state of the observer, the estimate of bias input, must be subtracted directly from the VCM input signal to cancel the effect of input bias. The controller equation for the linear state feedback is

$$u(k) = -K\bar{z}(k) = -k_1\bar{x}_1(k) - k_2\bar{x}_2(k) - \bar{x}_3(k). \qquad (2.57)$$

The first two component of the control equation is exactly the same as that of the linear part of PTOS. The block diagram of Figure 2.33 illustrates the realization of the PTOS with bias estimator.

Both integral control and bias estimation can eliminate the effect of input bias. The method with an integrator, uses an extra state in the feedback; this additional state is not estimated by the state estimator but generated by

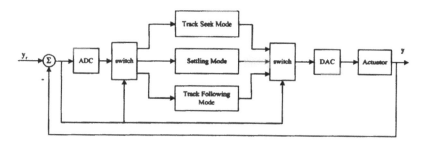

Figure 2.34: Schematic diagram of mode switching control.

integrating the error signal. Hence the order of the estimator remains the same as that of equation 2.51. For HDD, a second order estimator is required for estimating position and velocity. In that case, the estimator gain L is a 2×1 column vector and the feedback gain K is a 1×3 row vector. On the contrary, the estimator in the second method uses an augmented model increasing the order of the estimator by 1, but the feedback gain for the additional state of the estimator is fixed as 1. For the case of HDD, one needs to find a 3×1 column vector as the estimator gain L. And the feedback gain K, a 1×2 row vector, should be found using the plant model prior to augmentation. Several well defined and efficient algorithms are available for finding these gains [54], [152].

2.6.4 Mode Switching Control with Initial Value Compensation

The *Mode Switching Control* is an approach that uses several controllers with different structures and a switching function that decides changing from one controller to another depending on certain conditions in the plant and control states. This scheme, illustrated in the block diagram of Figure 2.34, can be used in HDD servomechanism to design the track seek control and the track following control independently.

The two modes of operation in HDD servomechanism requires different set of objectives to be met. The seek controller must be designed such that the actuator is moved from one track to another as fast as possible with due consideration to the power amplifier saturation. On the contrary, the track following mode is required to make the position error as small as possible. It is relatively easy to design and optimize the controllers one at a time without taking the other into consideration. The seek controller can be either a non-linear controller or a linear, two-degree-of-freedom controller to achieve a near time-optimal performance. Track-following controller, on the other hand, should be optimized to achieve a bandwidth as high as possible so that tracking error is minimized. A linear controller such as PID or H_∞ control can be used for precise head positioning during track-following. When the two controllers are designed independent of one another, continuity of the control signal is not guaranteed at the time of switching from one mode to the other.

A controller is usually designed with an assumption that the initial values of the states are zero. This assumption is not valid for the HDD servomechanism. When the states of the closed loop system move from one mode to the other, the initial states are not guaranteed to be zero-valued for the second controller. It may be pointed out that, in HDD servomechanism, switching always takes place from track-seek mode to track-following mode and not *vice versa*. Therefore, one can design the track-following controller using a technique that takes the non-zero initial conditions into consideration. The mode switching control with *Initial Value Compensation* (IVC) is a method that ensures smooth hand over between the two controllers [218]. The validity of the MSC-IVC design depends on two assumptions - (i) the track-following controller is a stable single-input single-output system, and (ii) the transfer functions of the plant and the controller are proper. The MSC-IVC design will be elaborated later in chapter 3.

Let the plant and controller be described by the following discrete-time state equations.

$$
\begin{aligned}
X_p(k+1) &= A_p X_p(k) + B_p u(k), \\
y(k) &= C_p X_p(k), \\
X_c(k+1) &= A_c X_c(k) + B_c(r(k) - y(k)), \\
u(k) &= C_c X_c(k) + D_c(r(k) - y(k)),
\end{aligned}
\tag{2.58}
$$

where X_p and X_c are the $m^{th} - order$ and $n^{th} - order$ state vectors of the plant and controller, respectively. The variables $u(k)$ is the control input to the plant, $y(k)$ is the output of the plant, and $r(k)$ is the reference. The matrices and vectors $A_p, B_p, C_p, A_c, B_c, C_c$ and D_c are of appropriate dimensions. The instant of mode switching is the sampling instant $k = 0$. During the track following mode, $r(k)$ is the reference of the track-center which is not measured directly. It is the error signal $e(k) = r(k) - y(k)$ that is available in the HDD servomechanism. In transform domain, the system of equation 2.58 can be written as

$$
Y(z) = \frac{N_p(z)}{D(z)} X_p(0) + \frac{N_c(z)}{D(z)} X_c(0)
\tag{2.59}
$$

where N_p, N_c and D are polynomials in z of appropriate dimensions. The states of the track-following controller can be initialized to desired initial values using a real coefficient matrix (K_{ivc}) such that $X_c(0) = K_{ivc} X_p(0)$. Then

$$
Y(z) = \frac{N_p(z) + K_{ivc} N_c(z)}{D(z)} X_p(0).
\tag{2.60}
$$

An appropriate selection of K_{ivc} can improve the transient characteristics followed by the mode-switching. The track following controller is designed to achieve better steady-state response. The poles in the track-following controller that are desirable for steady-state properties but not for transient can now be canceled by proper design of the matrix K_{ivc}.

2.6.5 Suppression of Residual Vibration

It explained earlier in this chapter that the lightly damped resonant modes of the head positioning actuator are inevitable. The problem is further aggravated by the present trend of making actuators lighter. Reduction of mass is an important consideration during designing the actuator as it makes movement of the read/write head faster using the available torque. Power consumption, an important issue in HDD for mobile and consumer electronics applications, is another factor that demands reduced moving mass. As a result, a light and sleek actuator always remains in the wish list of the designer.

Reduction of actuator mass however comes with a cost to pay — its vibration modes become more prominent. The achievable bandwidth of the servomechanism is upper bounded by the frequencies of these lightly damped modes. The mechanical vibrations are usually damped modes and, therefore, eventually decay to zero; but exciting these modes increases the seek time. No matter how fast the seek controller brings the read/write head to the destination track, operation of reading or writing can not be initiated unless the residual vibration is sufficiently low in amplitude. Suppression of the residual vibrations is an important step towards realizing a very high speed access mechanism.

Notch Filter or Band-reject Filter

A *notch filter* or *band-reject filter* in series with the VCM actuator can solve the problem of residual vibration. Such filter inhibits the band of frequencies around the resonant frequency from reaching the read/write head. If an analog filter is used, it is often integrated into the VCM driver circuit. But such filter, once designed, becomes a filter with fixed properties. Unfortunately, the actuators being mass produced, the frequencies of resonant modes vary from actuator to actuator. Resonant frequencies may also change due to many other factors, for example, changes in operating conditions, age of the components etc. To address this issue, a notch filter that can accommodate such variations in the resonant frequencies should be used. Digital filter is preferred over its analog counterpart as it comes with better flexibility to accommodate changes. Notch filters not only attenuates the frequencies around the frequency of resonance but also modifies the phase of the open loop transfer function. Care must be taken while including these filters in series with the plant. Design of notch filters for the HDD servomechanism is further explained in section 3.2.2.

Command Shaping

Use of notch filter improves the performance of the servomechanism during track-following, but it does not solve the problem when the resonances are excited by large jumps in the control signal from one polarity to another during the execution of any seek algorithm, e.g., the PTOS. Such excitation of the resonant modes is inevitable, but as long as they diminish before the end of seek, the tracking performance is not affected much. With continuing demand for lighter actuators, the resonant modes continue to be less and less damped and, therefore, it takes longer for the modes to decay. The problem of switching induced vibration is more severe for short seek as the seek algorithm ends before the vibration is reduced to negligible amplitude. These vibrations also extend the access time as it takes longer for the heads to settle on the target track.

The switching-induced vibration can be removed using a scheme called *Command Shaping* or *Input Shaping* [176]. In this method, the reference track is not changed abruptly at the beginning of the seek operation, but a filtered reference command is injected into the servo loop. This command generation scheme convolves the reference command with a sequence of impulses called the *input shaper* to create a command signal that cancels the vibration produced by the system to which it is applied. The concept of the command generation is explained next using a simple example.

If an impulse input is applied to an underdamped second order system at $t = 0$ then its response is a decaying sinusoid. After the application of the first impulse, if a second impulse, with suitable relative amplitude and phase with respect to the first, is applied, the oscillation due to the first impulse can be completely eliminated by the response to the second impulse. The relative amplitude and the phase of the second impulse can be determined considering the resultant output as the superposition of the responses to two impulses such that the condition for no residual vibration is satisfied. The relative amplitude and phase depend on the natural frequency and the damping coefficient of the resonant mode to be suppressed. Let us consider the simplest case with only one resonant mode. If the undamped natural frequency and the damping coefficient are ω_0 and ζ, respectively, then two impulses of magnitude $\frac{1}{1+K}$ and $\frac{K}{1+K}$ separated by ΔT satisfy the condition of zero-vibration for $\forall t > \Delta T$ where,

$$
\begin{aligned}
K &= e^{-\frac{\zeta \pi}{\sqrt{1-\zeta^2}}} \\
\Delta T &= \frac{\pi}{\omega_0 \sqrt{1-\zeta^2}}.
\end{aligned}
\tag{2.61}
$$

This sequence of impulses can be convolved to an arbitrary reference command to obtain vibration reduced output. For example, if the actual reference command is a step function, the convolved reference becomes a staircase signal. The responses of an under damped second order system to a step input and the corresponding staircase input are shown in Figure 2.35. The dotted lines

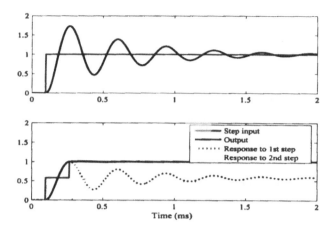

Figure 2.35: Cancelation of residual vibration by preshaping.

in the lower plot are the oscillatory responses to two step functions, and the solid line shows the resultant. The output of the system to the staircase input is vibration free after the second impulse is applied. It is obvious that values of the relative amplitudes of impulses and phase between them, i.e., the values of K and ΔT must be chosen exactly to achieve vibration-free response. Any change in these values fails to suppress the vibration completely. Accurate knowledge of the resonant mode is essential to make this method effective.

Command shaping method described above is very sensitive to parameter variation, and it is not suitable for application in HDD where the designer does not enjoy the freedom of selecting the sampling interval (T_S). The separation of impulses in the input shaper (ΔT) may not be an integer multiple of sampling interval (T_S). Use of *input shaper* with more impulses can resolve these issues.

Chapter 3

Design of Actuator Servo Controller

The head positioning servomechanism of an HDD uses the VCM actuator to move read-write head from one track to another and to regulate its position over the center of the track. Position feedback used by the servo loop is extracted from the readback signal obtained by the read head scanning the servo sector. Design objectives of the servo controller are fast movement of read/write head from one track to another during track seek and minimum variance regulation of the position of the head during track following. Various sources of disturbances and noise have detrimental influences on the performance of the servomechanism and the objectives of the servo controller must be fulfilled in presence of those influences. Moreover, a cost effective realization of the controller puts additional constraints on designer's choice of sophisticated and complex algorithms. To make such cost effective realization possible the designer must avoid a complex model of the process, leaving significant parts of plant dynamics unmodeled. The HDD servo controller is often implemented using fixed point digital signal processors (DSP), and low cost analog-to-digital converter (ADC) and digital-to-analog converter (DAC). This chapter explains issues related to design of digital controller, highlights factors to be considered while implementing such controllers, and provides different algorithms to solve problems specific to HDD servomechanism. Design considerations for track seek mode are already presented in chapter 2. Primarily the issues relevant to track-following controller are focused in this chapter, but some of these issues are equally important for the seek mode.

When the plant model, the disturbance model and design objectives are known, the first step towards the realization of a servomechanism is to select the structure of the controller. Since the objectives of the track seek mode are quite different from those of the track-following mode, two different controllers can easily be identified, each of which independently satisfies the requirements of one of the modes. Choice of time optimal or near time-optimal controller for the seek mode is explained in chapter 2. The PTOS ([54], [209]) discussed there is a solution that ensures smooth transition between a nonlinear seek controller and a linear state feedback controller for track-following. Since all states of the head positioning actuator are not available as measured variable, the states are estimated using an observer. Many well defined and efficient algorithms are available for calculating the parameters of state feedback controller and the observer. However, state feedback is only one of the many standard and well studied methods available for designing a linear controller. Traditionally, the industry finds itself more comfortable with PID(Proportional-Integral-Derivative) controller or PID-type compensators. One can realize the track-following servomechanism using one such controller. A simple feedback controller does not always provide the best solution to meet design objectives of the HDD servomechanism. Feed-forward control or piggy-back correction algorithms in addition to the usual feedback control are widely used for better performance of the HDD servomechanism.

3.1   Review of Design Methods

Design of controller for a plant with known model involves two steps — selecting the structure of the controller and determining the parameters such that the given robustness and perfomance objectives are fulfilled. While deciding the structure of the controller, one must consider different factors such as knowledge about the plant dynamics, type of reference signals, nature of disturbances and desired performances. For example, integral action should be included in the controller for a first order plant if it is desired to have zero steady state error in presence of constant input disturbance. There are several approaches that can be used to find the appropriate parameters once the structure is decided.

In a mechatronic system, the plant is a continuous-time system but the controller is usually implemented using a digital processor. In such case, two approaches can be taken to design and realize the controller — (1) by discretizing an analog prototype controller or (2) by direct digital design applied to a discretized model of the plant. The choice between these two approaches depends on the relative rate of sampling with respect to the desired closed-loop bandwidth. One particular concern in the design of discrete-time controller for continuous-time plant is the delay introduced by the discretization process and implementation of control algorithm. The inter-dependence of sampling frequency, desired bandwidth and choice of design approach is explained in the following.

3.1.1 Slow Dynamic Systems

If a continuous-time dynamic system is sampled such that the sampling rate is fast in comparison to the desired bandwidth then the discretization induced delay in the system can be considered negligible. Such a dynamic system is referred to as *slow dynamics system*. A commonly used design method for such system is as follows:

Step 1, Obtain, using identification or physical modeling, a suitable model $P(s)$ of the plant. Model the delay T_d due to digital control as $P_d(s) = e^{-T_d s}$ which can be expressed as $\frac{-0.5T_d s + 1}{0.5T_d s + 1}$;

Step 2, Design a continuous-time controller $C(s)$ for $P(s)P_d(s)$ using one of the many well-known design methods, such as LQR, LQG/LTR, H_2, H_∞, etc;

Step 3, Use transformation method such as the bilinear transformation

$$s = \frac{2}{T} \frac{z-1}{z+1} \tag{3.1}$$

to convert $C(s)$ to digital controller $C(z)$. Then the discrete-time controller $C(z)$ can be implemented using a μ-controller or DSP.

Conversion of continous-time poles into corresponding discrete-time poles using bilinear transformation is nonlinear and becomes significantly different for high frequency poles and zeros. However, this highly nonlinear mapping of the high frequencies in equation (3.1) can be corrected by applying a frequency pre-wrapping scheme before the bilinear transformation. This method replaces each s in the analog transfer function with $\frac{\omega_0}{\omega_p} s$, where ω_0 is the frequency to be matched in the digital transfer function and

$$\omega_p = \frac{2}{T} \tan \frac{\omega_0 T}{2}.$$

Such a mapping provides the matching of a single critical frequency between the analog domain and the digital domain. Bilinear transformation with frequency pre-wrapping provides a close approximation to the analog compensator and is widely used in practice.

An alternative discretization method is the matched pole-zero or matched z-transform. This method maps all poles and zeros of the compensator transfer function from the s-plane to the z-plane according to the relation

$$z = e^{sT} \tag{3.2}$$

where T is the sampling period. If there are more poles than zeros, additional zeros are added at $z = -1$. Also, the gain of the digital filter is adjusted to match the gain of the analog filter at some critical frequency such as the DC gain for a low pass filter. This method is somewhat heuristic and may not produce a desirable compensator.

Main disadvantage of designing controller based on an analog prototype is that the discrete compensator is only an approximation to the designed analog prototype. The analog prototype sets the upper bound on the effectiveness of the closed-loop response using the digital compensator.

3.1.2 Fast Dynamic Systems

Fast dynamic system refers to a system with a relatively slow sampling rate. Let us consider a continuous-time dynamic system described by the following

$$
\begin{aligned}
\dot{x} &= A_c x + B_c u, \\
y &= C_c x,
\end{aligned}
\tag{3.3}
$$

with zero-order hold (ZOH). Then its discretized model is as follows:

$$
\begin{aligned}
x(k+1) &= Ax(k) + Bu(k), \\
y(k) &= Cx(k),
\end{aligned}
\tag{3.4}
$$

where

$$
A = exp[A_c, T], B = \int_0^T exp[A_c, t]dt B_c.
$$

The designer can then use one of the many well defined methods for finding the discrete controller of the discrete-time model of the plant. Available design methods include shaping of the open loop transfer function, pole placement, minimizing certain norm or cost function etc [166], [27], [167].

3.1.3 Numerical Search to Find Controller Parameters

Various gains or coefficients of the controller must be found so that the desired performance criteria are met. To do this, it is a common practice to define a performance index or cost function C that includes in some way all the design criteria. The cost function may put different weightage on different design objectives. The coefficients or parameters of the controller must minimize the cost function. One possible way of finding the minimum of the cost function involves computation of the gradient of the cost function. As an alternative, numerical search methods that do not require gradient information can be utilized to search for the controller parameters directly in the closed-loop control system [101]. The size of the search space which is supposed to contain the optimum point and the relative weighting of various terms in the performance index are up to the designer to decide.

Among those non-gradient based optimization methods, Random Neighborhood Search (RNS) ([54]), Genetic Algorithm (GA) ([147], [62]), neural networks ([169]) and some other random optimization methods incorporated with statistical techniques ([147], [199]) are employed for their well known robustness property to the error of objective function and ability of global search. Simplex method ([204]) and Sequential Quadratic Programming (SQP) ([162]) have been used to find the optimal controller within a convex sub-region of performance surface.

One advantage of numerical search method is that the cost function C can be of any type, such as ITAE, the Integral of the Time multiplied by the Absolute value of the Error, or a combination of probability of the stability, phase margin, time responses, etc, and is not limited to the quadratic cost functions or desired pole locations which yield closed-form solutions. When the field of search is sufficiently wide, non-gradient based method do not get stuck, but a gradient algorithm may, for complex cost functions with a multitude of local minima. For the case of RNS algorithm it also has the advantage of being very simple to implement. Its main steps are given below:

1. The designer initiates the random search by defining the limits of the search space D.

2. A random number generator selects points d_k within D, where $k = 1, 2, \cdots, N_s$, is the number of search points.

3. The value of $J(d_k)$ is tested for each k, and the point giving the lowest value is taken to be the estimate of the global minimizer, d^*.

However, non-gradient based methods are typically slow and need many iterations to find the optimal or near optimal solution even in a relatively small region. A practical approach is to search in a wider space and successively zoom in to a smaller space which contains the optimal point.

If the performance surface is convex within the allowable region of controller parameters, response surface method [62] or gradient based method [72] can be used to find an optimal solution. Other application examples include FIR filter optimization using LMS method, iterative learning control for runout compensation, see [84] [143] [214], [160] and the references therein. There has been some study on whether the controller should adapt to the controlled output (which could be the true PES containing RRO and NRRO, or NRRO only), or measured output which is the measured PES that contains both RRO and NRRO and sensing noise [160][62].

The next few sections of this chapter provide an insight of the control problem for HDD servomechanism with the help of a few different design approaches. A simple PID type controller is first designed and its limitations to meet the requirements for high bandwidth are explained. Different methods are suggested to deal with the actuator resonances so that the bandwidth can be extended to meet design specifications. Given this basic control design, we then discuss the factors that limit the performance of the HDD servo system.

Application of optimal controller design is explained in section 3.4. Several advanced control strategies are explained in section 3.5, where HDD-specific issues such as residual vibration, RRO, low sampling frequency etc receive special attention. The specifications for HDD servomechanism are continuously becoming more and more stringent to meet the demands for increasing track density and diversification of applications. Application of a second actuator, lighter and faster than the VCM, is gradually becoming a necessity. Section 3.6 discusses pros and cons of different micro-actuators suitable for application in HDD. Different approaches to design controller for dual-stage actuators are presented in section 3.7.

3.2 PID-type Control

According to the Bode stability criterion (see references e.g. [55] or [144]), typical shape of the Bode magnitude plot of a compensated open-loop transfer function should have the following characteristics.

(a) Low-frequency band: high gain above 0 dB and decreases with increasing frequency at a rate of -20N dB/decade where N is an integer and $N \geq 2$;

(b) Crossover band: crosses the 0 dB with a slope of approximately -20 dB/deca to ensure stability;

(c) High-frequency band: low gain under 0 dB and decreases with increasing frequency at a rate of -20N dB/decade ($N \geq 2$).

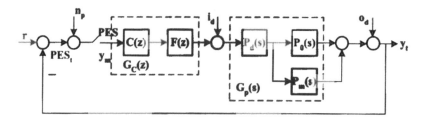

Figure 3.1: Block diagram representation of the control problem.

Any typical approach to design the controller for HDD head positioning servomechanism attempts to meet the above mentioned requirements and to achieve high servo bandwidth. This approach is explained in this chapter by illustrating the design of controller for a typical control problem shown in Figure 3.1. This illustration starts with a simple controller designed for the nominal model of the actuator. Then the limitations of this simple controller are explained, and solutions to overcome those problems are sought in the subsequent design approaches.

The transfer function of the HDD servomechanism plant can be described by the model [113]:

$$
\begin{aligned}
G_p(s) &= kP_d[P_0 + P_m], \\
&= k\frac{e^{-T_d s}}{T_{amp}s + 1}\left[\frac{r_0}{(s^2 + 2\zeta_{m0}\omega_{m0}s + \omega_{m0}^2)}\right. \\
&\quad \left. + \sum_{i=1}^{N_a}\frac{r_{mi}}{(s^2 + 2\zeta_{mi}\omega_{mi}s + \omega_{mi}^2)}\right],
\end{aligned}
\tag{3.5}
$$

where the loop gain k includes gains of various stages of the servo plant e.g. the DAC (Digital-to-Analog Converter) gain*, amplifier gain, torque gain, mass and position gain. The transfer function $P_d(s) = \frac{1}{T_{amp}s+1}e^{-T_d s}$ represents both the dynamics of power amplifier with time constant T_{amp} and the computational delay T_d. The rigid body model of the actuator coupled with linearized pivot friction is modeled as $P_0(s)$, where as $P_m(s) = \sum_{i=1}^{N_a}\frac{r_{mi}}{s^2+2\zeta_{mi}\omega_{mi}s+\omega_{mi}^2}$ represents N_a modes of mechanical resonances.

The computational delay in a typical HDD servomechanism is of the order of $T_d = 15$ μs. The phase lag of such a delay term is $2.7°$ at 500 Hz and $5.4°$ at 1 kHz. The phase lag of a typical amplifier with 40 kHz bandwidth is $0.72°$ at 500 Hz and $1.43°$ at 1 kHz. Hence these dynamics are ignored in the nominal model used for design of controller, but their effects must be taken into consideration while evaluating the performances of the designed closed loop system. For the sake of simplicity, only the first resonance mode is considered in the design examples presented in this chapter, i.e., $i = 1$ in the model of equation 3.5, and the pivot friction is ignored so that $P_0 = \frac{1}{s^2}$.

The series of design examples begins with a simple PID type controller followed by few compensators to be used with the basic PID type controller for enhancement of performance.

3.2.1 Basic PID-type Controller

In order to design the nominal controller, which is the first step in the process of designing practical servo controller, the actuator resonances, power amplifier time constant, and the delay caused by the digital control, shown in Figure 3.5, are ignored. A rigid body, double integrator model k/s^2 with -40 dB/dec slope is used to represent the nominal model of the servo plant. This simplification is reasonable when the desired open loop servo bandwidth is very low compared to the frequency of actuator resonance, power amplifier bandwidth, and the sampling frequency so that the ignored dynamics contribute negligible gain (approximately 0 dB) and phase (close to $0°$) for frequencies below the servo bandwidth. For example, the resonant mode at $\omega_{m1} = 2\pi \times 5000$ rad/s with a damping ratio of 0.02 adds only $-0.23°$ phase and 0.0101 dB gain at 500 Hz.

---

*the controller for modern HDD servomechanism is always implemented in discrete time

To change the slope of -40 dB/decade of the uncompensated nominal plant to -60 dB/decade slope for the compensated plant in the frequencies between ω_1 and ω_2, we need a lag compensator

$$C_g(s) = \frac{\frac{1}{\omega_2}s + 1}{\frac{1}{\omega_1}s + 1}, \tag{3.6}$$

where ω_2 is a few times higher than ω_1. Similarly to change the slope from -40 dB/decade to -20 dB/decade in the frequency range between ω_3 and ω_4, a lead section of

$$C_d(s) = \frac{\frac{1}{\omega_3}s + 1}{\frac{1}{\omega_4}s + 1}, \tag{3.7}$$

can be used, where ω_4 is a few times higher than ω_3. The open loop 0-dB crossover frequency f_v should be somewhere in the middle of ω_3 and ω_4. Because of 2 poles and 2 zeros in the combination of these two compensators, it results in 0 dB/decade slope in the high frequency. The required high frequency roll off -40 dB/decade at frequencies higher than ω_4 is achieved by the actuator model itself.

The combined lag-lead compensator $G_c(s)$ is written as

$$G_c(s) = k_c \frac{\left(\frac{1}{\omega_2}s + 1\right)\left(\frac{1}{\omega_3}s + 1\right)}{\left(\frac{1}{\omega_1}s + 1\right)\left(\frac{1}{\omega_4}s + 1\right)}, \tag{3.8}$$

$$k_c = \left| \frac{\left(\frac{1}{\omega_1}s + 1\right)\left(\frac{1}{\omega_4}s + 1\right)s^2}{\left(\frac{1}{\omega_2}s + 1\right)\left(\frac{1}{\omega_3}s + 1\right)k} \right|_{s = j2\pi f_v}, \tag{3.9}$$

which makes the open loop transfer function $G_c(s)G_p(s)$ crossing the 0 dB line at frequency f_v with a slope of -20 dB/decade.

The design can be further simplified by assigning pre-defined relations between frequencies of different poles and zeros of the compensator, and the desired cross-over frequency f_v. Let $\omega_3 = f_v/3$, $\omega_4 = 3f_v$, $\omega_1 = 20\pi$, $\omega_2 = f_v/5.2$. With these assumptions,

$$G_c(s) = k_c \frac{\left(\frac{5.2}{f_v}s + 1\right)\left(\frac{3}{f_v}s + 1\right)}{\left(\frac{1}{20\pi}s + 1\right)\left(\frac{1}{3f_v}s + 1\right)}, \tag{3.10}$$

$$k_c = \left| \frac{\left(\frac{1}{20\pi}s + 1\right)\left(\frac{1}{3f_v}s + 1\right)s^2}{\left(\frac{5.2}{f_v}s + 1\right)\left(\frac{3}{f_v}s + 1\right)k} \right|_{s = j2\pi f_v}, \tag{3.11}$$

This design results in an open loop bandwidth of f_v, 41° phase margin and infinite gain margin for the $\frac{k}{s^2}$ model. Figures 3.2-3.5 show the bode plots of different transfer functions with this nominal controller designed for an example system. These plots show the response for two cases — rigid body plant (solid line) and plant with resonance (dashed line). The models of the rigid body plant and the plant with flexible mode are

$$G_p(s) = \frac{2.6 \times 10^7 \times 2.3}{s^2 + 131.2s + 1.328 \times 10^5},$$ (3.12)

and

$$G_p(s) = 2.6 \times 10^7 \times [\frac{2.3}{s^2 + 131.2s + 1.328 \times 10^5} + \frac{-1}{s^2 + 2312s + 1.305 \times 10^9}]$$ (3.13)

respectively. The designed controller with one lag section and one lead section is

$$G_c(s) = \frac{1.8258 \times 10^{-8}(s + 1047)(s + 748)}{(s + 9425)(s + 62.83)}$$ (3.14)

with crossover frequency chosen to be $f_v = 500$ Hz or 3142 rad/s.

The bode plots of the following transfer functions are shown here and in the subsequent sections:

- Open loop transfer function: $L(s) = G_c(s)G_p(s)$,

- Controller transfer function: $G_c(s)$,

- Sensitivity transfer function: $S(s) = \frac{1}{1+G_c(s)G_p(s)}$,

- Complementary sensitivity transfer function: $T(s) = \frac{G_c(s)G_p(s)}{1+G_c(s)G_p(s)}$, and

- Shock transfer function: $S_h(s) = \frac{G_p(s)}{1+G_c(s)G_p(s)}$.

The step response of the closed-loop system is shown in Figure 3.6.

It is clearly evident from these figures that a simple lag-lead compensator is able to achieve desired shape of the open loop transfer function. The high frequency actuator resonance (compared with the open-loop servo bandwidth) does not adversely affect the system's behavior.

Remark 3.1: When ω_1 is sufficiently small, the lag-lead compensator (3.9) can be written as the PI-Lead compensator form:

$$G_c(s) = k_c \frac{T_{c1}s + 1}{s} \frac{T_{c2}s + 1}{T_{c3}s + 1}.$$ (3.15)

Reference [113] suggests $T_{c1} = T_{c2} = 1/(2\pi 250)$, and $T_{c3} = 1/(2\pi 10k)$ in one example.

It has been shown above that by using a simple lag-lead compensator, approximately 40° phase lead is achieved at the the crossover frequency f_v in the case of a double integrator plant. One may be tempted to increase the open loop crossover frequency f_v in order to achieve higher servo bandwidth. However, in reality, the effects due to the actuator resonances, sampling frequency

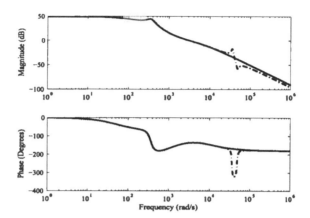

Figure 3.2: Open loop Bode plot with controller given by equation 3.14 with crossover frequency chosen $f_v = 500$ Hz $= 3142$ rad/s. Solid line: Rigid body plant of equation 3.12, and Dashed-line: Plant with resonant mode as modeled by equation 3.13.

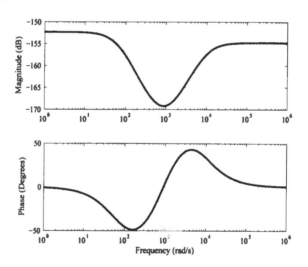

Figure 3.3: The Bode plot of the controller $G_c(s)$ corresponding to Figure 3.2.

and even power amplifier bandwidth which are ignored in equation 3.9 become prominent when servo bandwidth is increased.

The crossover frequency and thus the servo bandwidth can not be increased arbitrarily in presence of actuator resonances. The resonance modes of the actuator add phase lag and limit the achievable servo bandwidth. Methods

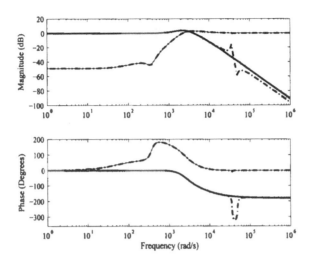

Figure 3.4: Sensitivity transfer function $S(s)$ and complementary sensitivity transfer function $T(s)$ corresponding to Figure 3.2.

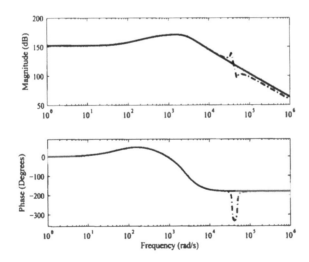

Figure 3.5: Shock transfer function $S_h(s)$ corresponding to Figure 3.2.

to overcome this limitation are explained in next few sections. Section 3.2.2 describes a method used to increase servo bandwidth by compensating for the resonance mode using notch filter, while sections 3.2.3 and 3.2.4 illustrate methods that do not compensate for the resonances but yet help to push the servo bandwidth higher. The capability of the closed loop to suppress vibration

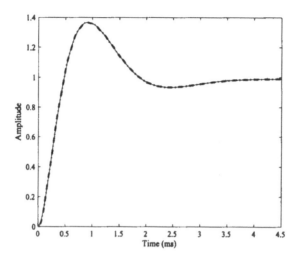

Figure 3.6: Closed-loop step response corresponding to Figure 3.2.

can also be enhanced by including the so called *peak filter* in the servo loop; this approach is explained in Section 3.2.5.

3.2.2   Cancelling Actuator Resonances using Notch Filter

It is shown in the previous section that PID-type simple compensator works well for a plant represented by rigid body dynamic model, but the phase and gain distortions due to the actuator resonances affect the overall phase and gain margins. One possible way to overcome this problem is to use a pre-compensator for the flexible modes such that the frequency response of the pre-compensated plant resembles that of a rigid actuator. This process is often called *gain stabilization* [54]. Typical method of gain stabilization involves use of notch filters to suppress the structural resonances [71]. Application of notch filter for gain stabilization is illustrated in this section using a plant model containing only one lightly damped resonant mode, but the idea can be extended to actuators with multiple flexible modes.

Let us consider a second order plant model

$$R(s) = \frac{\omega_n^2}{s^2 + 2\zeta_n\omega_n s + \omega_n^2}, \qquad (3.16)$$

where ω_n is the natural frequency of the resonant mode and ζ_n is the corresponding damping ratio. Resonances in actuators are usually damped, i.e., they eventually decay to zero; it means that the damping ratio is bounded by $0 < \zeta_n < 1$. The duration for which an excited resonance is sustained

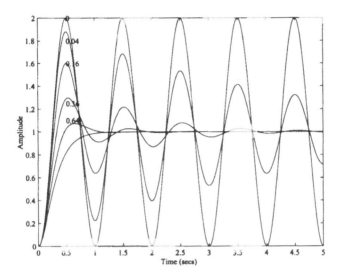

Figure 3.7: Step response of a second order model for different damping ratios.

depends on the value of the corresponding damping ratio. Smaller the damping ratio, longer it takes for the resonance to be diminished. The step responses of the second order plant of equation 3.16 are shown in **Figure 3.7** for different damping ratios. This figure shows responses for $\omega_n = 2\pi$ with $\zeta_n = 0, 0.04, 0.14, 0.36, 0.64$ and 1.0. As can be seen here, if the damping ratio is low, for example $\zeta_n \leq 0.1$ which is typical of a mechanical resonance, the step response is highly oscillatory.

If the controller is designed using a rigid body model of the plant ignoring the resonances, the response of the closed loop system shows oscillations whenever there is a change in reference command. Though these oscillations may eventually decay to zero, the performance of the servomechanism is hampered as it takes longer time for settling. Designing controller using rigid body model also ignores the gain and phase changes caused by the resonant mode dynamics, which may have more severe consequences on the stability. Effects of resonant modes can be minimized by adding a filter between the controller and the plant. Either a low pass filter or a notch filter can be used for this purpose. However, notch filter is prefered over low pass filter because of the cut-off in the magnitude response is sharper for notch filter and notch filter introduces lower phase in the loop. A simple notch filter that can be used to cancel the resonant frequency ω_{n1} is described by the transfer function

$$F(s) = \frac{s^2 + 2\zeta_{n1}\omega_{n1}s + \omega_{n1}^2}{s^2 + 2\zeta_{n2}\omega_{n2}s + \omega_{n2}^2}, \tag{3.17}$$

whose numerator cancels the lightly damped poles of the plant model (3.5). The denominator of the filter transfer function can be the stable numerators of equation 3.5, if available; otherwise one can add real pole, or well damped complex poles (e.g., complex poles with damping ratio of $\zeta_{n2} = 0.7$) at frequencies that is at or well above the resonance frequency so that the effects of the changes in phase and amplitude contributed by the denominator on the overall gain and phase margins become negligible.

With the filter $F(s)$ inlcuded in the loop, the open-loop transfer function $L(s)$, error rejection (or sensitivity) transfer function $S(s)$, complementary sensitivity transfer function $T(s)$, and shock transfer function $S_h(s)$ are:

$$
\begin{aligned}
L(s) &= G_c(s)F(s)G_p(s), \\
S(s) &= \frac{1}{1 + G_c(s)F(s)G_p(s)}, \\
T(s) &= \frac{G_c(s)F(s)G_p(s)}{1 + G_c(s)F(s)G_p(s)}, \\
S_h(s) &= \frac{G_p(s)}{1 + G_c(s)F(s)G_p(s)}.
\end{aligned}
\tag{3.18}
$$

For the example plant model shown in Figure 3.2, we have

$$
\begin{aligned}
G_p(s) &= \left(\frac{2.6 \times 10^7 \times 2.3}{s^2 + 131.2s + 1.328 \times 10^5} + \frac{-2.6 \times 10^7}{s^2 + 2312s + 1.305 \times 10^9} \right), \\
&= \frac{8.7880 \times 10^{14}}{(s^2 + 131.2s + 1.328 \times 10^5)} \frac{(s^2 + 3990s + 2.309 \times 10^9)}{(s^2 + 2312s + 1.305 \times 10^9)}. \tag{3.19}
\end{aligned}
$$

The numerator of the filter $F(s)$ are chosen such that its zeros cancel the poles of the flexible mode, i.e., the roots of $(s^2 + 2312s + 1.305 \times 10^9)$. For this actuator model there is a pair of stable zeros $(s^2 + 3990s + 2.309 \times 10^9)$. They are used as the poles of $F(s)$ to balance its zeros. Thus,

$$
F(s) = \frac{(s^2 + 2312s + 1.305 \times 10^9)}{(s^2 + 3990s + 2.309 \times 10^9)}. \tag{3.20}
$$

The transfer function of the compensated model, P_{comp}, is

$$
P_{comp}(s) = F(s)G_p(s) = \frac{8.7880 \times 10^{14}}{(s^2 + 131.2s + 1.328 \times 10^5)}. \tag{3.21}
$$

The Bode plot of the compensated actuator with controller of equation 3.11 in cascade with notch filter of equation 3.17 is shown in Figure 3.8. The crossover frequency f_v is chosen to be 1500 Hz or 9425 rad/s. It is clearly visible from this plot that the open loop transfer function looks identical to the case of rigid body model controlled by the lag-lead compensator (shown by solid line in the figure). Such a response is achieved because the notch filter of equation 3.20 is exactly the inverse of the model of resonant mode and cancels the resonance completely.

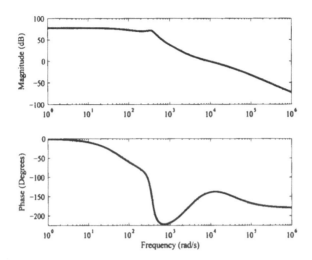

Figure 3.8: The Bode plot actuator with resonant mode controlled by equation (3.11) with notch filter equation (3.17).

The Bode plots of the controller, sensitivity transfer function $S(s)$, complementary sensitivity transfer function $T(s)$, and shock transfer function $S_h(s)$ are shown in Figures 3.9 - 3.11, and the closed loop step response is shown in Figure 3.12. It should be noted that the notch filter, being the exact inverse of the transfer function of the resonant mode, cancels completely the resonant mode of the plant and, therefore, the zero crossover of the open loop transfer function can be pushed to higher frequency.

Although this pre-compensator of notch filter produces desirable results for all responses related to the reference input, it fails to solve the problem for the response corresponding to the input distubance. One can easily see that the resonance peak in the open loop transfer is suppressed by the notch filter and the frequency response of the compensated actuator model resembles that of the rigid body model of the actuator coupled with the linearized pivot bearing model. Similar results are obtained for the sensitivity transfer function and the complementary sensitivity transfer function as they too resemble the responses of an actuator without resonance. Since the effect of resonance is suppressed, the step response of the closed loop corresponding to the reference input is smooth with no visible oscillation. However, the notch filter does not attenuate the H_∞-norm of the shock transfer function and a sharp peak is present in the bode plot of the shock transfer function. This is the transfer function between the input disturbance and the output of the closed loop system. Because of the sharp peak at the resonant frequency in the bode plot of the shock transfer function, any disturbance appearing at the input of the plant, i.e., the force disturbance, is not sufficiently attenuated and causes oscillations at the output

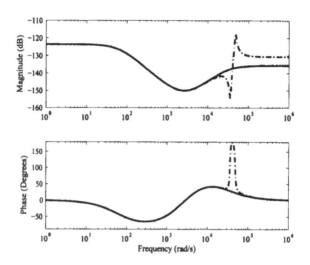

Figure 3.9: The Bode plot of the controller corresponding to Figure 3.8. Solid line: no resonance. Dashed-line: with resonance compensator.

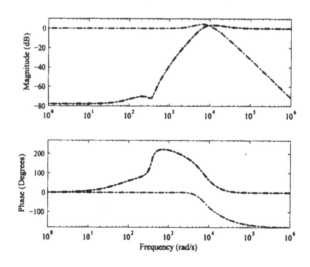

Figure 3.10: Sensitivity transfer function and complementary sensitivity transfer function corresponding to Figure 3.8. Solid line: no resonance. Dashed-line: with resonance compensator.

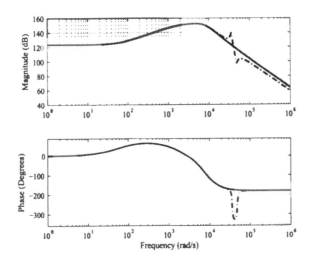

Figure 3.11: Shock transfer function corresponding to Figure 3.8. Solid line: no resonance. Dashed-line: with resonance compensator.

of the closed loop system.

Hard disk drives are mass produced systems and the actuators are also mass produced. So the physical properties of the actuators from the same batch of production may have slightly different frequencies and damping factors of their resonant modes. If a notch filter is designed to cancel one particular frequency of resonance, it may not suit well for all actuators in that batch causing imperfect cancelation of resonant modes for some. Such in-exact cancelation of the resonant modes may cause sustained oscillation in the closed-loop. Robustness of the compensator must be analyzed and examined carefully.

One possibile solution for this problem is to use a notch filter whose center frequency and width of the notch can be easily modified. From this point of view, digital notch filters are prefered over their analog counterparts. The parameters of a digital filter and, therefore, its properties can be changed easily. These filters are easily implementable in the firmware of the microprocessor used in digital control system or even using programmable digital hardware. One can substitute s in equation (3.17) with $s = \frac{2}{T_s} \frac{z-1}{z+1}$ as indicated in equation (3.1) to find the digital version of the notch filter. Many design softwares provide appropriate functions to convert an analog filter into a digital filter. For example, the command c2dm of $MATLAB^{TM}$ can be used to convert a continuous-time LTI (Linear Time Invariant) systems to equivalent discrete-time LTI system. One can use either a zero-order hold (ZOH), or a first-order hold (FOH) for plant discretization with this MATLAB function, and use the bilinear (Tustin) approximation to discretize the controller.

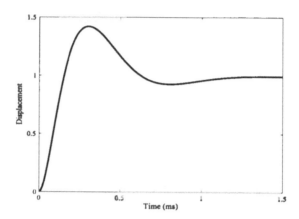

Figure 3.12: Closed-loop step response corresponding to Figure 3.8.

As mentioned in chapter 2, the sampling frequency used for HDD servo-mechanism is always a compromise between the demand for higher sampling frequency and limitations imposed by the rotational speed of the disks and the number of servo sectors used. If the Nyquist frequency for PES sampling is lower than the critical resonant frequency, a digital notch filter can not be designed using the same sampling frequency. In such situation, the notch filter as well as the controller can be discretized and implemented using a sampling frequency twice or more than the PES sampling frequency. This results in a multi-rate system [204], where the PES is sampled at the pre-defined sampling frequency but the compensator is implemented at a higher rate. Another advantage of the multi-rate implementation is the reduction of phase delay introduced by digital compensator. Additional phase delay caused by the implementation of a digital filter can be easily visualized with the help of a sinusoidal signal. If a continuous-time sine wave is sampled first and then it is reconstructed from the samples using a ZOH, the reconstructed continuous-time signal is a sine-like function with discontinuous step. The reconstructed signal has its fundamental frequency same as that of the input sine wave but is delayed from the input by half of the sampling interval [54]. In a multi-rate system, since the controller is implemented at a higher sampling frequency, the delay introduced by sampling is reduced.

The evaluation of the magnitude and phase of a discrete transfer function $H(z)$ can be done by evaluating the transfer function for different values of z lying on the unit circle, that is by substituting z by $e^{j\omega T}$ and then evaluating it for different values ω,

$$\text{magnitude} \quad = \quad |H(z)|_{e^{j\omega T}}, \tag{3.22}$$
$$\text{phase} \quad = \quad \angle H(z)|_{e^{j\omega T}}. \tag{3.23}$$

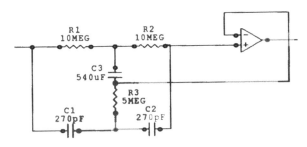

Figure 3.13: High Q notch filter circuit. $f_0 = \frac{1}{2\pi R1C1}$, $R1 = R2 = 2R3, C1 = C2 = \frac{C3}{2}$.

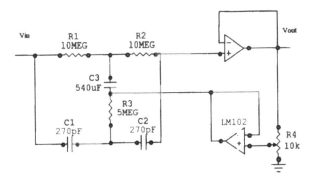

Figure 3.14: Adjustable Q notch filter circuit.

For a system where digital controller is not used or when it is required to use analog notch filter for other reasons, passive notch filters such as Bridged-T networks [152] or twin-T notch filter circuit or active notch filters shown in Figures 3.13 and 3.14 can be used. These active notch filters have relatively higher Q factors. Active notch filters can also be implemented using switched capacitor filters such as MF10, or LMF100[†].

We need several components such as resistors, capacitors, operational amplifiers etc dedicated to the realization of filter only when an analog filter is used. A discrete realization, however, does not require dedicated components. The components used to realize a discrete filter, e.g., ADC, processor etc can be shared by many other functions. For a mass produced product like HDD, reduction in the component count is one of the possible ways of reducing cost of production. In addition, the hardwared notch filters, once designed, can not be easily modified. Hence, in mass produced HDD's, where the frequency of resonance may vary up to 5% from one drive to another and in which feedback controllers are implemented using powerful digital signal processors,

[†]National Semiconductor, http://www.national.com/apnotes/apnotes_all_1.html

digital notch filters are more popular choice for tackling actuator resonance problems.

3.2.3   Cancelling Sensor Noise using Notch Filter

It is shown in the previous section how notch filters can be used to eliminate unwanted resonant oscillations of the actuator. Notch filter can also be used to eliminate sensor noise whose energy is concentrated in a narrow band of frequencies. When different signals are measured in a control system, the process of measurement often contribute to noise entering into the system. These measurement noises are usually random in nature containing wide band of frequencies. However, in many practical systems, the noise from a sensor either can be sinusoidal or has its energy concentrated in a narrow band of frequencies. Such noise has severe detrimental effects on the performance of the closed loop. A method useful for elimination of the effects of narrow band sensor noise is explained next.

As an illustrative example, let us assume that the noise contaminating the sensor output has a peak at f_n kHz. Such noise can be contributed by many practical issues, for example, the switching noise of a motor driver. The open-loop transfer function $L(s)$, error rejection (or sensitivity) transfer function $S(s)$, complementary sensitivity transfer function $T(s)$, and shock transfer function $S_h(s)$ are identical to those defined in the previous sections, given the plant model $G_p(s)$, controller $G_c(s)$ and filter $F(s)$.

Figures 3.15-3.19 show the responses of the plant model shown in Figure 3.2 controlled by the same lag-lead compensator as before but with an additional notch filter whose center frequency is 2.88 kHz and is derived from equation (3.17) by letting $\omega_{n1} = \omega_{n2} = 2\pi \times 2.88$ rad/sec, and setting the values of ζ_{n1} to a small positive number ($0.02 \rightarrow 0.2$) and ζ_{n2} close to 1. The controller is designed to obtain a crossover frequency f_v equal to 1000 Hz or 6284 rad/s.

The bode plots of the sensitivity transfer function and the complementary transfer function for this design are shown in Figure 3.17 (dashed line). The complementary transfer function shows a notch at the designed frequency, i.e., 2.88 kHz when the designed notch filter for eliminating sensor noise is included. These figures also show the bode plots when the notch filter cancelling sensor noise is not included (solid line). It is evident from the comparison between the two that the inclusion of the notch filter offers approximately 30 dB additional attenuation at the designed frequency of 2.88 kHz, i.e., the frequency of sensor noise. However, the output becomes more oscillatory for a step command in the reference input as illustrated by the closed loop step response shown in Figure 3.19.

Since the notch filter is used in this case for attenuating the sensing noise, the filter can be chosen such that its ferquency response is deeper and narrower than that of the notch filter used to attenuate actuator resonance.

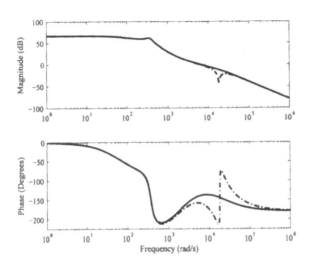

Figure 3.15: The Bode plot a rigid body actuator controlled by equation (3.11) with (dashed line) and without (solid line) notch filter.

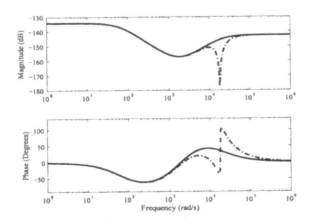

Figure 3.16: The Bode plot of the controller corresponding to Figure 3.15.

Similar to the case of adding notch filter for compensating actuator resonances, the stability and transient response of the closed-loop system need to be checked after adding the sensor noise eliminating notch filter. This is especially important when the notch frequency is close to the 0-dB cross-over frequency.

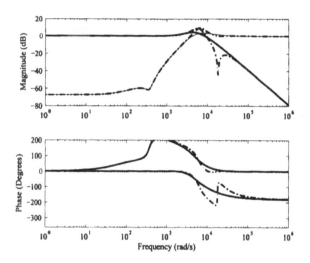

Figure 3.17: Sensitivity transfer function and complementary sensitivity transfer function corresponding to Figure 3.15.

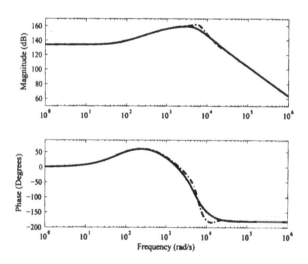

Figure 3.18: Shock transfer function corresponding to Figure 3.15.

3.2.4 Phase Stable Design

It is shown in section 3.2.2 that inclusion of a suitable notch filter suppresses the oscillation of the resonant modes caused by a step change in the command input. This is clearly visible in the step response of Figure 3.12. This phe-

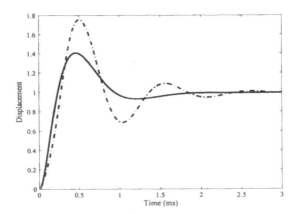

Figure 3.19: Closed-loop step response corresponding to Figure 3.15. Note that after using the notch filter, the step response is more oscillatory.

nomenon is also evident in the smooth plot of the complementary sensitivity transfer function shown in Figure 3.10. However, oscillation caused by a step change in input disturbance is not affected by inclusion of this filter. This can be easily concluded by observing the shock transfer function, which still shows peak at the frequency of actuator resonance (Figure 3.11). Therefore, any oscillation caused by a disturbance entering the loop at the input, such as windage induced vibrations, is not sufficiently suppressed at the frequencies of actuator's structural resonant modes. Furthermore, use of notch filter to suppress resonances leads to higher order controllers, requiring more computational power for realization and thus causing higher cost of implementation.

An aternative method of suppressing resonance with a fairly low order controller was proposed in [114] which was named as the phase stable design. We can explain the underlying concept using an example and let us consider again the plant model of equation 3.5. In the phase stable design, a low-order compensator $C(s)$ such as a PI-lead compensator which can take the form of equation 3.9 is used to provide phase lead in the region near the 0 dB crossover frequency and to lift the gain in low frequency range. While a notch filter reduces the open loop gain in the frequencies around the resonant frequency, in the phase stable design the filter

$$F(s) = F_{ps} = \frac{1}{T_{ps}s + 1}, \tag{3.24}$$

added to the controller $C(s)$ makes the phase and gain of the open loop transfer function about 360° and higher than 0 dB at the resonant frequency. This causes the sensitivity transfer function to have a notch which provides extra vibration suppression at the actuator resonant frequency. This design also provides attenuation of the peak in the shock transfer function near the resonant frequency more than that achieved in the notch filter based design.

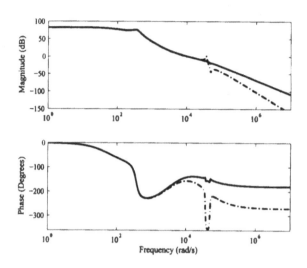

Figure 3.20: Open loop frequency response. Solid line: notch based design. Dashed-line: phase stable design.

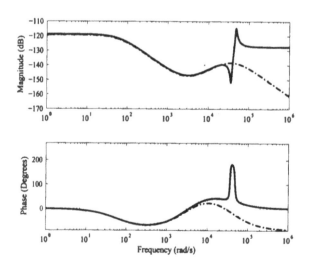

Figure 3.21: The Bode plot of the controller corresponding to Figure 3.20. Solid line: notch based design. Dashed-line: phase stable design.

Bode plots of the open loop transfer function $L(s)$, controller transfer function $C(s)$, senitivity transfer function $S(s)$ and shock transfer function $S_h(s)$

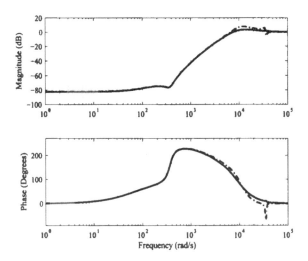

Figure 3.22: Sensitivity transfer function corresponding to Figure 3.20. Solid line: notch based design. Dashed-line: phase stable based design.

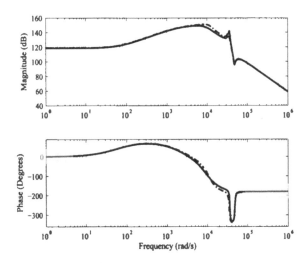

Figure 3.23: Shock transfer function corresponding to Figure 3.20. Solid line: notch based design. Dashed-line: phase stable based design.

are shown in Figures 3.20-3.23 for the phase stable design of compensator for the plant model used in previous sections. The controller is described by the equation 3.11 with phase stable compensator described by equation 3.24; the controller is designed to achieve crossover frequency $f_v = 1800$ Hz or 13310 rad/s and T_{ps} is set to $\frac{1}{2\pi 6000}$.

For the sake of comparison, the response with notch filter designed in the previous section is also shown in these figures. The Bode plot of the sensitivity transfer function (Figure 3.22) shows attenuation of approximately -4.45 dB at the resonance frequency of the actuator. On the contrary, the notch filter based design has an amplification by 2.1 dB at the resonant frequency. The bode plot of the shock transfer function (Figure 3.23) shows 136 dB gain at the resonant frequency (5.75 kHz) when phase stable design is used; but the gain is 142 dB when notch filter based design is used. The response at the output corresponding to a step change in the command reference is shown in Figure 3.24. The effectiveness of the phase-stable design is highlighted by perturbing the damping ratio ζ_{m1} of actuator resonant mode from 0.032 to 0.025.

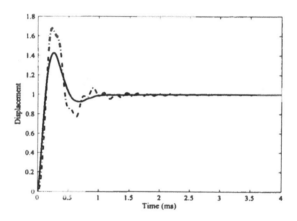

Figure 3.24: Closed-loop step response corresponding to Figure 3.20. Solid line: notch based design. Dashed-line: phase stable based design.

As can be seen from the Figures 3.22 and 3.23, although the order of the compensator with phase stable design is lower than that of the notch filter based design, it has better attenuation of vibration at the actuator resonant frequencies from both input disturbance and output disturbance.

Figure 3.25 shows the Nyquist plot of the notch based control design (solid line) and phase stable based control design (dashed-line). Note here that in the usual notch based design which is a gain stabilization, the gain of the open loop transfer function is lower than 0 dB. Hence, on the Nyquist plot, once the curve enters the unit circle, it remains inside. For the phase stable design, on the other hand, the gain of the open loop transfer function crosses the 0 dB line more than once. This is shown on the plot by the curve exiting the unit circle in the second quadrant and re-entering the unit circle in the first or 4th quadrant (Figure 3.25). Since there are more than one 0-dB crossover frequencies, there is a need to define the second phase margin and gain margin.

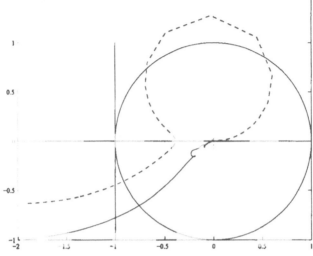

Figure 3.25: Phase stable design Nyquist plot. Solid line: notch based design. Dashed-line: phase stable based design.

Definition: The second phase margin of a Nyquist plot is an angle from the negative real axis to a vector from the origin to the point where the Nyquist plot crosses the unit circle from the inside to the outside [114].

For the example system we have above, the second phase margin is about 40°. In [114], it is recommended that the phase of an open loop characteristic could be secured within −360° ± 90° at the main resonance frequency, a phase margin could be designed at 30° or more, and a second phase margin could be secured at 40° or more.

Using this phase stable design approach, because the resonance mode is not compensated, there is a high suppression of the vibration at this frequency as shown in Figure 3.22. Notch filter based design is relatively flat for this frequency as can be seen in Figure 3.10. The phase-stable design, however, does not suppress the oscillation induced by step change in reference command. This issue can be resolved by using two degree-of-freedom (2-DOF) controller, discussed in section 3.5.1, where proper shaping of the reference input command is used to reduce the residual vibration of the actuator.

Like in all other design, the robustness of the compensator obtained using phase-stable design approach should be examined carefully. As have been discussed in most literature [90], the actuator resonant frequency may have a ±5% variation from part to part or due to changes in environmental conditions. The damping of the actuator resonant mode is also variable. These factors could make a nominally stable phase-stable design unstable. Actuator parameter calibration and/or adaptive implementation of the phase design might mitigate the robustness issue. Designing the actuator with certain phase property for achieving higher performance using simpler low order controllers is an interesting research topic. Interested readers may refer to [10], [122] and the references therein for further information.

3.2.5   Inserting a Peak Filter

Following methods of designing servo controller of HDD have been explained so far.

1. Design a PI controller using a nominal rigid body model of the actuator: The resulting controller produces satisfactory results as long as the frequencies of actuator resonant modes are a few order of magnitude higher than the servo bandwidth.

2. Design via gain stabilization: Gain stabilization is achieved by cancelling the resonant modes with a notch filter. It works well if the resonant properties of the actuator are well known and there exist a few prominant resonant modes. If the frequency and damping of resonances vary from actuator to actuator or if they are changed over time, the implemented filter must have the provision to adapt with changing parameters. The order of the compensator increases with increasing number of resonant modes having significant magnitude.

3. Design based on phase stabilization: This approach retains the actuator resonance but shapes its phase to maintain a high loop gain and hence increased servo bandwidth for vibration rejection.

We can expand the bandwidth using gain stabilization at the cost of reduced gain in certain frequencies. The phase stabilization method, on the contrary, helps to expand the servo bandwidth and retains the actuator's high gain at certain frequency providing vibration rejection at that frequency. In all these methods explained so far, the objective is to enhance the servo loop gain at desired frequencies or band of frequencies so that the sensitivity transfer function meets the requirements for good vibration rejection.

In HDD servomechanism, there exist both broad band and narrow band noise and vibrations. Sensor noise and windage induced vibrations are broad band, while the narrow band vibrations are contributed mainly by (1) structural vibrations of disks ([141], [140], [79]), (2) structural vibrations of spindle [175] and (3) actuator resonance. Moreover, imperfections of the shape of data tracks contribute to RRO whose frequency spectrum consists of frequencies which are integer multiple of the spinning frequency of disks. The servo loop must follow these variations of track.

According to the internal model principle [53] for dealing with external disturbance, a suitable copy of the exosystem should be included in each control

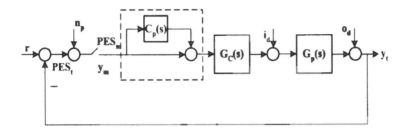

Figure 3.26: Control schemes for disturbance rejection using peak filter.

channel in such a way that the extended system is exponentially stabilizable via output feedback. A *peak filter* consisting of a pair of lightly damped poles can be used to model the narrow band disturbance, for example the RRO or narrow band NRRO, present in HDD servomechanism [48],[183],[229]. The peak filter transfer function is the inverse of a notch filter transfer function.

There are two possible ways to implement a controller that rejects narrow-band disturbances. In the first method, an internal model of the disturbance is embedded into the open loop to form an augmented model and the disturbance is eliminated via feedback control of the augmented model. The second approach, on the other hand, detects the disturbance and cancels it via active feedforward control [63]. The method involving internal disturbance model is discussed here. In the frequency domain, the compensator that includes an internal model of the narrow band disturbance exhibits a large peak at the disturbance frequency. That is why such compensator is commonly known as *peak filter*. Figure 3.26 shows the block diagram of the closed loop system using peak filter.

Consider the HDD servo control loop shown in Figure 3.26. The transfer functions of the plant and the nominal, appropriate and stabilizing controller are $G_p(s)$ and $G_c(s)$, respectively. A peak filter $C_p(s)$ with center frequency coinciding with the frequency ω_i of a narrow band disturbance represents the internal model of the disturbance. The input disturbance i_d may have periodic components. The reference signal, plant output, and the tracking error are represented in this figure by r, y and *pes*, respectively.

With the peak filter $C_p(s)$ included in the loop configuration shown in Figure 3.26, the open-loop transfer function $L(s)$ and the error rejection (or sensitivity) transfer function $S(s)$ are

$$
\begin{aligned}
L(s) &= (1 + C_p)G_cG_p, \\
S(s) &= \frac{1}{1 + G_pG_c(1 + C_p)}, \\
&= \frac{1}{1 + G_pG_c}\frac{1 + G_pG_c}{1 + (1 + G_pG_c)C_p}, \\
&\overset{\Delta}{=} S_nS_F.
\end{aligned}
\tag{3.25}
$$

where

$$S_0 = \frac{1}{1 + G_p G_c},$$ (3.26)

$$S_F = \frac{1}{1 + T_0 C_p},$$ (3.27)

$$T_0 = \frac{G_p G_c}{1 + G_p G_c}.$$ (3.28)

The transfer function S_O is automatically defined once the nominal controller $G_c(s)$ is chosen. It is easily deduced from the above equations that by using the controller structure of Figure 3.26, the narrow band disturbance compensator C_p can be designed to shape S_F so that overall sensitivity transfer function satisfy the desired performance [229].

Let us assume that the desired magnitudes of the peak and the baseline of the frequency response of the peak filter are M dB and N dB, respectively, as shown in Figure 3.27. Then a suitable peak filter can be constructed in the s-domain according to the following [213],

$$C_p(s) = \frac{s^2 + 2\zeta_1 \omega_p\, s + \omega_p^2}{s^2 + 2\zeta_2 \omega_p\, s + \omega_p^2}$$ (3.29)

with

$$\zeta_1 = \frac{\Delta^2 + 2\Delta}{2\,(1 + \Delta)}\,\sqrt{n^2 - 1}$$

$$\zeta_2 = \frac{\zeta_1}{m}$$ (3.30)

$$\theta = \tan^{-1}\frac{m - 1}{2\sqrt{m}}$$

where ω_p is the center frequency of the peak filter in rad/sec, ζ_1 and ζ_2 are the damping ratios with $\zeta_1 > \zeta_2$, Δ is the percentage of variation in center frequency, $m = 10^{M/20}$, and $n = 10^{N/20}$. Possible phase loss due to inclusion of the peak filter is estimated by θ. The inclusion of peak filter modifies both the gain and phase of the open loop transfer function.

This design method is now illustrated with the help of the same plant model used in previous sections. Let the desired magnitude at the peak and the baseline of the bode plot of the peak filter be M=40 dB at 360 Hz and N=2 dB. To keep the 0-dB crossover frequency same as in the earlier designs, the gain of G_c is reduced by half to offset the gain introduced by $C_p(s)$. The bode plots of L, and S are shown in Figures 3.28 and 3.29. The open loop frequency response for this design shows an increase in gain by approximately 40 dB at the disturbance frequency 360 Hz compared to the design that includes PID-type controller and notch filter. This additional gain is provided by the peak filter, which is effectively a bandpass filter. The increase in gain of the open

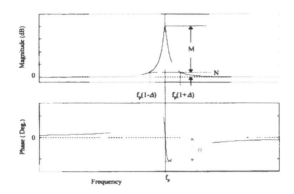

Figure 3.27: Frequency response of a peak filter.

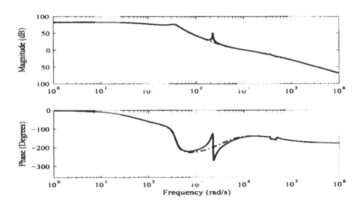

Figure 3.28: Open loop frequency response. Solid line: PID-type controller + notch filter + peak filter. Dashed-line: PID-type controller + notch filter.

loop transfer function at the frequency of disturbance means better rejection of this periodic disturbance. The bode plot of the sensitivity transfer function is shown in Figure 3.29 along with the sensitivity transfer function obtained using PID-type controller and notch filter.

The step response shown in Figure 3.30, however, exhibits more oscillatory behaviour for the controller with peak filter included compared with the case when no peak filter is used.

The discrete transfer function of the peak filter (3.29) can be obtained by substituting $s = \frac{2}{T}\frac{1-z^{-1}}{1+z^{-1}}$:

$$C_p(z) = \frac{A + Bz^{-1} + Cz^{-2}}{1 + Ez^{-1} + Fz^{-2}}, \tag{3.31}$$

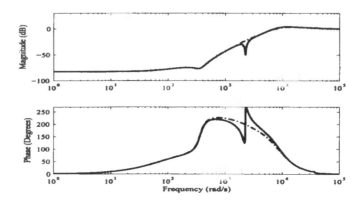

Figure 3.29: Sensitivity transfer function frequency response. Solid line: PID-type + notch filter + peak filter. Dashed-line: PID-type + notch filter.

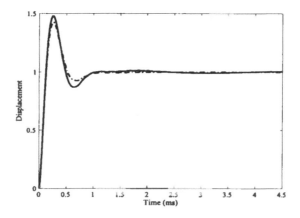

Figure 3.30: Step response. Solid line: PID-type + notch filter + peak filter. Dashed-line: PID-type + notch filter.

where

$$
\begin{aligned}
D &= 1 + b_1 + b_2, \\
A &= (1 + a_1 + a2)/D, \\
B &= (-2 + 2a_2)/D, \\
C &= (1 - a_1 + a_2)/D, \\
E &= (-2 + 2b_2)/D, \\
F &= (1 - b_1 + b_2)/D
\end{aligned}
$$

and

$$
\begin{aligned}
a_1 &= 2\pi\zeta_a f^*, \\
a_2 &= (\pi f^*)^2, \\
b_1 &= 2\pi\zeta_b f^*, \\
b_2 &= (\pi f^*)^2,
\end{aligned}
$$

and $f^* = f_0/f_s$ for sampling frequency f_s.

Using the bias free structure, the internal AC and DC gains can both be lower compared with those obtained direct form implementations [183].

The peak filter can be used for disturbances in the low frequency band such as for runout compensation [183][48] or the disturbances in higher frequency band similar to the phase stable design [229]. It provides additional vibration suppression of few dB for the sinusoidal disturbances of sufficiently long time. One drawback of this approach is the inferior transient response of the closed loop system. With such a scheme in use, the transient response of the system becomes oscillatory because of the phase distortion induced by the peak filter. While responding to an external periodic disturbance, such as an RRO, it takes few cycles before the filter can reproduce the disturbance internally so that it can be cancelled. This is not what one would like to see when the system is responding to the reference command. Time taken by the transient of the peak filter's response to decay is the main factor contributing to the oscillatory output. This can be reduced by letting the filter run even during the track seek process; interested readers may refer to [183] for further explanation on this solution. An alternative solution suggested in [229] selects filter zeros such that the phase lag contributed by the peak filter is minimized.

3.2.6 Summary: Application of Different Filters

Over the years, the demand for higher bandwidth of the HDD servomechanism continued. It was about 300 Hz in the middle of 1990s and increased to over 1 kHz by 2001 [220]. This has been possible by multi-pronged developments including better mechanics with higher bandwidth, faster microprocessor that allows implementation of complex higher order controller with less computational delay, and better servo modeling and control designs. Higher bandwidth

Table 3.3: Applications of various filters

Problem	Solution	Remark
HDD actuator	Lag-lead compensation	Lag section to elevate low frequency gain. Lead section for achieving -20 dB/decade slope for the open loop transfer function in the range of frequencies around the 0-dB crossover for stability.
Actuator resonance	Notch filter	Smooth output in response to step change in reference command. Shock transfer function may still show a peak indicating inadequate suppression of oscillation induced by input disturbance. Relatively robust to actuator resonance variation compared with phase stable design.
Actuator resonance	Phase stable design	Shock transfer function has lower peak. Good for reducing oscillation induced by narrow band input disturbance at resonant frequency. Response to reference command shows oscillatory output, but can be mitigated via input shaping. Not robust to changes in actuator resonance.
Narrow band noise	Notch filter	Need to check stability and transient response.
Narrow band disturbance	Peak filter	Loop gain elevated at center of peak filter. Transient response of the loop degraded. Need to check the stability.

results in a lower sensitivity transfer function, and hence rejects the process disturbances more effectively making more accurate tracking control possible.

Though a simple lag-lead controller can be used as the nominal controller for HDD servomechanism, such design results in low servo bandwidth and therefore poor disturbance rejection capability. Increase in bandwidth is possible with proper compensation of actuator resonances and various narrow-band and broad-band noise and disturbances. These compensation techniques often include special purpose filters. Table 3.3 summarizes these basic control filters used in the HDD servo loop.

In a practical servomechanism, a combination of the above mentioned filters is used. Selection of parameters involves more complex procedure when these filters are combined. Optimal control theory or on-line optimization is useful tool for finding the parameters for such complex compensator. The application of optimal control is discussed in section 3.4, but before that, the fundamental limits of a servo control system are discussed in the next section.

3.3 Factors Limiting Servo Performance

It is generally expected to have the performance of servo system as high as possible. However, the performance can not be improved to any arbitrary level. Therefore, knowing the bounds on the limit of performance in a control system is of great interest to the designer.

Let us consider the read-write head positioning servomechanism shown in Figure 3.31 with actuator model $G_p(s)$, controller transfer function $G_c(s)$, input and output disturbances $I_D(s)$ and $O_D(s)$, and noise $N_P(s)$. Then true

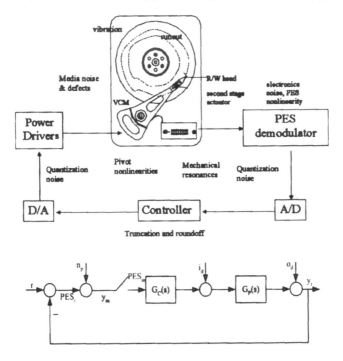

Figure 3.31: HDD servo loop block diagram.

PES ($PES_t(s)$) is,

$$PES_t(s) = -G_p(s)S(s)I_D(s) - S(s)O_D(s) - T(s)N_P(s), \qquad (3.32)$$

where the sensitivity transfer function $S(s)$ and the complementary sensitivity transfer functions $T(s)$ are defined by

$$S(s) = \frac{1}{1 + G_p(s)G_c(s)}, \qquad (3.33)$$

$$T(s) = \frac{G_p(s)G_c(s)}{1 + G_p(s)G_c(s)}. \tag{3.34}$$

Assuming that $I_D(s)$, $O_D(s)$ and $N_P(s)$ are independent of each other, the power spectrum of true PES S_{PESt} is:

$$
\begin{aligned}
S_{PESt}(s) &= |G_p(s)S(s)|^2|I_D(s)|^2 + |S(s)|^2|O_D(s)|^2 + |T(s)|^2|N_P(s)|^2, \\
&= (|G_p(s)|I_D(s) + O_D(s)|)^2|S(s)|^2 + |T(s)|^2|N_P(s)|^2, \\
&= (\tilde{O}_D(s)|)^2|S(s)|^2 + |T(s)|^2|N_P(s)|^2. \tag{3.35}
\end{aligned}
$$

The power spectrum of measured PES S_{PESm} is:

$$
\begin{aligned}
S_{PESm}(s) &= |G_p(s)S(s)|^2|I_D(s)|^2 + |S(s)|^2|O_D(s)|^2 + |S(s)|^2|N_P(s)|^2, \\
&= |G_p(s)S(s)|^2|I_D(s)|^2 + |S(s)|^2(|O_D(s)|^2 + |N_P(s)|^2). \tag{3.36}
\end{aligned}
$$

Achievement of position accuracy required for higher TPI has traditionally been realized by scaling. A simple way to deal with NRRO effects is to reduce them from the source. Technical advancements in the design of components as well as in the design of the HDD mechanics have resulted in significant reduction of vibrations induced by electromagnetic force [31], disk-spindle pack imbalance [86], ball or fluid bearings [31], [227], disk platter [26], [141]. These improvements effectively reduce the magnitude of output disturbance O_D. Improvement of the pattern of airflow inside the drive enclosure by modifying the mechanical structures such as base casting [39], squeeze air bearing plate [42], and shape of actuator [154] can help to reduce both I_D and O_D.

Use of alternative material and different structure for actuator assembly [9] and for disk substrate [79] can reduce structural vibrations, mostly manifested as actuator resonance and disk vibration. Resonance of actuator can also be reduced by actively stiffening its flexible modes using instrumented suspension [90], [89] or using active damping with sensor buried on the arm, and passively stiffening the flexible modes via change of actuation structure [76], using alternative actuator material, reducing weight, and even simply by increasing the position error signal (PES) sampling frequency [207].

Given the mechanical system for actuation with all possible component level improvements stated above, the objective of servo controller design is to attenuate the effects of vibrations via various loop shaping. Expanding servo bandwidth via phase stable design and gain stable design has been discussed in previous sections. Peak filters can be used to increase the servo loop gain. However, these improvements can not be pushed arbitrarily further. Limits of acheivable performance are explained in the following sections for factors such as

1. compromising noise and disturbance,

2. water bed effect,

3. actuator uncertainty, and

4. sampling of PES and others.

3.3.1 Limitation of $S + T = 1$

The target of the track following controller $G_c(s)$ is to maintain the PES to a minimum level in the presence of different forms of disturbance and noise. In other words, the servo controller should be designed such that the variance of $PES(s)$ is minimized.

Performance of a closed loop system can be easily assesed with the help of its sensitivity transfer function $S(s)$ and complementary sensitivity transfer function $T(s)$. The complementary sensitivity function defines the systems response to the reference input as well as to the measurement noise, where as the sensitivity transfer function measures the system's capability in rejecting the effects of vibration and other disturbances. If for any frequency ω_0, the complementary transfer function $T(j\omega_0) = 1$ then the magnitude of the output sinusoid is equal to the magnitude of the reference input which is a sinusoid of frequency ω_0. However, $T(j\omega_0) = 1$ also implies that if the noise has a frequency component ω_0 then the output of the closed loop system contains a sinusoid of the same frequency with magnitude equal to that of the noise. Similarly, $S(j\omega_0) = 1$ implies that the magnitude of the output sinusoid is equal to the magnitude of the disturbance sinusoid of frequency ω_0.

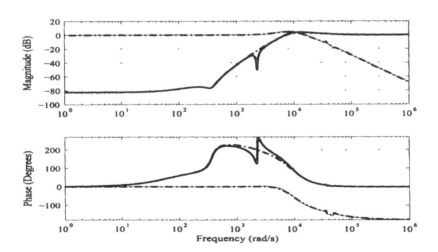

Figure 3.32: Frequency response of S (solid line) and T (dashed-line) for the flexible actuator controled using PID type + notch + peak filter discussed in Section 3.2.5.

It is easily deduced from the definitions of sensitivity and complementary sensitivity functions that $T(s)+S(s) = 1$. It is obvious from this identity that it is not possible to achieve simultaneously the objectives of rejecting disturbance and keeping the influence of noise on PES low. Increased servo bandwidth means low magnitude of $S(j\omega)$ for wider range of frequencies ensuring better rejection of disturbances ($I_D(s)$ and $O_D(s)$ in Figure 3.31). But it also implies higher magnitude of $T(j\omega)$ and, as a result, more effect of measurement noise on PES. To reduce the measurement noise in PES, one can choose to reduce the servo bandwidth such that $T(j\omega)$ has smaller magnitude in high frequencies. But this increases the magnitude of $S(j\omega)$ in those frequencies causing less effective rejection of disturbances, and the main contribution to PES comes from various disturbances.

The challenge of designing a tracking controller is thus to pick the structure and parameters of the controller that balance the impact of disturbance sources and measurement noise. This must be achieved inspite of the limitations of actuator bandwidth and PES sampling frequency, and without using unreasonable knowledge of the disturbance (or vibration) models and noise models.

High bandwidth actuator has always been considered as necessity for achieving higher positioning accuracy. This is especially true when the sensing noise level is low, or the disturbance is concentrated mainly in the low frequency band. It is easily deduced from equation 3.36 that increasing the bandwidth always reduces the *measured* PES. However, depending on the spectrum of noise, a lower bandwidth servomechanism may achieve higher positioning accuracy compared to that obtained by simply pushing the servo bandwidth higher [121]. When the sensing noise level is high, one must differentiate between the measured PES and the true PES while designing and optimizing the servo controller.

The MEMS actuators reported so far in the published literature show constant magnitude response up to 40 kHz, beyond which, the resonant modes appear. Using the rule of thumb that the bandwidth is limited to 1/4 of the high frequency actuator resonance, these MEMS actuators make it possible to extend servo bandwidth to minimum 10 kHz, leaving all mechanical vibrations under 10 kHz compensated by the MEMs actuator. The MEMS actuators have very limited range of movement and, therefore, must be used together with the VCM making it a dual-stage actuator. The single stage actuators (just the VCM actuator) can support servo bandwidth up to 2 kHz only. However, the single stage actuator costs less than the dual stage and is widely used in the HDD industry. Moreover, the 10 kHz servo bandwidth supported by MEMS actuator is not achievable with the current state of the HDD technologies as the PES sampling frequency available so far does not support such bandwidth. Designing a servo controller which is bandwidth restricted and yet effective in rejecting vibrations is of prime interest in the HDD industry.

For better rejection of vibration via feedback control, the loop gain $G_p(s)G_c($ should be elevated to higher magnitude at the frequencies where the disturbance power is concentrated so that the sensitivity transfer function $S(s)$ has notch at those frequencies. If the loop gain is increased in the band of frequencies where the disturbance spectrum is concentrated, the vibration is more effectively rejected though the servo loop bandwidth is not increased. Inclusion of peak filter is an example of this approach. The sensitivity and complementary sensitivity functions shown in Figure 3.32 underscores the effectiveness of adding a peak filter with center frequency at 360 Hz. The peak filter improves vibration suppression at its central frequency, but amplifies vibration at other frequencies. This limitation can be understood better with the help of *waterbed effect* discussed later.

Bode plot measures a system's steady state responses when the input is pure sinusoids of different frequencies. When we analyze the performance of the closed loop system using the Bode plots, it should be kept in mind that it takes sufficiently long time for the closed-loop to attenuate the pure sinusoid disturbances. If the peak filter is designed for narrow band NRROs (and thus not pure sinusoid of sufficiently long duration), the vibration reduction may not be as good as what is expected from the analysis of the Bode plot.

As the recording density continues to increase, each bit consists of less number of magnetic grains. This causes a possible decrease in the SNR (Signal-to-Noise Ratio) of the read back signal. The position feedback in HDD servomechanism is generated from the read back waveform produced by the servo patterns on the disks. Low SNR of the readback waveform thus has a detrimental effect on the performance of the HDD servomechanism. The spectrum of the noise in PES generation lies in the higher range of frequencies. Increased bandwidth of the servo loop means higher magnitude of measurement noise in the PES. Therefore, the PES noise puts a limit on the achievable bandwidth. Study of low noise PES generation is important so that the servo system can effectively utilize the high bandwidth servo to achieve high tracking accuracy.

3.3.2 Waterbed Effect

The continuous time Bode's Integral Theorem is given as follows:

Let $L(s)$ be a stable open-loop transfer function of a continuous-time, single-input-single-output (SISO), linear time-invariant (LTI). Then the sensitivity function is $S(s) = 1/(1 + L(s))$. When the closed-loop system is stable and $k_s = \lim_{s \to \infty} sL(s)$, then

$$\frac{1}{\pi} \int_0^\infty \ln |S(j\omega)| d\omega = -\frac{1}{2} k_s. \tag{3.37}$$

When the relative degree of $L(s)$ is no less than 2, $k_s = 0$, then

$$\frac{1}{\pi} \int_0^\infty \ln |S(j\omega)| d\omega = 0. \tag{3.38}$$

If $|S(j\omega)| < 1$ ($\ln|S(j\omega)| < 0$) over some frequency interval, then the above mentioned relation implies that $|S(j\omega)| > 1$ ($\ln|S(j\omega)| > 0$) at other frequencies.

This result suggests that it is not possible to achieve an arbitrary sensitivity reduction (i.e., $|S| < 1$) at all points on the imaginary axis. If $|S(j\omega)|$ is smaller than one in a particular range of frequencies, then it must be greater than one in another range of frequencies as illustrated in Figure 3.33. This phenomenon is known as the *waterbed effect*.

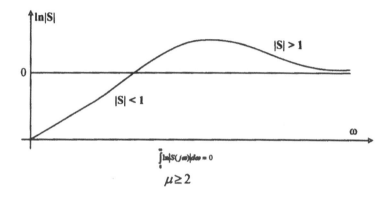

Figure 3.33: Illustration of the waterbed effect.

The waterbed effect is apparent in continuous time systems when the relative degree of the open-loop transfer function is greater than or equal to two. The movement of the actuator used in all position servo loops is governed by the Newton's laws of motion, i.e., the actuator generates the driving force that creates the acceleration, integration of the acceleration is equal to the velocity of motion, and further integration of velocity is the displacement of the actuator. So the actuator has a relative degree of at least 2, and therefore, the waterbed effect is inevitable. Low sensitivity hump control designs tend to lower the relative degree of the open loop in a wide range of frequency [46] [159].

When a peak filter is included in the servo loop, it increases the loop gain in the vicinity of the peak filter's center frequency and reduces the vibration at those frequencies. This large increase in the magnitude of the loop gain results in smaller gain or even attenuation in other frequencies because of the loop's integral is a constant. Though the peak filter helps to improve suppression of vibration for frequencies around the peak filter's center frequency, the sensitivity is increased at some other frequencies, as shown in the example of Figure 3.29. However, regardless of the waterbed effect, the effectiveness of the increase in loop gain at frequencies of abundant disturbances is still significant when the compromised frequencies have relatively less vibrations.

If integral action in the controller is not required and if a pure velocity or acceleration feedback is available by adding additional sensors, then k_s might be greater than 0, resulting in more attenuation of vibration than amplification with a continuous time controller. Additionally, we note that Clegg integrator has 39 degree phase lag instead of the usual 90 degree phase lag. As a result, it provides the necessary integral action but does not add the amount of phase lag found in the usual linear controls. Hence it is possible to avoid the waterbed effect. Nevertheless, with limited sampling frequency, a digital servo loop will have more vibration amplification than a continuous time servo loop of the same bandwidth.

3.3.3 Bandwidth Limitations

Assuming a typical servo loop shape (Section 3.2) and stability margin requirements, the two major factors thet limit the achievable servo bandwidth are (1) the uncertainties related to the actuator, and (2) various delays in the control system.

The phase margin of a control system defines the extent of additional phase lag that can be tolerated before the stability is lost, and the gain margin defines the amount of increase in gain that makes the loop unstable. So the servo loop can be pushed to the point that enough margin is left for the actuator's phase and gain uncertainty. Since the control systems are typically designed to have 6 dB gain margin, the servo bandwidth attainable is limited to the frequency where the uncertainty of actuator gain is equal to 6 dB. In general, a rule of thumb states that a servo bandwidth of 1/4 of the critical resonant frequency (beyond which, the frequency response will show uncertain behavior at different excitation level) can be achieved [144].

The PES sampling frequency and the achievable servo bandwidth are interrelated. It is generally preferred to have PES sampling frequency roughly 10 times the open loop servo bandwidth or higher for effective suppression of vibration. The sampling frequency depends on the number of servo sectors per revolution which, in turn, is limited by the space on the disk allocated for servo bursts. Increasing the rotating speed of disk increases the sampling frequency, but it also increases the level of internal vibrations and therefore demands for better servo design, which in turn requires a even higher PES sampling frequency. Similar to the limitations on control performance due to the sampling frequency, the computation delay and more importantly the delay introduced by the ADC limit the achievable performance by adding extra phase lag. These delays must be kept less than a fraction of the PES sampling period.

The success in achieving the objective of the HDD servo control to meet the requirements on tracking accuracy and response time demanded by the system depends on many factors including the limitations of actuator's performance (plant uncertainty), lack of accurate disturbance model, and insufficient feedback information due to limited sampling frequency. It is obvious that no single solution exists that can tackle all these limitations, and optimization plays an important role in the design of the servo controller. The next section provides an optimal control scheme which results in a system with the highest TPI possible in presence of all the NRRO sources. The limits on the servo control loop's performance described so far can be broken using advanced techniques such as sensor assisted feedforward control of disk's vertical vibration [63].

3.4   Optimal Control

Optimal control is a well known design technique in the control community. In this approach, the controller design problem is first formulated as the problem of optimizing certain norm of a pre-selected function (objective function) that includes design specifications and description of noise and disturbances. The controller is then identified such that the chosen norm of the objective function is minimized. In this section, the problem of designing HDD track-following servo is formulated as a standard H_2-optimal control problem. The objective of finding the minimum TMR budget is treated as an equivalent problem of minimizing the H_2 norm of the corresponding transfer function. TMR or *Track Mis-Registration* is an important metric that determines the track density (*Tracks per inch* or TPI) and, therefore, the achievable areal density of the HDD.

As discussed before, the TMR during track following is defined as $3\sigma_{pest}$ where

$$\sigma_{pest} = \sqrt{\frac{1}{n}\sum_{i=0}^{n-1} y_{pest}(i)^2}. \tag{3.39}$$

Here n in equation (3.39) is the number of samples of the true PES. For a given system with all the disturbances and noise described earlier, one must minimize the value of $3\sigma_{pest}$ in order to achieve the highest track density [83].

We can associate σ_{pest} with the problem of designing the track-following controller by considering the H_2 norm and H_2-optimal control. The H_2 norm of a system can be interpreted as the RMS value of the output when the system is driven by independent zero mean white noise with unit power spectral densities. In a hard disk drive servo as shown in Figure 3.34, all the disturbance sources can be viewed as colored noise generated by filtering independent white noises.

Let the true PES be the output of the system and the transfer function from the vector of three independent white noise sources $w = [w_i', w_o', w_n']'$ to the true PES y_{pest} be defined as T_{zw}. Then the H_2 norm of the transfer function T_{zw} is defined as

$$\|T_{zw}\|_2 = \sqrt{\frac{1}{n}\sum_{i=0}^{n-1} y_{pest}(i)^2}, \tag{3.40}$$

when n is sufficiently large. As a result, the problem of controller design for accurate tracking, which is to minimize $\|T_{zw}\|_2$, can be solved using H_2 optimal control design method when an accurate noise and vibration model is available. The solution formulae for both continuous-time and discrete-time control cases are given next.

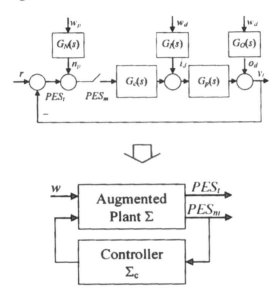

Figure 3.34: H_2 output feedback problem for an HDD servo system considering noise and disturbance model [128].

3.4.1 H_2 Optimal Control: Continuous-time Case

Consider a stabilizable and detectable linear time-invariant system Σ with a proper output feedback controller Σ_c, shown in Fig 3.34 [166], where,

$$\Sigma: \begin{cases} \dot{x} = A\,x + B\,u + E\,w, \\ y = C_1\,x \qquad\quad + D_1\,w, \\ z = C_2\,x + D_2\,u, \end{cases} \tag{3.41}$$

with $x \in \Re^n$ being the state, $u \in \Re^n$ the control input, $w \in \Re^l$ the disturbance input, $y \in \Re^p$ the measurement output, and $z \in \Re^q$ the output to be controlled.

The problem of H_2 optimal control is equivalent to finding an internally stabilizing proper controller such that the H_2 norm of the resulting closed-loop transfer matrix $T_{zw}(\Sigma \times \Sigma_c)$ is minimized. A proper controller Σ_c is said to be an H_2 optimal controller if it internally stabilizes Σ and $\| T_{zw}(\Sigma \times \Sigma_c) \|_2 = \gamma_2^*$.

If the following conditions are satisfied:

1. D_2 is injective, i.e., D_2 is of maximal column rank,

2. the subsystem (A, B, C_2, D_2) has no invariant zeros on the imaginary axis,

3. D_1 is surjective, i.e., D_1 is of maximal row rank,

4. the subsystem (A, E, C_1, D_1) has no invariant zeros on the imaginary axis,

then the H_2 optimal control problem is said to be regular, and the output feedback controller is given by:

$$\Sigma_c : \begin{cases} \dot{v} = (A + BF + KC_1) \ v \ - \ K \ y, \\ u = \qquad F \qquad v, \end{cases} \tag{3.42}$$

where

$$F = -(D_2^T D_2)^{-1}(D_2^T C_2 + B^T P), \tag{3.43}$$
$$K = -(QC_1^T + ED_1^T)(D_1 D_1^T)^{-1}, \tag{3.44}$$

and $P \geq 0$ and $Q \geq 0$ are respectively the solutions of the following algebraic Riccati equations (ARE),

$$A^T P + PA + C_2^T C_2 - (PB + C_2^T D_2)(D_2^T D_2)^{-1}(D_2^T C_2 + B^T P) = 0, \tag{3.45}$$

$$QA^T + AQ + EE^T - (QC_1^T + ED_1^T)(D_1 D_1^T)^{-1}(D_1 E^T + C_1 Q) = 0. \tag{3.46}$$

Moreover, the infimum γ_2^* of the H_2 norm of the closed-loop transfer matrix $T_{zw}(\Sigma \times \Sigma_c)$ is given by

$$\gamma_2^* := inf\{\| T_{zw}(\Sigma \times \Sigma_c) \|_2 \,|\Sigma_c \text{ internally stabilizes } \Sigma\},$$
$$= \{trace(E^T PE) + trace[(A^T P + PA + C_2^T C_2)Q]\}^{1/2}. \tag{3.47}$$

When the problem is singular, the so-called perturbation approach can be used by adding some small values to z, and redefining a new D and E,

$$\tilde{z} := \begin{bmatrix} z \\ \epsilon x \\ \epsilon u \end{bmatrix} = \begin{bmatrix} C_2 \\ \epsilon I \\ 0 \end{bmatrix} x + \begin{bmatrix} D_2 \\ 0 \\ \epsilon I \end{bmatrix} u, \tag{3.48}$$

$$\tilde{E} = \begin{bmatrix} E & \epsilon I & 0 \end{bmatrix}, \tag{3.49}$$
$$\tilde{D}_1 = \begin{bmatrix} D_1 & 0 & \epsilon I \end{bmatrix}, \tag{3.50}$$

where $\epsilon > 0$. The constructed perturbed system

$$
\tilde{\Sigma}: \begin{cases} \dot{x} = A\,x + B\,u + \tilde{E}\,w, \\ y = C_1\,x \qquad\qquad + \tilde{D}_1\,w, \\ z = \tilde{C}_2\,x + \tilde{D}_2\,u, \end{cases} \tag{3.51}
$$

is hence regular.

For the hard disk drive servo control design, we can formulate the following H_2-optimal problem [128],

$$
\begin{cases} x &= (x_a'\ x_i'\ x_o'\ x_n')', \\ w &= (w_i'\ w_o'\ w_n')', \\ A &= \begin{pmatrix} A_a & 0 & 0 & 0 \\ 0 & A_i & 0 & 0 \\ 0 & 0 & A_o & 0 \\ 0 & 0 & 0 & A_n \end{pmatrix}, \\ B &= (B_a'\ 0\ 0\ 0)', \\ E &= \begin{pmatrix} B_a & 0 & 0 \\ B_i & 0 & 0 \\ 0 & B_o & 0 \\ 0 & 0 & B_n \end{pmatrix}, \\ C_1 &= (C_a\ C_i\ C_o\ C_n), \\ D_1 &= (D_i\ D_o\ D_n), \\ C_2 &= (C_a\ C_i\ C_o\ 0), \\ D_2 &= 0. \end{cases} \tag{3.52}
$$

Here the vector $x = (x_a', x_i', x_o', x_n')'$ represents the state variables from the actuator $G_p(s)$, input disturbance filter $G_I(s)$, output disturbance filter $G_O(s)$ and measurement filter $G_N(s)$ in Figure 3.34. The disturbance vector $w = (w_i', w_o', w_n')'$ includes the white noise sources that drive the process disturbance filters and the measurement noise filter. $\Sigma_a(A_a, B_a, C_a, D_a)$, $\Sigma_i(A_i, B_i, C_i, D_i)$, $\Sigma_o(A_o, B_o, C_o, D_o)$ and $\Sigma_n(A_n, B_n, C_n, D_n)$ denote the disk drive actuator, the input disturbance filter, the output disturbance filter and the measurement noise filter, respectively, and all these transfer functions are assumed strictly proper.

Substituting the model parameters for HDD actuator, disturbance and sensing noise parameters into equation 3.52 which in turn is substituted into the Riccati equations 3.45 and 3.46, we can obtain an output feedback H_2 optimal control equation 3.42 which achieves the theoretically highest tracking accuracy.

3.4.2 H_2 Optimal Control: Discrete-time Case

Consider a stabilizable and detectable linear time-invariant system Σ with a proper output feedback controller Σ_c, show in Fig 3.34 where,

$$\Sigma: \begin{cases} x(k+1) &= Ax(k) + B_1 w(k) + B_2 u(k), \\ y(k) &= C_1 x(k) + D_{11} w(k), \\ z(k) &= C_2 x(k) + D_{21} w(k) + D_{22} u(k), \end{cases} \quad (3.53)$$

with $x \in \Re^n$ the state, $u \in \Re^n$ the control input, $w \in \Re^l$ the disturbance input, $y \in \Re^p$ the measured output (the measured PES in case of HDD), and $z \in \Re^q$ the output to be controlled (the true PES in this case).

The controller in the form

$$\Sigma_c: \begin{cases} x_c(k+1) &= A_c x_c(k) + B_c y(k), \\ u(k) &= C_c x_c(k) + D_c y(k), \end{cases} \quad (3.54)$$

such that $\|\Phi_{zw}\|_2^2 < \mu$ are parameterized by LMI (Linear Matrix Inequality) [41]:

$$\begin{cases} trace(W) < \mu, \\ \begin{bmatrix} W & C_2 X + D_{22} L & C_2 + D_{22} R C_1 \\ * & X + X' - P & I + S' - J \\ * & * & Y + Y' - H \end{bmatrix} > 0, \\ \begin{bmatrix} P & J & AX + B_2 L & A + B_2 R C_1 & B_1 + B_2 R D_{11} \\ * & H & Q & A + U C_1 & Y B_1 + U D_{11} \\ * & * & X - X' - P & I'_S - J & 0 \\ * & * & * & Y + Y' - H & 0 \\ * & * & * & * & I \end{bmatrix} > 0, \end{cases} \quad (3.55)$$

where

$$\begin{cases} D_c &= R, \\ C_c &= (L - R C_1 X)\Lambda^{-1}, \\ B_c &= \Xi^{-1}(U - Y B_2 R), \\ A_c &= \Xi^{-1}[Q - Y(A + B_2 D_C C_1)X - \Xi B_c C_1 X - Y B_2 C_c \Lambda]\Lambda^{-1}, \end{cases} \quad (3.56)$$

and Ξ and Λ are nonsingular with $\Xi \Lambda = S - YX$.

For the case of disk drive, augmenting the plant mode with process disturbance models and measurement noises, we have [44][45]:

$$
\left\{
\begin{array}{rcl}
A &=& \begin{bmatrix} A_a & B_a C_i & 0 & 0 \\ 0 & A_i & 0 & 0 \\ 0 & 0 & A_o & 0 \\ 0 & 0 & 0 & A_n \end{bmatrix}, \\[2em]
B_1 &=& \begin{bmatrix} B_a D_i & 0 & 0 \\ B_i & 0 & 0 \\ 0 & B_o & 0 \\ 0 & 0 & B_n \end{bmatrix}, B_2 = \begin{bmatrix} B_a \\ 0 \\ 0 \\ 0 \end{bmatrix}, \\[2em]
C_1 &=& [\, C_a \quad 0, \quad C_o \quad C_n \,], \\
D_{11} &=& [\, 0 \quad D_o, \quad D_n \,], \\
C_2 &=& [\, C_a \quad 0, \quad C_o \quad 0 \,], \\
D_{21} &=& 0, \\
D_{22} &=& 0.
\end{array}
\right.
\tag{3.57}
$$

Using LMI toolbox in MATLAB, matrices X, L, Y, F, Q, R, S and J can be found by solving the matrix inequalities 3.56 to minimize $trace(W)$. Hence an optimal solution in the form of 3.54 can be obtained.

3.4.3 An Application Example

Let us consider the actuator and disturbance models given below [44]:

$$
\begin{aligned}
G_p &= \frac{2.861 \times 10^{21}}{(s^2 + 50.27s + 1.579 \times 10^4)(s^2 + 816.8s + 1.668 \times 10^9)}, \\[1em]
G_I &= \frac{1.3916 \times 10^{-5}(s + 575.8)(s + 575.6)(s^2 + 0.04389s + 161.6)}{(s^2 + 315.5s + 8.178 \times 10^4)(s^2 + 315.4s + 8.178 \times 10^4)}, \\[1em]
G_N &= \frac{1.1695(s + 1.431 \times 10^4)(s + 766.2)(s^2 + 8609s + 4.672 \times 10^7)}{(s + 4630)(s + 1538)(s^2 + 4451s + 1.507 \times 10^7)}, \\[1em]
G_o &= \frac{0.7016(s + 1.271 \times 10^4)^2(s^2 + 6.125 \times 10^{-5}s + 4.373 \times 10^8)}{(s + 708.4)^2(s^2 + 0.0001317s + 4.376 \times 10^8)}
\end{aligned}
\tag{3.58}
$$

w_i $(i = 1, 2, 3)$ are independent white noises with variance 1.

The optimal control obtained through the LMI approach is

$$
G_c(z) = \frac{1 \times 10^{-5}(4.289z^5 - 3.541z^4 - 7.898z^3 + 6.485z^2 + 3.745z - 3.073)}{z^5 - 0.3219z^4 - 1.37z^3 + 0.2231z^2 + 0.4031z + 0.06635}.
$$

The frequency responses of the controller and open loop transfer functions are shown in Figures 3.35 and 3.36.

The plant output disturbance is composed of two components. One is the repeatable run-out (RRO), which is attributed to disk shift and written in position error. The RRO is phase-locked to the spindle rotation. The other is the non-repeatable runout (NRRO), which is not phase-locked to the spindle rotation, and is attributed to spindle, disk and actuator assembly resonances.

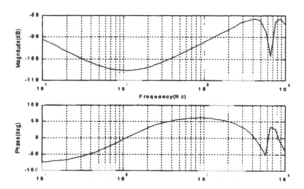

Figure 3.35: H_2 optimal controller frequency response.

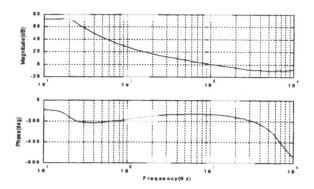

Figure 3.36: Compensated open-looped transfer function frequency response.

Self-induced NRRO imposes fundamental limitations on the viability of very high-TPI drives while RRO components can be compensated for using simple methods such as look-up table. Hence, while calculating the feedback controllers above, only the NRRO component in the output disturbance is considered.

Moreover it should be noted that the H_2-optimal control attempts to minimize the TMR without considering any other specifications such as low frequency gain for counter-reacting the bias force, which typically needs an integral action. Also, the resultant controller might be of very high bandwidth which the actuator may not necessarily be compatible to the actuator. However, the resulting controller provides a good reference of the tracking accuracy that is achievable for more practical control designs.

In general, higher order models can represent the disturbance more accurately. However, considering higher order models for different disturbance sources ends up with a controller, which is very high order and requires more powerful microprocessors for implementation. The numerical stability is an important issue when solving higher order Riccati equations and attention is required to this problem if more accurate disturbance model is chosen. If the resultant controller is of very high order, controller's performance robustness also need to be examined. A realizable and practical controller can be obtained from the designed high order controller through reduction of controller order. But the model order reduction process may offset the benefits of using optimal control. On the other hand, using lower order disturbance models during the design phase makes it easier to find an optimal solution. Discretization of either the controller or the plant model given the limited PES sampling frequency leads to an approximation of the true transfer function, which degrades the optimality of the controllers obtained.

Designs of PID-type controller cascaded with various filters discussed in Section 3.2 appear to be ad-hoc as they follow the designer's understanding of the system to achieve desirable performance through shaping the servo loop manually. On the other hand, the optimal controller design methods, such as the H_2-optimal control discussed in this section, the H_∞-optimal control [28], [58], LQG/LTR control [24], [205], RPT [59] etc provide closed form optimal solution, but the control objectives must be translated first into one of the few types of performance index. The conflict between robustness and performance, the availability of accurate models for plant, noise and disturbance, and the accuracy with which the performance index reflects the actual design problem are some of the critical practical considerations for design based on optimal-control.

The problem of increasing the tracking accuracy can be translated into a problem of finding the minimal H_2 norm of a feedback control system with noise of variance one as input. The H_2 output feedback control that achieves the highest tracking accuracy given the noise and disturbance model can be found by solving Riccati equations or LMIs. The resultant controller is of high order.

All the design approaches presented so far assumes that all the design objectives are to be achieved using feedback controller. It is not easy to meet all requirements using feedback control or single degree-of-freedom controller of low order dynamics. Both the traditional loop shaping and the optimal control design are subject to various control limitations. This shortcoming can be overcome by including feedforward control for known output trajectory such as output overshoot, initial value response, RRO and so on and leaving the task of rejecting noise and disturbance to the feedback controller.

3.5 Advanced Topics

There are some techniques that can be used to improve the transient response as well as the steady state performance of the closed loop servomechanism.

Since the vibration of the actuator is a big hurdle in achieving good performance, elimination of structural vibration is a very important issue. We can use a feedforward control to shape the reference command, which is otherwise a step function, to inhibit actuator vibration. The settling performance of the HDD servomechanism is affected by non-smooth handover between track seek and track-following controllers. Application of initial value compensation (IVC) eliminates the undesirable transient response induced by initial value. Similarly an RRO compensator can be used in addition to the feedback controller so that performance of the closed loop is improved. Two other methods, use of multirate control and multi-sensing servo, give additional freedom to the designer in his/her effort to achieve improved performance.

3.5.1 Input Command Shaping

The methods described in the previous sections address primarily the issues related to accurate track following controller. The step responses are examined for all design examples, but the main purpose of these examples is to see how oscillatory the system is when subjected to a change in one of the inputs. These linear feedback controllers are not suitable for large reference commands expected in track seek mode of operation as the fixed gain controllers makes the VCM driver saturate. As a result, the output goes through excessive overshoot and undershoot. Use of nonlinear control for track seek, such as PTOS described in chapter 2, can eliminate such problems. However, the linear part of the PTOS is simple state feedback and the tracking performance is not very satisfactory particulalrly in presence of external disturbances such as RRO and NRRO. The controllers explained in this chapter so far are better suited for tackling the issues of track following, but transition between such controller and PTOS with smooth handover becomes a challenging issue.

It is possible to meet the contradicting objectives of track seek and track following satisfactorily if the controller structure has two degree-of-freedom. In such case, the track following issues are handled by the linear feedback controller such as a PID-type control law in cascade with suitable filter for enhancement of performance. Since this feedback control is not suitable for large changes in external signals, the command input should not be allowed to enter the loop directly. In stead, shaping the command input using a two-degree-of-freedom structure can be used.

Let us consider the closed-loop system shown in Figure 3.37. In this figure, G_c and G_p represent the transfer functions of the controller and the plant, respectively. The overall output of the system is,

$$y = \frac{G_c G_p}{1 + G_c G_p} r.$$
(3.59)

Let the command signal be r'. Assuming that the feedback controller G_c has been designed for accurate track following, we can design an input shaper I_s

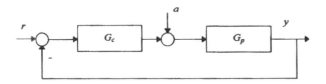

Figure 3.37: Feedback control system.

such that

$$y = \frac{G_c G_p}{1 + G_c G_p} I_s r',$$
$$= G_{clp} I_s r', \tag{3.60}$$

has a desirable step response. Here G_{clp} is the closed-loop transfer function.

The advantages of this two degree-of-freedom structure are,

1. I_s can be selected such that slow dynamics in the transfer function from r' to y is removed,

2. feedback controller G_c can be designed to produce relatively fast closed-loop poles. However, feedback control does not shape the zeros of the closed loop transfer function. The input shaper I_s can be designed to relocate the zeros and, in addition, possibly the poles, and

3. I_s can be designed to filter out the high frequency components in the command signal r'.

Example 1: When the closed-loop transfer function represented by

$$G_{clp}(z) = \frac{z^{-d} N_{clp}^+(z^{-1}) N_{clp}^-(z^{-1})}{D_{clp}(z^{-1})} \tag{3.61}$$

where $N_{clp}^-(z^{-1})$ represents all the zeros outside the unit circle and $N_{clp}^+(z^{-1})$ those inside the unit circle, I_s can use the form of ZPET controller [124]:

$$I_s = \frac{z^d D_{clp}(z^{-1}) N_{clp}^-(z)}{[N_{clp}^-(1)]^2 N_{clp}^+(z^{-1})}. \tag{3.62}$$

This results in

$$y = \frac{N_{clp}^-(z^{-1}) N_{clp}^-(z)}{[N_{clp}^-(1)]^2} r'. \tag{3.63}$$

In this case, the performance is limited by the non-minimum phase zeros of the closed-loop system if the control signal is not saturated.

Example 2: Consider the system shown in Figure 3.37. Suppose the plant is a double integrator i.e. $G_p(s) = \frac{1}{s^2}$ and the controller is an ideal PD controller $G_c(s) = K_p(1 + T_d s)$. The open-loop transfer function is

$$G_c(s)G_p(s) = \frac{1}{s^2}K_p(1 + T_d s). \qquad (3.64)$$

And the closed-loop transfer function from r to y:

$$\frac{G_c(s)G_p(s)}{G_c(s)G_p(s) + 1} = \frac{K_p(1 + T_d s)\frac{1}{s^2}}{1 + K_p(1 + T_d s)\frac{1}{s^2}}$$

$$= \frac{K_p T_d s + K_p}{s^2 + K_p T_d s + K_p}. \qquad (3.65)$$

The natural frequency of the closed loop is $\omega = \sqrt{K_p}$ and damping ratio is $\zeta = \frac{\sqrt{K_p}T_d}{2}$. Figure 3.38 shows the step response of this closed loop.

We can use input shaping controller I_s to improve the performance during step response. Let

$$I_s = \frac{s^2 + K_p T_d s + K_p}{K_p T_d s + K_p}\frac{1}{T_s s + 1}, \qquad (3.66)$$

where T_s is a desirable time constant, the command input response from r' ($r = I_s r$) to y through r now becomes $\frac{1}{T_s s + 1}$. T_s can be selected according the the limitations on control current. No change has been made in the parameters of the feedback controller K_p and T_d to achieve the desirable step response of a first order system.

Step responses of $\frac{K_p(1 + T_d s)}{s^2 + K_p(1 + T_d s)}$ for $K_p = 1$ with different values of T_d are shown in Figure 3.38. When T_d increases, overshoot decreases but can not be eliminated even if the damping ratio of the closed-loop system is above 1. However, when the input shaper $I_s = \frac{s^2 + 0.5s + 1}{0.05s^2 + 0.6s + 1}$, designed using equation 3.66 with $T_s = 0.1$ and $T_d = 0.5$, is used, step response shows rapid change without overshoot regardless of the feedback loop's overshoot and damping. This example also illustrates the fact that the closed-loop step response is affected not only by the closed-loop poles but also by the zeros.

It is easy to verify that the closed-loop transfer function from a to y is:

$$\frac{G_p(s)}{G_p(s)G_c(s) + 1} = \frac{1}{s^2 + K_p T_d s + K_p}. \qquad (3.67)$$

One can easily select T_d such that $\zeta = \frac{\sqrt{K_p}T_d}{2} > 1$ hence no overshoot in step response from a to y.

Comparison between equations 3.65 and 3.67 reveals that even if the open-loop transfer functions are the same, the closed-loop behaviours may be different because different transmission zeros. It is well known that state feedback control can arbitrarily place the closed-loop poles for controllable plants but

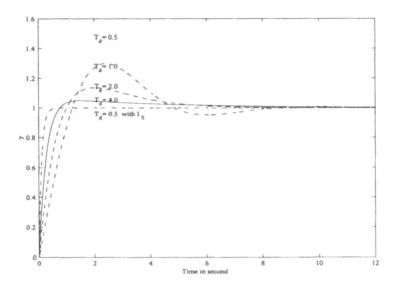

Figure 3.38: Step response of $\frac{K_p(1+T_d s)}{s^2+K_p(1+T_d s)}$ for $K_p = 1$ with different values of T_d.

not the zeros. Hence feedforward control and/or input shaping can be designed to affect the tranmission zeros as well as poles to achieve the desired step responses. In other words, we *design the feedback controller to achieve accurate track following via suppression of vibration and sensing noise, and design the feedforward control and/or input shaping to achieve desirable step responses.* The later is not a feedback control design problem and hence the stability issue is of less concern.

One special way of designing command input shaper for lightly damped systems is to apply two successive step changes instead of one such that the oscillations from the two step changes cancel each other. This method was first reported by N. Singer and W. Seering to eliminate the ringing effects in lightly damped system [176]. Hai T. Ho presented in [82] the application of this method in HDD fast servo bang-bang control. Figure 3.39 gives an illustrative example of the outcome. For this example, the plant transfer function is $P(s) = \frac{\omega_n^2}{s^2+2\zeta\omega_n+\omega_n^2}$. Command shaping is realized using an FIR filter $b_0 + b_1 z^{-1}$ with delay time $T_{fir} = \frac{\pi}{\omega_n\sqrt{1-\zeta^2}}$ where $b_0 = \frac{1}{1+K}, b_1 = \frac{K}{1+K}$, and $K = e^{-\frac{\zeta\pi}{\sqrt{1-\zeta^2}}}$. The simulation is carried out with $\omega_n = 1$, $\zeta = 0.2$, $T_{fir} = 3.2064$, $b_0 = 0.6550$, $b_1 = 0.3450$, $K = 0.5266$. It is clearly seen that the oscillations from the two substeps cancel each other. The combined step response is no longer oscillatory even though the plant model is lightly damped.

Besides the PTOS explained in chapter 2 and the two degree-of-freedom controller with command shaping, fast step response without overshoot can

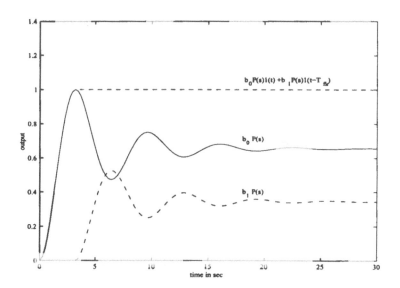

Figure 3.39: Step response of a second order model with flexible modes controlled by a input shaper.

also be achieved in presence of control saturation using time varying feedback control gain. See for reference, the workds presented in [30], [230], [127] and the references therein.

3.5.2 Initial Value Compensation

In order to meet the stringent specifications of both track following and track seek, it is better to design two controllers independently so that each controller can be optimally tuned to address specific issues. While designing the track following controller, the initial states of the plant and controller are usually assumed to be zero. However, when the control is transferred from the seek mode to track-following mode, the final states at the end of the seek controller are not necessarily zero. If proper care is not taken, continuity of the control signal at the time of mode switching is not ensured. This is not desirable as it excites the resonant modes of the actuator and, as a consequence, time taken for the head to settle on the track is extended. Possible solution lies in the compensation of the initial states included in the track following mode.

Let us consider again the plant G_p and controller G_c shown in Fig. 3.37 where the command input signal $r = 0$, the output of the system with non-zero initial states is given by:

$$y = \frac{N_p}{D} X_p(0) + \frac{N_c}{D} X_c(0) + \frac{N_a}{D} a \tag{3.68}$$

where N_a/D is the transfer function between signal injection point a and y,

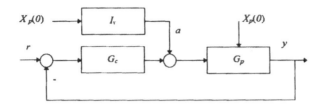

Figure 3.40: IVC via injecting a signal.

and $X_c(0) = 0$. Thus

$$y = \left[\frac{N_p}{D} + \frac{N_a}{D} \frac{n}{d} \right] X_p(0). \tag{3.69}$$

The objective of the initial value compensator is to find $I_v = \frac{n}{d}$, as opposed to changing the closed-loop system characteristic equation D, such that $\frac{N_p}{D} + \frac{N_a}{D} \frac{n}{d}$ has a more desirable dynamics than $\frac{N_p}{D}$.

According to Yamaguchi et al [219], any N_a can be represented as

$$N_a = N_a^u N_a'$$

where N_a' contains only the stable zeros and N_a^u contains only unstable zeros. Let the desired poles of the transfer function between the initial states $X_p(0)$ and y described by equation 3.69 be located at ζ_i $(i = 1, 2, ...l)$, and let

$$d_m = (z - \zeta_1)(z - \zeta_2)...(z - \zeta_l),$$

and

$$d = d_m d' = d_m N_a', \tag{3.70}$$

where d' contains only stable roots which we choose to be N_a'. Then

$$
\begin{aligned}
y &= \left[\frac{N_p d' d_m + N_a n}{D d'} \right] \frac{1}{d_m} X_p(0) \\
&= \left[\frac{N_p d_m + N_a^u n}{D} \right] \frac{1}{d_m} X_p(0). \tag{3.71}
\end{aligned}
$$

Now we select n such that the roots of $N_p d_m + N_a^u n$ in equation 3.71 include all the roots of D, which are $\lambda_i, i = 1, 2, ...m + n_d$, and $\lambda_{i+1} = 0$ for continuous time or $\lambda_{i+1} = 1$ for discretize time model. Then the transient response of $y(t)$ is dominated by the desired poles ζ_i $(i = 1, 2, ...l)$. Hence it is necessary to find $n = [n_1, n_2, ...n_m]$ in the form of

$$n_i = a_{i,q} z^q + a_{i,q-1} z^{q-1} + ... + a_{i,1} z + a_{i,0}, \tag{3.72}$$

so that the following equation is satisfied:

$$N_p(\lambda_j) d_m(\lambda_j) + N_a^u(\lambda_j) n_i(\lambda_j) = 0, i = 1, ..., m; j = 1, ..., m + n_d + 1. \tag{3.73}$$

Substituting equation 3.72 into equation 3.73 yields

$$
\begin{bmatrix}
\lambda_1^q & \lambda_1^{q-1} & \cdots & \lambda_1 & 1 \\
\lambda_2^q & \lambda_2^{q-1} & \cdots & \lambda_2 & 1 \\
& & \vdots & & \\
\lambda_{m+n}^q & \lambda_{m+n}^{q-1}, & \cdots & \lambda_{m+n} & 1
\end{bmatrix}
\begin{bmatrix}
a_{i,q} \\
a_{i,q-1} \\
\vdots \\
a_{i,0}
\end{bmatrix}
$$

$$
=
\begin{bmatrix}
-N_a^u(\lambda_1)^{-1} N_p(\lambda_1) d_m(\lambda_1) \\
-N_a^u(\lambda_2)^{-1} N_p(\lambda_2) d_m(\lambda_2) \\
\vdots \\
-N_a^u(\lambda_{m+n})^{-1} N_p(\lambda_{m+n}) d_m(\lambda_{m+n})
\end{bmatrix} . \tag{3.74}
$$

If $q + l = m + n_d + 1$ and if D does not have repeated roots, then the first term in the above equation is a non-singular square matrix, and n_i and n can be easily calculated by solving equation (3.74). Furthermore, d has been specified by equation (3.70). Hence we have the $I_V = \frac{n}{d}$.

Example: Let the actuator transfer function be $G_p(s) = \frac{1 \times 10^4}{s^2}$, which is controlled by a PD controller

$$
G_c(s) = \frac{0.014s + 1}{0.00014s + 1}. \tag{3.75}
$$

Following the above equations,

$$
S = \frac{s^3 + 714.3s^2}{s^3 + 714.3s^2 + 10^5 s + 7.143 \times 10^6},
$$

$$
T = \frac{100000s + 7.143 \times 10^6}{s^3 + 714.3s^2 + 10^5 s + 7.143 \times 10^6}. \tag{3.76}
$$

The poles of the closed-loop system are: -848.97, $-75.516 + 77.95i$, $-75.516 - 77.95i$.

We wish to inject the IVC signal from the reference signal point r. Let the transfer function from the initial position value to plant output considering I_v:

$$
y = (S + I_v T) y(0). \tag{3.77}
$$

Solution 1:

Following the equations given above, we have $N_a = (s^2(s + 714.2857))$. The roots of N are $\lambda_i = 100 \times [-5.5802, -0.7813 + 0.8183i, -0.7813 - 0.8183i]$. d_m is to be 20 times faster than these poles which can be determined. Substituting into equation 3.74 we have $a_i = 9.7843 \times 10^{-6}, 6.6852 \times 10^{-3}, 3.1529 \times 10^{-2}, 0$.

Hence

$$
I_v = \frac{n}{d} = \frac{5.591 \times 10^5 s^3 + 3.82 \times 10^8 s^2 + 1.802 \times 10^9 s}{s^4 + 1.304 \times 10^4 s^3 + 2.625 \times 10^7 s^2 + 5.895 \times 10^{10} s + 4.082 \times 10^{12}}. \tag{3.78}
$$

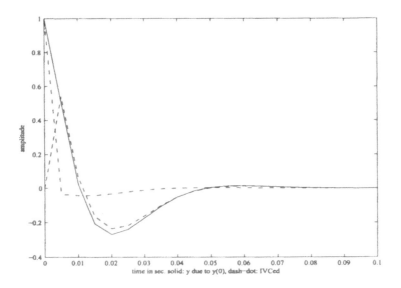

Figure 3.41: Simulation result of the IVC.

The response of the system is shown in Figure 3.41.

Solution 2:

In view of equation (3.77), let

$$I_v = -\frac{S(s)}{T(s)} = -\frac{(s + 1000)7.143 \times 10^{-6}s^2}{s + 71.43}, \tag{3.79}$$

so that the effect of initial values is totally canceled out. However, this transfer function is not causal. We can make $I_v(s)$ causal by including additional (fast) poles. Then I_v can be expressed as

$$\begin{aligned} I_v(s) &= -\frac{S(s)}{T(s)} \frac{1}{\text{additional (fast) poles}} \\ &= -\frac{10^{-6}s^3 + 0.007143s^2}{s + 71.43} \frac{1}{\text{additional (fast) poles}}. \end{aligned} \tag{3.80}$$

Taking I_v into consideration, the transfer function from the initial position value to plant output is,

$$\frac{y}{y(0)} = S + I_v T = S(1 - \frac{1}{\text{additional (fast) poles}}). \tag{3.81}$$

By selecting the additional poles to be a few times faster than those of $S(s)$ with a damping ratio close to 1, the effect of $y(0)$ on y is dominated by these fast poles and diminishes quickly.

In the present example, choosing the additional poles to be 20 times faster than the fastest of $S(s)$:

$$I_v(s) = -\frac{(s+1000)7.143 \times 10^{-6}s^2}{s+71.43} \frac{1}{2.122 \times 10^{-7}s^2 + 0.0003686s + 1}. \quad (3.82)$$

Simulation results shown in Figure 3.42 includes the plant's responses due to initial value (solid line), due to the compensator (dashed line), and the combined response. It is clearly evident from this simulation that the IVC makes the initial value response decay very fast, and it is achieved without any change in the feedback controller.

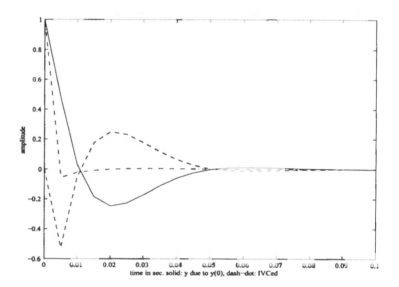

Figure 3.42: Simulation result of the IVC.

3.5.3   RRO Compensation

Ideally the shape of a track written on the disk surfaces is expected to be circular. However, due to the presence of nonrepeatable runout (NRRO) sources such as disk-spindle pack vibration, disk and slider related vibrations, sensing noise etc in the servo track writing process, the tracks created on the disks are not perfectly circular. In addition, any misalignment of the center of the spindle motor axis and the geometric center of the servo tracks introduces non-circularity in the tracks. Even the distortion of disk caused by clamping contributes to non-circularity of tracks.

Imperfections of track are illustrated in the Figure 3.43. The non-circularity of the tracks manifests as repeatable runout (RRO) during the track following controller's operation. This is a disturbance signal on top of the NRRO present in the head-disk assembly and spindle-disk assembly.

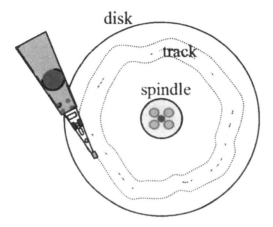

Figure 3.43: A schematic diagram of disk runout.

The PES signal extracted from the servo bursts written on the tracks contains both RRO and NRRO components. The RRO is synchronized with the disk's rotation but the NRRO component is not. Synchronous averaging of the PES signal can separate the repeatable components from the PES signal.

If there are N servo sectors in an HDD, we get N samples per revolution when PES is measured. For the sake of averaging, we need to measure PES for several revolutions. If measurement is performed for M revolutions then the measured PES y is a $M \times N$ matrix. Let the fundamental frequency of RRO signal be ω. The repeatable component $R_y(n)$ of PES and the nonrepeatable component $N_y(n)$ can then be obtained according to the following:

$$R_y(n) = \frac{1}{M} \sum_{m=1}^{M} y(n,m), \quad n = 1 \cdots N, \tag{3.83}$$

$$N_y(n,m) = y(n,m) - R_y(n), \quad n = 1 \cdots N, m = 1 \cdots M. \tag{3.84}$$

Figure 3.44 shows a sample of PES obtained from a track of an HDD with 62 sectors and rotational speed of 3325.65 RPM. The power spectral density of the PES is shown in Figure 3.45.

Since the RRO and NRRO are added to head position to form the measured position error signal (PES), only the frequency components that fall within the servo bandwidth can be attenuated by the basic servo loop that we have discussed previously.

When the servo system is to follow the track center that contains large amount of runout, the actuator has to move more frequently in both direction and therefore uses higher servo power. Reducing the amount of RRO compo-

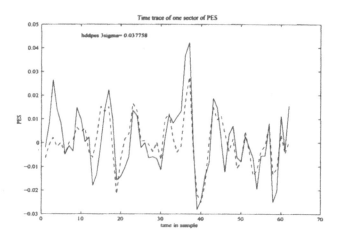

Figure 3.44: RRO (solid line) and one sector of PES (dashed line) time trace.

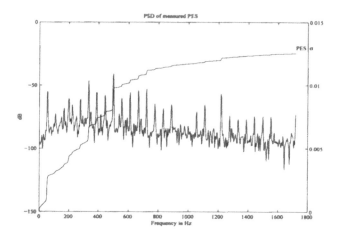

Figure 3.45: PES power spectrum density.

nent written on the disk, especially the higher frequency components of RRO, is essential for relaxing the burden on servo control. This must be taken care of during the process of servo track writing, which is discussed in chapter 5.

Since RRO contains narrow band vibrations at frequencies which are integer multiples of the spindle frequency, a common approach to cancel the effects of RRO is to use peak filter discussed in section 3.2.5. However, such approach alters the servo loop gain and one must check the acceptability of the stability and transient response if the peak filter is added in the loop.

Fortunately, the RRO is phase locked to spindle rotation and, therefore, can be modeled easily using simple experiment. Once a model of the RRO is available, it can be compensated for using feedforward control. There are different methods to implement this correction of RRO such as

1. calculating the signal required to be injected at each frequency of RRO and storing the information in a lookup table which can be used for feedforward compensation, or

2. using an Adaptive Feedforward Control (AFC) scheme [206].

These methods are explained in the following sub-sections. Another method, that includes simultaneous multiple frequency RRO compensation using periodic signal generator with a delay term, is explained later in this section.

Compensation of RRO using Inverse Signal

Since RRO is a repeatable signal, it can be decomposed as a sum of a series of sine and cosine waves, which can be represented in the following formats:

$$
\begin{aligned}
R_y(n) &= \sum_{i=1}^{L} A_i \sin(\frac{2\pi i(n-1)}{N} + \phi_i), \\
&= \sum_{i=1}^{L} A_i e^{-j\phi_i}, \\
&= \sum_{i=1}^{L} [a_i \sin(\frac{2\pi i(n-1)}{N}) + b_i \cos(\frac{2\pi i(n-1)}{N})], \\
&= \begin{bmatrix} \sin\frac{2\pi(n-1)}{N}, \cdots, \sin\frac{2\pi\frac{N}{2}(n-1)}{N} \\ \cos\frac{2\pi(n-1)}{N}, \cdots, \cos\frac{2\pi\frac{N}{2}(n-1)}{N} \end{bmatrix} \\
&\quad \cdot \begin{bmatrix} a_1, \cdots, a_{N/2} \\ b_1, \cdots, b_{N/2} \end{bmatrix}^T.
\end{aligned}
\tag{3.85}
$$

$L = 0.5 * F_s/rpm/60 = N/2$ is number of frequencies which is half of the number of sectors, Δt is the sampling period. Let $\theta_a = \begin{bmatrix} a_1, \cdots, a_{N/2} \\ b_1, \cdots, b_{N/2} \end{bmatrix}^T$, expanding the above equations we have

$$
\begin{aligned}
R_y(1) &= \begin{bmatrix} \sin\frac{2\pi(1-1)}{N}, \cdots, \sin\frac{2\pi\frac{N}{2}(1-1)}{N} \\ \cos\frac{2\pi(1-1)}{N}, \cdots, \cos\frac{2\pi\frac{N}{2}(1-1)}{N} \end{bmatrix} \theta_a, \\
R_y(2) &= \begin{bmatrix} \sin\frac{2\pi(2-1)}{N}, \cdots, \sin\frac{2\pi\frac{N}{2}(2-1)}{N} \\ \cos\frac{2\pi(2-1)}{N}, \cdots, \cos\frac{2\pi\frac{N}{2}(2-1)}{N} \end{bmatrix} \theta_a,
\end{aligned}
\tag{3.86}
$$

$$
\vdots \quad \vdots
$$

$$
R_y(N) = \begin{bmatrix} \sin\frac{2\pi(N-1)}{N}, \cdots, \sin\frac{2\pi\frac{N}{2}(N-1)}{N} \\ \cos\frac{2\pi(N-1)}{N}, \cdots, \cos\frac{2\pi\frac{N}{2}(N-1)}{N} \end{bmatrix} \theta_a.
\tag{3.87}
$$

We can rewrite the above equation in matrix form as:

$$\mathbf{R}_y = \mathbf{\Phi}\theta_a, \tag{3.88}$$

then we have

$$
\begin{aligned}
\theta_a &= (\mathbf{\Phi}^T\mathbf{\Phi})^{-1}\mathbf{\Phi}^T\mathbf{R}_y, \\
&\overline{\overline{\delta}} \quad \mathbf{\Phi}_c\mathbf{R}_y. \tag{3.89}
\end{aligned}
$$

Hence the amplitude of each sine and cosine wave can be obtained. It should be pointed out here that,

1. if one is interested to find the amplitudes of a few selected harmonics, then the corresponding columns of the $\mathbf{\Phi}$ matrix in the above equation can be retained omitting the remaining columns, and

2. $R_y(k)$ in the above equation is the RRO measured from the closed-loop and is not the realtime PES signal which contains the NRRO signal.

Let us assume that the transfer functions of the feedback controller and the plant are G_c and G_p, respectively, then the counter signal to be injected at a of Figure 3.40 to cancel RRO of frequency if_0 should be

$$a_{fi} = A_i e^{j\phi_i}\left|\frac{1+G_cG_p}{G_p}\right|_{f=if_0} = \frac{A_i}{B_i}e^{j(\phi_i-\theta_i)}, \tag{3.90}$$

where B_i and θ_i are the magnitude and phase of the transfer function between a and y at frequency if_0. The above process is essentially means inverting the RRO signal with respect to a suitable transfer function to have a counter signal for RRO cancelation.

The Bode plot from a to y for an example system is shown in Figure 3.23. The bode plot is evaluated using sweept sine excitation of the model and then measuring the output signal's amplitude ratio and phase delay with respect to the input signal at each frequency.

Using the equations given above, we find for the signal shown in Figure 3.44, $a_i = [2.1608, 0.5680, 1.1550, -0.4569, \cdots] \times 10^{-3}$, $b_i = [-2.0524, -0.0731, -0.5640, \cdots] \times 10^{-3}$.

Then we can calculate using equation 3.90 the control signal required to compensate for these RRO components. The spectrum of the resultant PES with all the RRO components taken out is shown in Figure 3.46. Comparing Figure 3.46 with Figure 3.45, there is a 33% reduction in the standard deviation of PES when all components of RRO are cancelled.

Implementation of this method requires measurement of the RRO signal, which is then used to calculate the control input that can corrects the phase and amplitude of RRO at certain frequencies via plant inversion. This can as

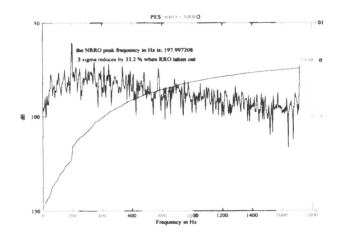

Figure 3.46: PES NRRO power spectrum density.

well be realized by first creating a look up table using equation 3.90 and then injecting the appropriate signal instead of implementing equations 3.84, 3.89, and 3.90.

It is obvious from the discussions above that an inverse of the shock transfer function is calculated at the runout frequencies. In general, inverting a plant model at all frequencies for canceling the output PES can not be realized as the process is a causal system and inverting the model results in a non-causal model. It involves pure differentiation of the output signal which is not feasible. Moreover, even if a causal approximation of the inverse of the model is used, the noise in PES will be significantly amplified. Additionally, the high frequency disturbances will not be canceled due to the phase error in the approximate inverse.

RRO Compensation using Adaptive Feedforward Cancelation

Refering to the block diagram shown in Figure 3.47 where O is the feedback control system, the equivalent periodic disturbance $d_i(\omega_i)$ is represented by[206]

$$d_i(\omega_i) = a_i \cos \omega_i t + b_i \sin \omega_i t. \tag{3.91}$$

The *adaptive feedforward control* (AFC) attempts to reproduce the disturbance by estimating a_i and b_i in an adaptive manner. The disturbance d_i can be exactly canceled when the estimates of the disturbance coefficients are such that,

$$\hat{a}_i(t) = a_i(t), \tag{3.92}$$

$$\hat{b}_i(t) = b_i(t). \tag{3.93}$$

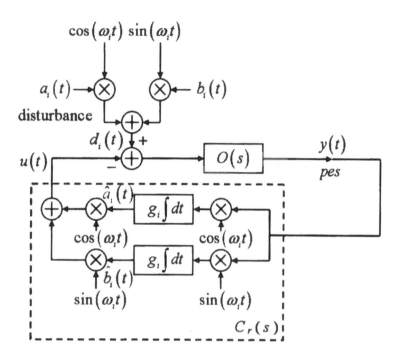

Figure 3.47: Basic AFC scheme for RRO compensation at single frequency ω_i.

The continuous-time adaptive control to adjust the estimates of $a_i(t)$ and $b_i(t)$ consists the following update laws as shown in Figure 3.47,

$$\frac{d}{dt}\hat{a}_i(t) = g_i y(t) \cos(\omega_i t), \qquad (3.94)$$

$$\frac{d}{dt}\hat{b}_i(t) = g_i y(t) \sin(\omega_i t), \qquad (3.95)$$

where $y(t)$ is the position error signal (PES), $g_i > 0$ is the adaptation gain and ω_i is the desired compensation frequency.

Based on Laplace transform analysis, Bodson et al [19] showed that the adaptive control scheme of equations 3.94 and 3.95 is equivalent to the scheme based on internal model principle

$$C_r(s) = g_i \frac{s}{s^2 + \omega_i^2}, \qquad (3.96)$$

in the sense that given the same disturbance $d_i(t)$, the responses $y(t)$ are identical for zero initial conditions. Hence the stability of the AFC system, for both single and multiple frequency runout compensation cases, can be verified by checking the stability of the closed-loop system consisting of O and C_r. Furthermore, the adaptive control is stable for all g if O is positive real; when

O has positive real part in low excitation frequency range, the system is stable for sufficiently small adaptive gain g_i. At excitation frequency where O has a negative real part, the adaptation gain can be a negative small value. Note that very often O is the closed-loop servo system with the baseline servo control while C_r or AFC is the compensator added for canceling the RROs.

The discrete-time representation of the adaptive control and transfer function equivalent are as follows,

$$\hat{a}_i[k] = \hat{a}_i[k-1] + g_i y[k] \cos(\omega_i T_k), \tag{3.97}$$

$$\hat{b}_i[k] = \hat{b}_i[k-1] + g_i y[k] \sin(\omega_i T_k), \tag{3.98}$$

$$C_r(z) = g_i \left\{ \frac{z^2 - \cos(\omega_i T_k)z}{z^2 - 2\cos(\omega_i T_k)z + 1} \right\}, \tag{3.99}$$

where T_k is the sampling period.

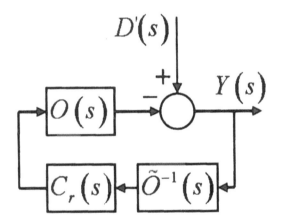

Figure 3.48: Simplified block diagram for analyzing the RRO compensation effectiveness. $\tilde{O}^{-1}$ is the approximate inverse of the plant O for improving the effectiveness of the RRO compensation.

Now, to enhance the effectiveness of RRO compensation for the known closed-loop servo system $O(s)$, consider an alternative equivalent RRO disturbance $D'(s)$ as shown in Figure 3.48 where $\tilde{O}^{-1}$ is the approximate inverse of O for reasons to be discussed below. The transfer function from $D'(s)$ to $Y(s)$ can be written by:

$$R(s) = \frac{1}{1 + C_r(s)\tilde{O}^{-1}(s)O(s)}. \tag{3.100}$$

When $R(j\omega T) = 1$, the loop gain at runout frequency ω equals to 1 meaning there is no runout compensation. When $R(j\omega T) < 1$ or > 1, the closed-loop will attenuate or amplify the corresponding frequencies respectively.

To enhance the stability of AFC algorithm as well as to avoid amplification of other RRO components, a suitable $\tilde{O}^{-1}(z)$ should satisfy: $R(z) < 1$ at other harmonics and $O(z)\tilde{O}^{-1}(z)$ must be stable. As $C_r(e^{j\omega_i T})$ is pure imaginary, the objective is to find a suitable $\tilde{O}^{-1}(z)$ that shapes the plant such that $O(z)\tilde{O}^{-1}(z)$ is close to SPR in a wide frequency range so that $R(e^{j\omega T}) < 1$. If the equivalent plant $O(z)$ is not a zero phase system, pole/zero cancelation and phase cancelation can be used to design $\tilde{O}^{-1}(z)$ to shape the plant such that $O(z)\tilde{O}^{-1}(z)$ is close to a zero-phase system [224].

Without loss of generality, let the equivalent plant $O(z)$ be expressed as

$$O(z) = \frac{N(z)}{D(z)} = \frac{b_0 z^m + b_1 z^{m-1} + \cdots + b_m}{z^n + a_1 z^{n-1} + \cdots + a_n}. \tag{3.101}$$

Because the equivalent plant is the closed-loop system under nominal controller, all the roots of $D(z)$ are inside the unit circle in the Z-plane, and the roots of $N(z)$ can be either inside, on or outside the unit circle.

If the equivalent plant $O(z)$ is a minimum phase system, i.e., all the roots of $N(z)$ are inside the unit circle, then all the poles and zeros of the plant are cancelable. Therefor we can choose:

$$\tilde{O}^{-1}(z) = \frac{D(z)}{z^d N(z)}, \tag{3.102}$$

where $d = n - m$. After that all the poles and zeros of $O(z)$ are canceled by $\tilde{O}^{-1}(z)$.

If the plant is a non-minimum phase system, and suppose that there is no zero on the unit circle, then the numerator polynomial $N(z)$ can be factored into two parts such that,

$$N(z) = N^s(z)N^u(z), \tag{3.103}$$

where $N^s(z)$ includes stable zeros which are cancelable, and $N^u(z)$ includes zeros which are not inside the unit circle. Then $\tilde{O}^{-1}(z)$ can be designed as

$$\tilde{O}^{-1}(z) = \frac{D(z)N^{u^*}(z)}{z^{d+2u}N^s(z)}, \tag{3.104}$$

where u is the order of $N^{u^*}(z)$. $N^{u^*}(z)$ can be designed according to Butterworth transforms. If $N^u(z)$ is represented as,

$$N^u(z) = d_0 + d_1 z + \cdots + d_u z^u, \tag{3.105}$$

then $N^{u^*}(z)$ can be designed as,

$$N^{u^*}(z) = d_u + d_{u-1}z + \cdots + d_0 z^u. \tag{3.106}$$

Note that $\frac{N^{u^*}(z)}{z^u}$ is the complex conjugate of $N^u(z)$ when $z = e^{j\omega T_s}$. Therefore, $N^u(z)\frac{N^{u^*}(z)}{z^u}$ is positive real.

When $\tilde{O}^{-1}$ is used, the phase of the shaped plant is $1/z^{d+u}$ and is near zero phase in a wider low frequency range compared with the case without $\tilde{O}^{-1}$. Such a modification can improve the robustness and convergence rate of the AFC scheme. Additionally, such a modification might lower the sensitivity transfer function hump from D' to y and prevent significant amplification of other RRO harmonics or NRRO signal when canceling the selected RRO harmonics.

We note here that by using (3.104), C_r with a suitable gain g_i might be able to generate a notch at the desired frequency and at the same time have attenuation at frequencies other than those of C_r's center frequency [224]. Nevertheless, such a system is still governed by the Bode Integral theorem, and hence there will be amplification of RRO and NRRO at some other frequencies.

RRO Compensation via Periodic Signal Generator using Delay Terms

In the above designs, compensating each RRO frequency requires a second order controller. If we want to compensate for more frequencies of RRO, the order of the compensator increases to twice the number of frequencies to the compensated for. Adopting the same "plug-in" structure as shown in Figure 3.49, periodic signal generators (PSGs) with a simple delay term in a feedback loop can be used to generate the internal model for disturbance [73] [137].

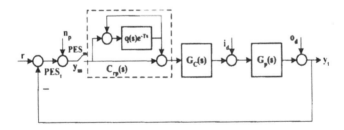

Figure 3.49: "Plug-in" repetitive compensation using periodic signal generator-continuous time case [145].

Hara *et al* has proved in [73] that exponential stabilization is not achievable for such repetitive control systems with strictly proper transfer functions. However, when a low-pass filter is used in conjuction with the delay section, the internal model is able to generate the signal with certain cancelation of the disturbances [73] [145]. Figure 3.50 shows an example of a scheme with PSG based control. Discrete time version of the such a scheme can be found in [156] and the references therein.

In the case of digital control, we can assume that the plant model shown

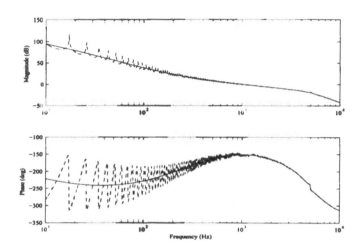

Figure 3.50: Frequency response of a control system using PSG. Solid line: with controller $C(s)$ and without the C_{rp}, Dashed-line, with controller $C(s)$ and without the C_{rp}.

in Figure 3.49 is expressed as

$$G_p(z^{-1}) = \frac{z^{-k}B(z^{-1})}{A(z^{-1})},$$ (3.107)

where k is the number of delays in the plant. Following [107], the controller with a periodic signal generator is given by:

$$G_c(z^{-1}) = K_r \frac{z^{-N+k}q(z^{-1})B^u(z^{-1})}{(1 - q(z^{-1})z^{-N})B^s(z^{-1})b},$$ (3.108)

where K_r is the repetitive control gain, N is the number of discrete-time samples of the periodical disturbance per revolution, $B^u(z^{-1})$ is the non-minimum phase zeros (non-cancelable part of the numerator), $B^s(z^{-1})$ is the minimum phase zeros (cancelable part of the numerator),

$$q(z^{-1}) = \frac{z + 2 + z^{-1}}{4},$$ (3.109)

and $b = [B^u(1)]^2$. It is an inverse model of the plant, modified for unstable zeros, and the remainder of the controller places poles on the unit circle at the harmonics of the fundamental frequency. The low-pass filter $q(z^{-1})$ brings the poles inside the unit circle and sacrifices high-frequency regulation, in order to improve robustness to the unmodeled dynamics and to guarantee stability [34], [168].

The formulation of the RRO and a few methods to cancel the effects of RRO on the performance of the servomechanism are explained in this section. Based on the known RRO trajectory, inverting the RRO signal against the transfer function allows us to inject a signal into the loop such that the RRO is cancelled. Such method is based on pre-calculation of the reference signal and is implemented in feedforward manner. This method does not affect closed loop stability. The disadvantage of this method is that the values of the compensation signal need to be stored in a look-up-table (LUT) after pre-calculation; when there is a change in the RRO profile due to motor aging, disk slip etc, the LUT need to be re-calculated and updated. The RRO compensation system can be designed to follow the servo-burst-defined track centers. Alternatively, knowing that the RRO reflects the deviation of the servo-burst-defined track centers from perfect circles, we can redefine the actual data track centers which are not coincident with the servo-burst-defined track centers by not following the RRO signal in PES. For the later scheme, interested readers may refer to the zero-acceleration-path or ZAP scheme reported in [33].

Feedback control based peak filters of limited peak gains, for both single and multiple peaks, as well as delay-generated multiple frequency peaks, allow partial cancelation of RRO signals. Peak filters of infinity peak gain (and their equivalent automatic feedforward control or AFC) allow complete cancelations of a specific RRO frequencies. All these schemes affect the control loop gain and hence affect the system stability.

In addition to the accuracy of the RRO compensation scheme, time taken by effective compensation is a critical issue as it affects the access performance of the HDD. When the head moves from one track to another, the time taken to arrive at the desired data block is anywhere between the seek-settle time plus zero and seek-settle time plus the time for one full revolution. If, for example, an AFC scheme takes one full revolution to converge in order to cancel the necessary percentage of a target RRO component, then all the data read/write actions need to wait for the full rotational period in addition to the average latency of half of the rotational period. However he AFC scheme, which typically takes 10 ms or more to converge, can be used to learn in real time or offline the RRO pattern which varies slowly. The RRO pattern learnt can then be used to update the RRO cancelation LUT which is used for realtime RRO correction.

3.5.4 Multirate Control

A multi-rate system is a discrete time system in which more than one sampling rate is used with different sampling rates for different sections [105, 116]. For example, in many practical systems, the output is sampled at a rate of f_O Hz where as the input is updated at a rate $f_i \neq f_O$ Hz [204]. Multi-rate system has been studied for control systems with different time-scale loops, and different sampling rate and control update rate. While designing controller for such systems, it can be first converted to an equivalent single rate system, either at the slowest common base rate or at the fastest common rate. Then the conventional single rate control design methodologies can be applied. Successive loop closure (SLC), pole placement, the singular perturbation method, the Linear Quadratic Gaussian (LQG) method, parameter optimization methods [13] are few of these methods that have been used for multi-rate systems.

As recounted in [5], multi-rate control can increase servo bandwidth when the PES sampling rate is relatively low compared to the open-loop crossover frequency. Multi-rate design is also useful in designing sharp notch filters to combat the actuator resonances by discretizing the notch filters at faster rates than the PES sampling rate [204]. Design of controller with multi-rate discretization of state space model is discussed in this section. This design follows the one presented in [56] where a perfect tracking control based on multirate feedforward control is discussed.

Let us consider a continuous-time nth-order single input single output (SISO) plant $P_c(s)$ described by

$$\begin{aligned} \dot{x}(t) &= A_c x(t) + B_c u(t), \\ y(t) &= C_c x(t) + D_c u(t). \end{aligned} \tag{3.110}$$

The discrete-time model of the plant $P[z_s]$ discretized by single rate sampling period T_y ($= T_u$) becomes

$$\begin{aligned} x[k+1] &= A_s x[k] + B_s u[k], \tag{3.111} \\ y[k] &= C_s x[k] + D_s u[k], \tag{3.112} \end{aligned}$$

where $x[k] = x(kT_y)$, $z_s \triangleq e^{sT_y}$ and $A_s = e^{A_c \frac{T_f}{N}}$, $B_s = \int_0^{T_f/N} e^{A_c \tau} B_c d\tau$.

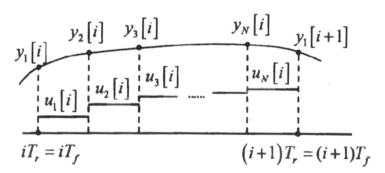

Figure 3.51: Intersample of multirate sampling.

The discrete-time plant $P[z]$ discretized by generalized multirate sampling control as shown in Figure 3.51 can be represented by

$$\begin{aligned} x[i+1] &= Ax[i] + Bu[i], \tag{3.113} \\ y[i] &= Cx[i] + Du[i], \tag{3.114} \end{aligned}$$

where $x[i] = x(iT_f)$, $z \overset{\Delta}{=} e^{sT_f}$, and according to sampled-data control theory, the multirate input and output vectors u and y are defined as,

$$
\begin{aligned}
u[i] &\overset{\Delta}{=} [u_1[i], \cdots, u_N[i]]^T, \\
&= [u(kT_y), u((k+1)T_y), \cdots, u((k+N-1)T_y)]^T, \quad (3.115) \\
y[i] &\overset{\Delta}{=} [y_1[i], \cdots, y_N[i]]^T, \\
&= [y(kT_y), y((k+1)T_y), \cdots, y((k+N-1)T_y)]^T, \quad (3.116)
\end{aligned}
$$

and matrices A, B, C, D can be calculated by

$$
\left[\begin{array}{c|c} A & B \\ \hline C & D \end{array} \right] \overset{\Delta}{=}
\left[\begin{array}{c|cccc}
A_s^N & A_s^{N-1}B_s & A_s^{N-2}B_s & \cdots & B_s \\
\hline
C_s & D_s & 0 & \cdots & 0 \\
C_sA_s & C_sB_s & D_s & \cdots & 0 \\
\vdots & \vdots & \vdots & \cdots & \vdots \\
C_sA_s^{N-1} & C_sA_s^{N-2}B_s & C_sA_s^{N-3}B_s & \cdots & D_s
\end{array} \right], \quad (3.117)
$$

where $P[z_s] = A_s, B_s, C_s, D_s$ is the plant discretized by the zero order hold on $T_y(= T_u)$ and $z_s \overset{\Delta}{=} e^{sT_y}$. $A_s = e^{A_c \frac{T_f}{N}}, B_s = \int_0^{T_f/N} e^{A_c\tau}B_c d\tau$.

To verify the above, we have

$$
\begin{aligned}
x[k+1] &= A_s x[k] + B_s u[k], & (3.118) \\
x[k+2] &= A_s x[k+1] + B_s u[k+1], \\
&= A_s^2 x[k] + A_s B_s u[k] + B_s u[k+1], & (3.119) \\
x[k+3] &= A_s x[k+2] + B_s u[k+2], \\
&= A_s^3 x[k] + A_s^2 B_s u[k] + A_s B_s u[k+1] + B_s u[k+2], & (3.120)
\end{aligned}
$$

$$\vdots$$

$$
\begin{aligned}
x[k+N] &= A_s^N x[k] + A_s^{N-1}B_s u[k] + \cdots \\
&\quad + A_s B_s u[k+N-2] + B_s u[k+N-1], & (3.121) \\
y[k] &= C_s x[k] + D_s u[k], & (3.122) \\
y[k+1] &= C_s x[k+1] + D_s u[k+1], \\
&= C_s A_s x[k] + C_s B_s u[k] + D_s u[k+1], & (3.123) \\
y[k+2] &= C_s x[k+2] + D_s u[k+2], \\
&= C_s A_s^2 x[k] + C_s A_s B_s u[k] + C_s B_s u[k+1] + D_s u[k+2], \\
& & (3.124)
\end{aligned}
$$

$$\vdots \qquad\qquad (3.125)$$

$$
\begin{aligned}
y[k+N-1] &= C_s A_s^{N-1} x[k] + C_s A_s^{N-2}B_s u[k] + \\
&\quad \cdots + C_s B_s u[k+N-2] + D_s u[k+N-1]. & (3.126)
\end{aligned}
$$

In state space form, the above can be written as

$$
\left[\begin{array}{c} x[k+N] \\ \hline y[k] \\ y[k+1] \\ \cdots \\ y[k+N-1] \end{array}\right] = \left[\begin{array}{c|cccc} A_s^N & A_s^{N-1}B_s & A_s^{N-2}B_s & \cdots & B_s \\ \hline C_s & D_s & 0 & \cdots & 0 \\ C_sA_s & C_sB_s & D_s & \cdots & 0 \\ \vdots & \vdots & \vdots & \cdots & \vdots \\ C_sA_s^{N-1} & C_sA_s^{N-2}B_s & C_sA_s^{N-3}B_s & \cdots & D_s \end{array}\right],
$$

$$
\triangleq \left[\begin{array}{c|c} A & B \\ \hline C & D \end{array}\right]. \tag{3.127}
$$

After multi-rate discretization, the system becomes an n-dimensional plant with N inputs and N outputs, with D being a square matrix of full rank. Thus we can get the system's inverse state-space model $\tilde{O}^{-1}(z) = \{A, B, C, D\}^{-1}$ directly:

$$
\left[\begin{array}{c|c} A & B \\ \hline C & D \end{array}\right]^{-1} = \left[\begin{array}{c|c} A - BD^{-1}C & BD^{-1} \\ \hline -D^{-1}C & D^{-1} \end{array}\right]. \tag{3.128}
$$

Recall that in Section 3.5.3 we discussed that by adding the plant inverse in the AFC scheme can work in a wider frequency range. We shall explain the advantage of using the multi-rate inverse scheme with the help of an example.

Example

Let us consider a first order SISO system described by a transfer function

$$
O(s) = \frac{s - 10}{s + 23}. \tag{3.129}
$$

Its single rate discrete form with sampling frequency 15 kHz is,

$$
O_1(z) = \frac{0.9989z - 0.9996}{z - 0.9985}. \tag{3.130}
$$

When discretized using dual rate sampling, its state space representation becomes,

$$
O_2(z) = \left[\begin{array}{c|c} A & B \\ \hline C & D \end{array}\right] = \left[\begin{array}{c|cc} 0.9985 & 0.0002664 & 0.0002666 \\ \hline -4.125 & 1 & 0 \\ -4.122 & -0.0011 & 1 \end{array}\right]. \tag{3.131}
$$

The dual rate inverse model is,

$$
\tilde{O}_2^{-1}(z) = \left[\begin{array}{c|c} A & B \\ \hline C & D \end{array}\right]^{-1} = \left[\begin{array}{c|cc} 1.001 & 0.0002667 & 0.0002666 \\ \hline 4.125 & 1 & 0 \\ -4.126 & 0.0011 & 1 \end{array}\right]. \tag{3.132}
$$

Because there is one unstable zero of the plant, the phase of the shaped plant $O(z)\tilde{O}^{-1}(z)$ is the same as $\frac{1}{z}$ when using $\tilde{O}^{-1}(z)$ is obtained using the pole/zero and phase cancelation scheme,.

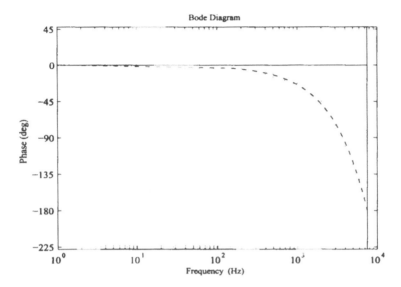

Figure 3.52: Phase of shaped plant $O(z)\tilde{O}^{-1}(z)$. Solid line: pseudo inverse scheme; dash-dot line: dual rate inverse scheme.

Figure 3.52 compares the phase of the shaped plant $O(z)\tilde{O}^{-1}(z)$ when using the two inverse schemes. When using the pseudo inverse scheme, the phase drops down exponentially when the frequency increases while zero phase is assured over the whole frequency range when using multirate inverse scheme. The multirate scheme does not have this problem of phase drop.

3.5.5 Multisensing Servo

Vibration of disk and suspension, repeatable runout (RRO), windage (shown as i_d in Figure 3.31) etc are the major contributing factors to track misregistration (TMR) in an HDD servo system [47]. According to the discussions presented so far, peak filters can be used to deal with some narrow band disturbances and, in general, high bandwidth servo for broad band disturbances.

Another source of disturbance in HDD servomechanism comes from disk flutter which refers to the vertical motion of disk contributed mainly by spindle and air-disk interaction. Although disk flutter is a vertical motion, it can have an in-plane component that can cause shift in the track center. Thus it has direct impact on the position error signal (PES) [66]. Such vibration is not phase locked to the disk rotation and, therefore, is a contributor to nonrepeatable runout (NRRO) [66] [185]. Compensation methods for RRO has no influence on this source of disturbance.

But the energy of disk vibration is concentrated in certain frequencies, and therefore, one common approach to suppress these vibrations is to increase the servo loop gain at the disturbance frequencies by inserting a narrow band filter. This method can be applied successfully in single-stage actuator (e.g. [48], [168], [184], and the references therein) as well as for dual-stage actuator [225], [113]. Use of dual-stage actuator, which is discussed in details in section 3.7, enables higher bandwidth of the servo system. With the help of this increased gain in designated frequencies, the runout signals can be rejected more effectively. However, significant amount of disk flutter energy lies in the frequencies around 500 Hz and above while the bandwidth of the servo systems nowadays ranges from a few hundred Hz to 2 kHz. Inserting a narrow band filter to suppress disk flutter, therefore, affects the stability of the loop. Besides, such feedback control systems typically show poor settling performance [48]. The prolonged settling caused by the insertion of narrow band filter can be reduced by initializing the states of controller to make the transient of the filter minimal [184], [213]. If it is necessary to insert multiple narrow band filters then the initialization of the controller becomes very complicated, if not impossible.

Modifying the feedback control for addressing the issues of disturbances of specific nature is relatively difficult. However, mitigating with the basic feedback servo loop by including additional sensors and hence applying the concept of multi-sensing servo (e.g., instrumented suspension [89], active damping of actuator vibration [90], and acceleration feedback [112]) is a viable option for achieving better rejection of NRRO. Multi-sensing can improve the bandwidth by up to a few hundred hertz only and hence the improvement in vibration rejection is limited.

The concept of multisensing servo has been widely studied by HDD servo engineers and researchers [99], [161], [2], [126], and [64]. The vertical motion of disks can be modeled as equivalent off-track motion of the head slider [185], [66]. These developments inspire a new approach using feedforward control for reducing TMR induced by disk flutter. In order to establish the applicability of this approach, we first show that the disk's vertical vibration signal picked from the slider has a fairly good correlation with the track misregistration at the frequencies of disk flutter. Once this correlation is established, we shall show how a feedforward controller can be designed to cancel the effects of disk flutter. The feedforward controller is an approximated differential element whose input is the vertical component of the velocity of disk. It was proven through experiment that the feedforward controller reduced the TMR induced by the first four modes of disk flutter by an average of about 56%. One potential application of this method is the high TPI servo track writers (STW), where this approach can reduce the written-in RRO. This, in turn, will have positive influence on the performance of the HDD servomechanism. Another method presented in [110] overcomes the disk flutter problem by using a flexure that is optimized to cause less off-track error at locations where the disk vibration is maximum, e.g., the OD. Compared to this method, the feedforward controller given in the following paragraphs works for wider operating conditions. An experimental setup for measurement of disk flutter is presented first, followed by an analysis on the correlation between the disk flutter and TMR.

Measurement of Disk Flutter

Figure 3.53 shows an experimental setup for measuring the vertical vibration of the disks. A 3.5" glass disk of 1.27 mm thickness is spun at the speed of 4800 RPM by a fluid dynamic bearing (FDB) spindle motor. The system uses the micro positioner[‡] PA-2000 to position a slider with 0.42 μm wide write head, which is used to perform servo track writing. A clock head is also available to provide the necessary timing signal during the servo writing process. Approximately 0.3 μm wide servo tracks are created by erasing part of the servo bursts. The read sensor has a width of about 0.25 μm. A Laser Doppler Vibrometer (LDV[§]) is used to measure the slider's vertical vibration by shining a spot on the slider with a perpendicular laser beam. The LDV measures the velocity of the slider and derives the displacement from there. A position error signal (PES) processing channel is also used [61] to measure the displacement of the slider parallel to the disk surface.

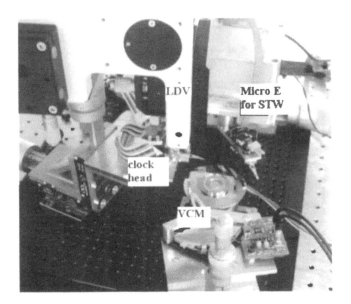

Figure 3.53: Experimental setup.

The model of the actuator is identified by injecting a swept sine signal into the voice coil motor (VCM) and measuring the frequency response using a Hewlett Packard dynamic signal analyzer (DSA) 35670A. The model of the

---

[‡]MicroE Systems Inc.
[§]Polytec GmbH, model OFV512.

VCM actuator is:

$$G_p(s) = \frac{1.715 \times 10^{12}}{s^2 + 153.9s + 4.836 \times 10^4} \frac{1.668 \times 10^9}{s^2 + 816.8s + 1.668 \times 10^9}. \tag{3.133}$$

The discretized feedback control law is:

$$G_c(z) = \frac{.107z^6 + .181z^5 - .033z^4 - .221z^3 - .092z^2 + .067z^1 + .046}{z^6 + 2.376z^5 + 1.463z^4 - .916z^3 - 1.626z^2 - .692z^1 - .056}, \tag{3.134}$$

with a sampling frequency of 12.64 kHz.

Figure 3.54 shows the open-loop transfer function of the servo system. The phase margin is 52 degrees, the gain margin is about 8 dB, and the cross frequency is about 1 kHz. Figure 3.55 shows the shock transfer functions measured in the setup (solid line) and derived from simulation model (dashed line).

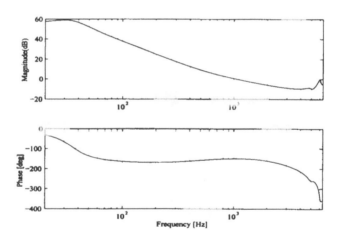

Figure 3.54: Open-loop frequency response of servo system.

Due to the design of the air bearing surface, the slider flies above the rotating disk surface. The flying height is about 10 nm and variation in flying height is also in the nanometer scale regardless of the vertical disk flutter which is in μm scale [132]. Therefore the measured vertical displacement of the slider is a fairly accurate representation of vertical movement of the disk. It is observed that the variation of the slider in vertical direction has a repeatable component synchronous to the spindle rotation measured by the clock signal. Therefore, the measurement output from the LDV with the synchronized averaged value removed from it can be considered as the disk flutter signal.

Figures 3.56 and 3.57 show the power spectral densities of the measured disk flutter and the nonrepeatable components of the PES, respectively. Figure 3.56 clearly shows the modes of disk flutter, especially for those at 504 Hz, 598 Hz, 648 Hz, and 696 Hz. These are the four dominant modes in disk flutter. All

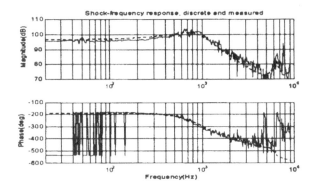

Figure 3.55: Shock transfer function of servo system.

the four modes also appear in Figures 3.57. However, the modes above 2 kHz seen in Figure 3.56 do not appear in the PES spectrum. These modes may be the vibration modes of the suspension in the vertical direction.

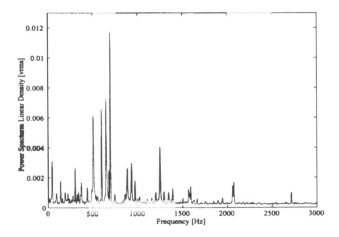

Figure 3.56: Linear spectrum of disk flutter (1 Vrms = 0.11 μm).

Correlation between Disk Flutter and PES

There exist coincidences of the amplitude peaks between the power spectrum of PES and the power spectrum of disk flutter. Let us now compare the phase relationship between them. Figure 3.58 shows the time domain signals for the vertical vibration of disk and the NRRO component of the PES.

It appears that the NRRO component of PES and the vertical vibration of disk have the same fundamental frequency. We can, therefore, establish a

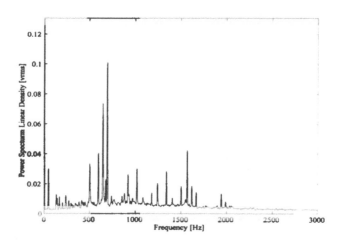

Figure 3.57: Linear spectrum of PES.

transfer function relationship between the vertical vibration velocity and the PES signal. Instead of using an external excitation source when measuring the system's frequency response, we can use disk flutter velocity (with the repeatable part removed) as the input signal and the PES as the output of the system to be modeled. Table 3.4 shows the amplitude and phase differences, obtained from the DSA, between the dominant modes of PES and the disk flutter measured.

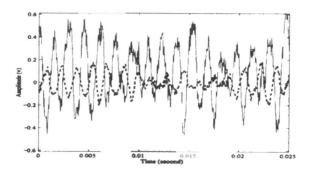

Figure 3.58: Time domain signal of disk vertical vibration velocity (dashed-line) and the PES (solid line).

Figure 3.59 shows the closed-loop system sensitivity transfer function multiplied by an integrator (solid curve) and the points, marked with asterisks (*), corresponding to Table 3.4. From this figure, we see good match between the phases and also between the magnitudes of the first four dominant modes.

These results confirm that the disk vertical vibration is proportionally reflected in the in-plane positioning error. As such, we can modify the standard

Table 3.4: The relationship between the corresponding modes of disk flutter and the PES

Frequency in Hz	Phase in degree	Amplitude in dB
504	65.0	-78.5
598	53.6	-74.1
648	44.2	-75.0
694	37.1	-72.3
882	10.7	-81.6
932	-2.2	-80.6
1018	-19.9	-68.0
1248	-45.8	-84.1

servo control block diagram Figure 3.31 to Figure 3.60 where the effect of the vertical motion of disk is shown as disk's motion entering the VCM output through a scaling factor $K = constant$.

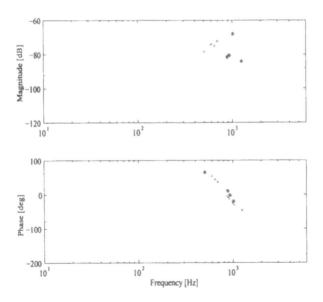

Figure 3.59: The frequency response from disk flutter measured (velocity output of LDV) to PES.

Feedforward Control Design

Although it is appealing to determine the feedforward compensator via adaptive control or self tuning control ([161] [2]), the method based on manipulating

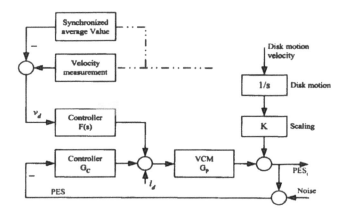

Figure 3.60: Feedforward compensation scheme for disk flutter.

the transfer function is simple to design and easy to implement. From Figure 3.60, the feedforward compensator $F(s)$ whose output is added to the normal feedback controller $G_c(s)$ can be obtained by letting:

$$S_h(s)F(s) = S(z)\frac{1}{s}K, \qquad (3.135)$$

where

$$S(s) = \frac{1}{1 + G_P(s)G_c(s)}, \qquad (3.136)$$

is the sensitivity transfer function and

$$S_h(s) = \frac{G_p(s)}{1 + G_p(s)G_c(s)}, \qquad (3.137)$$

is the shock transfer function. Substituting equations 3.136 and 3.137 into equation 3.135, we immediately have

$$F(s) = \frac{S(s)}{S_h(s)}\frac{1}{s}K = \frac{P_{den}(s)}{P_{num}(s)}\frac{1}{s}K, \qquad (3.138)$$

provided $F(s)$ is realizable.

For the system described above, $G_{(s)} = \frac{P_{num}(s)}{P_{den}(s)}$ has a relative degree of 2. Therefore, is a first order low pass filter added to the right hand side of equation 3.138 makes $F(s)$ realizable. A feedforward compensator can be designed as

$$F(s) = \frac{1.4596 \times 10^{-8}s}{s + 2\pi \times 4000}\frac{2\pi \times 12000}{s + 2\pi \times 12000},$$

which contains a second low pass filter. The feedforward controller is discretized at 25.28 kHz which is two times as fast as the feedback controller

$G_c(z)$. This yields a $F(z) = \frac{(z^2-1) \times 5.838 \times 10^{-9}}{z^2 - 0.1342z - 0.06662}$. Figure 3.61 shows the frequency response of $F(z)$ which is very much like a differentiator below 2 kHz. In this case, $F(s)$ can not totally cancel out the effects of the disk vertical motion because of the additional phase delay introduced by the low pass filters.

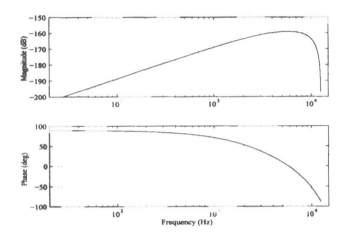

Figure 3.61: Frequency response of the feedforward controller.

Effect of the Feedforward Control

Figure 3.62 shows the PES spectrum after disk flutter is compensated using $F(z)$. We see in Figure 3.62 that the first mode is almost disappeared, while the remaining 3 dominant modes are decreased in amplitude, though they are still present. This is due to the phase lag of low-pass filter and phase lag of digitization which can be seen in Figure 3.61.

We can conclude by comparing Figure 3.62 with Figure 3.57 which has the same feedback controller $G_c(z)$ but without the feedforward control $F(z)$, that there is no obvious amplification of the modes above 1 kHz. The shape of the baseline in PES spectrum has not change as well. This is due to the fact that the feedforward compensation has no effect on the original feedback servo loop. Furthermore, the amplitudes of the four dominant modes due to disk flutter were significantly reduced. Table 3.5 summarizes the comparison results. As can be seen from the table, the amplitudes of vibration modes decreased by about 16% to 68%. The first four modes are attenuated by an average of 56%.

It is verified in this section that, at a number of disk resonance frequencies, the disk flutter induced PES has a fixed phase and amplitude relationship with the disk vertical vibration. Therefore, disk flutter induced TMR can be compensated for by adding a feedforward control loop without modifying the basic servo control loop. A simple, practical differentiator type feedforward controller with the vertical direction vibration velocity measured using an LDV

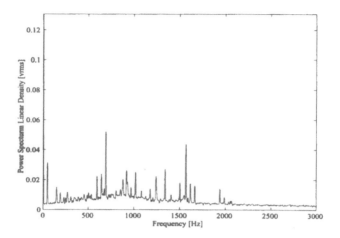

Figure 3.62: Linear spectrum of PES after disk vertical vibration feedforward compensation.

Table 3.5: The amplitude of dominant modes in PES spectrum before and after compensation.

Frequency (Hz)	Magnitude before compensation (mVrms)	Magnitude after compensation (mVrms)	Reduction %
504	33.1	11.2	66
598	40.3	22.5	44
648	73.9	23.9	68
696	100.9	52.0	48
1018	30.0	25.1	16

as input to the basic servo loop and reduced the TMR induced by the first four disk flutter modes by 56%.

The method presented here requires a vertical vibration sensor, and therefore, increases the cost when implemented in a commercial HDD. However, this technique can be used in applications whereby precision positioning is required and the extra sensor can be reused. For example, the method can be used to improve the positioning accuracy in the testing equipments such as servo track writers or spinstands to reduce the written-in runout. Development of various technologies such as MEMS will eventually bring the cost down and measurement of the vertical vibration will be a practical solution that cen be applied in HDD. Additionally, the same instrumented suspension for expanding the servo bandwidth and reducing the suspension vibrations (all in the in-plane direction) might also be used for measuring the vertical direction vibration for the feedforward compensation discussed in this section.

3.6 Emergence of Dual-Stage Actuator

The design methods discussed so far in this chapter attempts to improve the performance of the closed loop head positioning servomechanism by working around the limitations and shortcomings that come with the system built using conventional VCM actuator. The torque producing VCM is located at one end of the actuator arm while the read-write head slider sits at the other end. The mechanical structure lying between the point of actuation (VCM) and the point to be controlled (head) is not perfectly rigid and gives rise to different modes of vibration that affect the positioning accuracy. The suspension arm and the gimbal used to attach slider to the suspension contribute to these vibrations. These vibrations as well as the nonlinearity and stiffness of pivot bearing can not be neglected when it is desired to have the servo bandwidth as high as possible to suppress the effects of disturbances and noise.

Removing the bottleneck factors at the source is always a better approach to accomplish the task of achieving high servo bandwidth. Over the years, continuous efforts have been made to improve the performance of the VCM actuator in various ways such as improved mechanical design, use of new materials to make actuator and suspension arms, using high bandwidth voice coil motor (VCM) [9], alternative design of arm and suspension, application of active damping etc. The bandwidth of the conventional single-stage actuator has been above 1 kHz since 2000.

Another approach, investigated in parallel to improve the performance, adds a second actuator placed conveniently between the pivot and slider in a piggy back fashion. Such actuator, known as the dual-stage actuator, still uses the conventional VCM actuator for long-range seek while the secondary stage located closer to the slider can improve tracking performance during track following. The actuation mechanism for the secondary stage can be classified as electro static, electromagnetic, or piezo electric. Depending on where the secondary stage actuator is located, the dual-stage actuator can be one of the following,

1. Actuated suspension [50] [149] [50],

2. Actuated slider [173] [179] [51] [148] and

3. Actuated head [134] [57] [136].

When the actuator is located nearer to the read/write head, i.e., either actuating the suspension or actuating the slider or actuating the read/write head, the load to be moved by the secondary actuator is reduced. The plant model for the secondary actuator has less low frequency flexible modes, and higher servo control bandwidth can be achieved.

3.6.1   Actuated Suspension

The piezoelectric material undergoes a dimensional change when voltage is applied. It can have expansion or contraction or shear when appropriate voltage is applied. This dimensional change of the piezo materials can be exploited to produce precise actuation. Commonly available piezoelectric materials such as quartz, tourmaline, Rochelle salt etc. show very small piezoelectric effect. Polycrystaline ferroelectric ceramic materials such as Lead Zirconate Titanate (PZT) possess significantly high actuation sensitivity. These materials are widely used to make piezoelectric actuator.

The actuation property of the PZT material can be exploited to move the suspension or the slider used in HDD leading to dual-stage actuation. The actuated suspension, often called milli-actuator due to its size, has a simple structure and is relatively easier to implement.

Figure 3.63 shows the schematic of a suspension based actuator [149] [88]. Two parallel piezoelectric actuators, each constructed with a reinforcement element sandwiched between two pieces of piezoelectric plates, are located in between the base plate and spring beam. The reinforcement element takes the shape of a meander-line spring and is designed to reinforce the shock resistance capability of the milli-actuator while without constraining its stroke output. Two stiff ribs across the base plate and milli-actuator part are designed to further increase the vertical rigidity of the milli-actuators.

Two piezoelectric actuators are driven in such a way that when one expands the other contracts, making the load beam turn around the flexible hinge, causing an amplified displacement generated at the end where the read/write head is situated. The length of the suspension between this PZT actuator and the head slider provides a mechanical amplification. The deflection of the slider is larger compared to the actual deflection the PZT actuator experiences. The suspension beam's spring rate can be adjusted by partial-etching without significantly affecting the resonance of the suspension.

The actuated suspensions are typically designed to generate a displacement in the range of about 1 μm at the R/W head. It can be modeled as:

$$P_M(s) = k \prod_{i=1}^{n} \frac{k_i(s)}{s^2 + 2\zeta_i\omega_i + \omega_i^2}. \tag{3.139}$$

A set of possible parameters is shown in Table 3.6 with its frequency response shown in Figure 3.64.

Other designs such as using a pocket in the suspension to hold a slanted PZT for reducing the torsional mode have also been proposed [178] [196].

A narrower, lighter, and, if sprint rate is acceptable, shorter design of the suspension is widely adopted in the industry to increase the resonance frequency of the suspension regardless of being actuated or not. Nevertheless, due to the presence of the suspension resonance, actuated suspensions have

Table 3.6: Parameters of a PZT actuated suspension model

		$k = 5 \times 10^5$	
$\zeta_1 = 0.05$	$\omega_1 = 2\pi \times 4600$	$k_1(s) = 1$	
$\zeta_2 = 0.02$	$\omega_2 = 2\pi \times 8300$	$k_2(s) = s^2 + 2 \times 0.02s + (2\pi \times 8100)^2$	

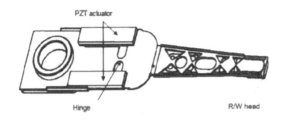

Figure 3.63: Illustration of actuated suspension.

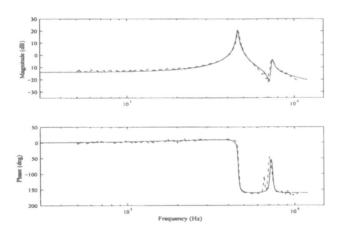

Figure 3.64: PZT micro-actuator frequency response.

only been able to support an open-loop bandwidth of about 3 kHz so far [212].

3.6.2 Actuated Slider

In order to bring the microactuator closer to the point of control, it can be assembled with the *head gimbal assembly* (HGA). One such design is illustrated by the schematic shown in Figure 3.65. In this configuration, the microactuator carries the slider and rotates it as and when required to generate a rotational motion of the read-write head.

The design of electrostatic microactuator driving the slider described in [195] uses a single crystal silicon electrostatic comb drive microactuator mounted between the slider and suspension. It drives the slider on which a magnetic head element is attached. Figure 3.66 shows a quarter of the actuator which consists of a movable rotor connected to an anchored central column via electrical conductive silicon spring beams, and a stator connected to the silicon substrate by a bus bar which electrically isolates the stator from the rotor. The stator and rotor of the microactuator are suspended and directly processed from a highly N-doped single crystal silicon substrate. The dimension of the microactuator is 1.4 mm × 1.4 mm × 0.18 mm.

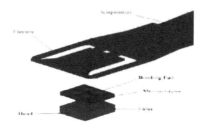

Figure 3.65: Schematic of microactuator driving slider.

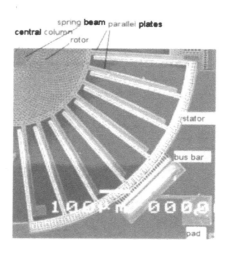

Figure 3.66: View of one quarter of the microactuator.

The parallel plate pairs attached to the stator and rotor generate electrostatic force. For small rotational angles θ, these plate pairs may be modeled as parallel plate capacitors separated by gaps. The torque T with a driving voltage V to half of the structure is given by

$$T \propto R\epsilon A \left(\frac{V}{x_o - R\theta} \right)^2, \tag{3.140}$$

where R is the distance from the centroid of the plate to the center of rotation of the rotor, x_n is the nominal capacitive gap with zero rotation, A is the area of each plate and ϵ is the permittivity of air. The comb drive structure, originally proposed by Tang [194], increases the area and hence generates a larger force or torque for rotary actuator.

Because the torque is inversely proportional to the square of the gap of the parallel plate pair, typically a small gap is desirable provided the structure can be manufactured. Deep Reactive Ion Etching (DRIE) technology [111] enabled the fabrication of high aspect ratio MEMS structures and especially the comb drive structure. Typically, the microactuator has electrically isolated microstructures with an aspect ratio of 20:1 with a finger width of 2 μm and gap width of 2 μm. Novel mechanical designs that are supported by the micro fabrication process have been studied to achieve larger deflections, low driving voltages, better area-efficiency and improved lateral stability.

The actuator can be driven with differential driving scheme to linearize the quadratic voltage nonlinearity of the electrostatic force. The slider, driven by the rotor for fine positioning of the magnetic head, is bonded on the top of the rotor using ultraviolet (UV) curable adhesive [57], [51], [195], [25], [85], [197].

The transfer function of a MEMS actuated slider is modeled as a second-order system

$$P_M(s) = \frac{k}{s^2 + 2\zeta\omega s + \omega^2}, \qquad (3.141)$$

with a typical set of parameter as $\omega = 2\pi \times 1500$, $\zeta = 0.03$ and $k = 2.5 \times \omega^2$.

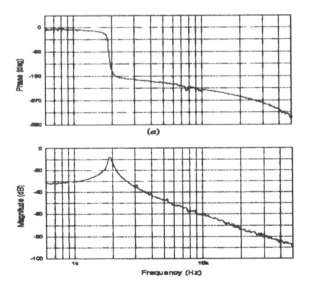

Figure 3.67: MEMS micro-actuator frequency response.

Although the electrostatic actuator has a low resonant frequency at around 1.5 kHz, it has a very *smooth* and *clean* transfer function similar to a single spring mass system up to above 40 kHz. Such microactuator can be considered as having little uncertainty except uncertainty in the gain of its model. The displacement range of such an actuator is about 1 μm with a typical driving voltage of less than 40 Volts.

When a PZT actuator is used to drive the slider or head, the resonant frequency can move up to above 12 kHz [179] with a displacement range of about 0.5 μm, or even above 25 kHz with a displacement range of about 0.2 μm [134]. The later design is able to support a servo bandwidth of above 4 kHz.

3.6.3   Actuated Head

Instead of driving the entire slider, one can choose to use a microactuator to drive a small part of the slider that holds the read-write head. This scheme has the lightest moving mass among all possible placements of the secondary microactuator. A silicon electrostatic microactuator fabricated on the trailing edge of an AlTiC slider [136] is shown in Figure 3.68. Figure 3.69 shows a plan view of this microactuator design.

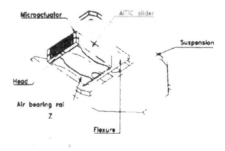

Figure 3.68: Schematic of microactuator-slider assembly attached on suspension.

The microactuator consists of stationary structures and movable structures with a solid head plate. All the movable structures are suspended by two springs. The read/write element is attached to the head plate of the movable structure via four flexure of wires with pads on the slider which are stationary. Comb drives are used to generate the electrostatic force. When voltages are applied across one pair of the metal pads, the movable structures together with the head plate and the head element are driven by electrostatic attraction. Only half of the comb finger pairs are used to generate the attractive force; the remaining pairs can be used as capacitive sensors during this period to measure displacement that can be used for feedback control. Applying alter-

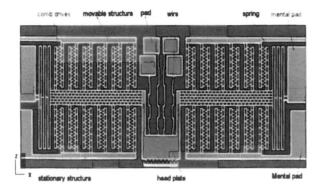

Figure 3.69: Plan view of a microactuator design.

nating voltages on each pair of the metal pads can drive the movable structure in both directions.

In the straight beam suspension design, the Y direction stiffness k_y of a suspension beam is expressed as

$$k_y = \frac{Eh^3w}{l^3}, \qquad (3.142)$$

where E is the Young's modulus, h is the height of the spring, w is the spring width and l is the spring length along the Z direction. On the other hand, the X direction (operational direction) stiffness k_x is expressed as

$$k_x = \frac{Ehw^3}{l^3}. \qquad (3.143)$$

The stiffness ratio is

$$\frac{k_y}{k_x} = \left(\frac{h}{w}\right)^2. \qquad (3.144)$$

The ratio between the stiffness in the Y direction and the stiffness in the X direction must be maximized to make the structure very stiff in the Y direction (normal to XZ plane). If the spring is designed with beam width of 2 μm in the X direction and 60 μm in the Y direction, then the aspect ratio (h/w) of the spring is 30, and the beam is very stiff in the Y direction. The aspect ratio is somehow limited by the fabrication process.

The electrostatic driving force F is expressed as

$$F = \frac{\epsilon hnV^2}{g}, \qquad (3.145)$$

where ϵ is the permittivity of air, h is the height of the structure, n is the number of the electrode pairs in the comb drives, V is the driving voltage, and g is the gap width of the comb-drive fingers.

The head-microactuator-slider assembly is fabricated in an integrated fabrication process described below.

1. Releasing slots on back of the silicon wafer are etched by DRIE. There are two etching steps, one for releasing movable parts (including springs and head plate) and the other for reducing the height of the wire structures. Different heights of the structures can thus be formed from the backside of the wafer.

2. The patterned silicon wafer is bonded to a slider wafer using adhesive bonding. The thickness of the SU-8 bonding layer is approximately 10 μm.

3. Magnetic heads are fabricated together with its microactuator processes on the front-side of the silicon wafer. Silicon dioxide (SiO_2)layer is used as hard etching mask to fabricate microactuator structures.

4. The silicon-AlTiC bonded wafers are diced into bars and air-bearing rails are fabricated on the dicing surface which is perpendicular to the wafer surface.

5. DRIE is used to release the movable part of the microactuator.

The frequency response of such an electrostatic microactuator driving the head is shown in Figure 3.70. The response is measured on a probe station [136]. Figure 3.71 highlights the movable part that has the 1st resonant frequency at 15.4 kHz in the XZ plane in an FEM model.

3.6.4 Microactuator for Controlling Head-Media Spacing

With the increase of recording density of HDD to above 100 Gb/in^2, a constant flying height of approximately 10 nm or smaller is required. Variations in flying height are contributed by (1) manufacturing tolerances, (2) environmental changes such as atmospheric pressure change, and (3), slider localized heating by writing current induced slider crown and chamber variation, and many more. Adjustment of the spacing between the read/write head and the recording media, i.e., the height (z-direction) of the read/write head with respect to the slider can be effectuated using a microactuator integrated in the slider. Such a possibility was studied in tandem with the study of lowering the slider flying height [223].

One such design adopts the configuration of an active head slider with PZT unimorph cantilever structure which can be fabricated monolithically using silicon micro machining. The active head sliders were fabricated using micro machining process, which includes spin-coating of the sol-gel PZT, ABS etching, and deep RIE for forming cantilevers of about 50 μm thickness. The usual pads on the slider generate high pressure to support the load. The

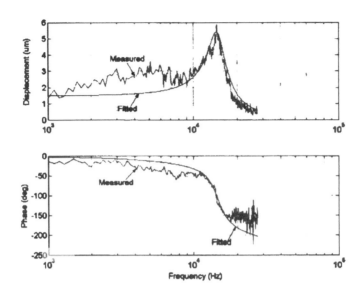

Figure 3.70: In plane frequency response with a resonant peak at 14.7 kHz.

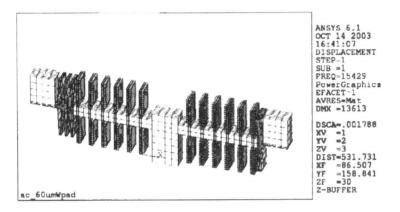

Figure 3.71: FEM model showing the movable part that has a 1st resonant frequency at 15.4 kHz in the XZ plane.

pad with the PZT z-direction actuator that carries the head generates little pressure even if the head approaches a disk surface. The recording element can be raised and lowered on demand by applying a control voltage. Change in flying height of some 70 nm was obtained at the cantilever tip with the control voltage of 4 V [188], [189], [118], [119]. References [192] and [193] report the fabrication process for the PZT thin film.

We note that the writing current induced variation in slider crown and chamber affects the flying height and can conversely be used as a thermal actuator to control the flying height. Additionally, the electrostatic force between the slider and the disk can also be used to adjust the flying height [182].

A brief introduction to the piezo-electric or electrostatic microactuator is presented in this section. These actuators can be used to move the suspension or the slider or the read/write head for fine head positioning of the read/write element and also for active control of the spacing between head and disk. These actuators typically require a driving voltage of less than 40 Volts, generates a displacement of around 1 μm for the head positioning control, and offer a "clean" transfer function which is very useful for reducing the microactuator plant model uncertainty and hence expanding the servo bandwidth.

In the case of the actuated suspension design, the resonant frequency of the microactuator is limited by the suspension resonance which is about 10 kHz at this moment. Using the rule of thumb that open loop servo bandwidth is about 1/3 of the critical resonant frequency which has been the case for the dual-stage servos reported so far, actuated suspension can support a servo bandwidth of about 3.3 kHz. The electrostatic actuated slider or head, on the other hand, have frequency responses that resemble a spring-mass system to about 50 kHz.

Given a suitable PES sampling, it is easy for such an actuator to support a servo bandwidth 10 kHz and above, which can effectively reject the vibrations due to disk, spindle and HSA to achieve accurate positioning. As such, actuated slider or head is the actuator of choice from point of servo performance. Accordingly, low overhead, accurate and very fast PES sampling is required for such a microactuator. Nevertheless, in addition to the higher bandwidth feature they offer, the microactuators must be reliable and low cost in order to be useful in practical systems.

Addition of a secondary stage actuator brings in significant changes in the design of controller as it turns out to be a double-input-single-output (DISO) problem. Examination of the controllability matrix reveals that the system is controllable from both inputs. However, care must be taken to coordinate the movements of the two actuators in order to make use of the microactuator for high bandwidth control.

The next section discusses the issues related to the design of controller for dual-stage servo for achieving higher servo bandwidth. All examples of dual-stage servo controller illustrated there use actuated suspension as the plant model since it is more readily available.

3.7   Control of Dual-Stage Actuator

Though the motion of the read/write head slider can be generated by energizing either the VCM or the secondary actuator, the only available measurement in the dual-actuated HDD servomechanism is the displacement of the slider.

Designing the controller for a dual-stage actuator is a problem of designing controller for a dual-input single-input (DISO) system. If the relative displacement of the secondary stage with respect to the VCM can be somehow measured, it would make the actuator a dual-input dual-output (DIDO) system and the controllers for the two actuators can be designed independently. Various methods of designing controller for dual-actuated HDD servomechanism have been reported in the published literature. Commonly used configurations are,

- parallel loop (Figure 3.72),

- master-slave, and

- decoupled master-slave (DMS, Figure 3.90) ([173], [65]).

The controller parameters for each of these structures can be found either by classical loop shaping method or using optimal control in state space. Some very simple methods for parameter selection are introduced in the next few pages to illustrate the design considerations for dual-stage servo system. This design problem has been researched by many researchers and designers, and they can be found in published literatures such as [173], [212], [65], [125], [123], [113], [62], [106], [128], and the references therein. Interested readers may refer to these works to get better understanding of each of these approaches.

3.7.1 Control Design Specifications

Following the the shape of the magnitude of a compensated open-loop transfer function given in Section 3.2, a typical design specification for dual-stage HDD servomechanism is summarized in Table 3.7.

Table 3.7: Specifications for Dual-stage Control Design Example

Open-loop bandwidth f_o:	>2000 Hz
Disturbance attenuation:	>40 dB below 100 Hz
Phase margin (PM):	> 40 degree
Gain margin (GM):	> 6 dB
Rise time:	< 0.2 ms
Overshoot:	< 20%
Peak in the sensitivity transfer function:	< 10 dB

In the dual-stage actuator, the primary stage, i.e., the VCM actuator has larger inertia and lower bandwidth but provides a larger range of movement. The secondary stage, on the other hand, is lighter, has greater bandwidth but supports a very small range of displacement. Because of these contradictory properties, efforts must be distributed properly between the two actuators. According to the design published in [173], [106], and [67], the following guidelines should be adhered to while designing the dual-stage servo controller:

- The VCM responds to low frequency components of error while the secondary stage responds to high frequency error signals.

- Hand-off: Hand-off frequency is the frequency where the magnitudes of the compensated VCM branch and the compensated secondary actuator branch are equal. The phase difference between the outputs of the secondary stage and that of the VCM should be $\leq 120°$ at the hand-off frequency to avoid destructive interference. Since the resultant displacement of the slider is sum of displacements contributed by two actuators, they cancel each other when they are out of phase. Typical hand-off frequency reported so far is about 400 Hz.

- The gain of the VCM should exceed the secondary stage's gain by 20 dB for frequencies below 60 Hz to avoid saturation of the secondary stage actuator.

The basic concepts of shaping the responses of two actuators are explained here using few simple examples of dual-stage servo system with PZT actuated suspension. For simplicity, when introducing various design concepts, we assume that the resonances of both VCM and PZT actuator are compensated using filters in series so that the VCM can be represented approximately by the simple model of k/s^2 whereas the PZT actuator is described by a pure gain g_m. We use P_V, P_M, C_V, C_M to denote the transfer functions of the VCM, the secondary actuator, and their respective controllers.

3.7.2   Parallel Structure

The parallel structure dual-stage control is shown in Figure 3.72. The open loop transfer function $O_p(s)$, closed-loop transfer function $T_p(s)$, sensitivity transfer function $S_p(s)$ are given as [173]

$$O_p = C_V P_V + C_M P_M, \tag{3.146}$$

$$T_p = \frac{C_V P_V + C_M P_M}{1 + C_V P_V + C_M P_M}, \tag{3.147}$$

$$S_p = \frac{1}{1 + C_V P_V + C_M P_M}, \tag{3.148}$$

respectively if the actuators are not saturated.

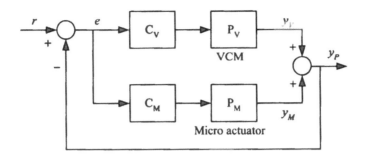

Figure 3.72: Configuration of parallel structure.

PID-type Controllers for Parallel Structure

PID type controllers are the most commonly found in practice. As the first example, we consider PID controller for both VCM and microactuator. As elaborated in the following design, this approach gives a stable system although some aspects of hand-off are not so desirable.

VCM Controller

Let us assume that the VCM resonance is compensated using a notch filter so that the compensated model can be regarded as a double integrator k/s^2. Let the desirable crossover frequency for the VCM actuator be f_V. Then the lag-lead compensator is:

$$C_{VP}(s) = k_c \frac{(1/\omega_2 s + 1)(1/\omega_3 s + 1)}{(1/\omega_1 s + 1)(1/\omega_4 s + 1)}, \tag{3.149}$$

with ω_2 higher than ω_1 by a factor 5, ω_4 higher than ω_3 by a factor of 5 or more, and f_V higher than ω_3 by a factor of 2-3 or more but lower than ω_4 by a factor of 2-3 or more. As discussed in the case of single stage actuator, such a scheme can give us a lag-lead compensator for the double integrator type plant with 35° or more phase margin and minimum of 6 dB gain margin.

Microactuator Controller

Let the microactuator gain be g_m and the microactuator loop crossover frequency be f_m. Obviously f_m should be a few order of magnitude higher than f_V as the microactuator is expected to respond to high frequency components of PES. We use a lag filter with the corner frequency of $1/\beta$ (e.g. 1/4) of f_m:

$$C_m = k_m \frac{1}{\frac{\beta}{2\pi f_m}s + 1}, \tag{3.150}$$

with $|g_m C_m(s)|_{s=j2\pi f_m} = 1$, such that the microactuator loop crosses 0-dB line with -20 dB/dec slope.

Since the microactuator loop works in parallel with the VCM loop which has a -60 dB, -40 dB, -20 dB and -40 dB/dec slope, the low frequency gain follows the VCM loop. For frequencies above f_V, the gain of the VCM is below 0-dB where as the microactuator path has a higher gain. The parallel loop thus crosses the 0-dB line at a -20 dB/dec slope following the microactuator path. The combined loop has a phase delay of less than 120° following the microactuator path, giving a phase margin of more than 60°.

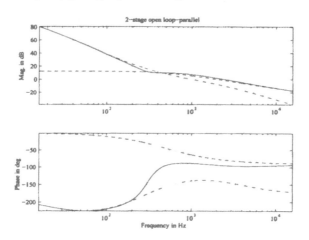

Figure 3.73: Open loop transfer function. Solid line: combined loop, dash-dot line: VCM path, dashed line: PZT path.

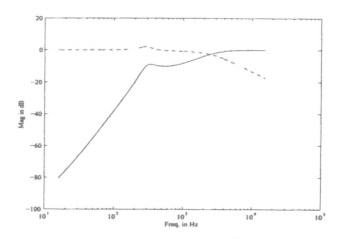

Figure 3.74: S (solid) and T (dashed line) for parallel structure.

Now let the desired crossover frequency of the microactuator loop be f_m = 2 kHz, and the same for the VCM path be f_V = 1 kHz. The gain and phase delay of the microactuator loop at 2 kHz are 1 and 76°, respectively. The gain and phase of the VCM loop are 0.4133 and $-139°$, respectively at 2000 Hz. These two loops work in parallel and generate a loop gain of 1.2419 and phase delay of about $93°$ at f_m. At the 0-dB crossover frequency, which is approximately 2020 Hz, the phase is about $80°$ above $-180°$.

However, one can notice that the frequency at which the magnitude responses of the two paths cross each other, the overall gain drops to a value lower than the gain of either of the two loops. It indicates that the two actuators are conflicting with each other at the hand-off frequency. This is because of the phase difference between the two loops at the hand-off frequency. To reduce the phase difference between the microactuator output and VCM output at the hand-off frequency, it is necessary to change the loop gains by reselecting the controller parameters.

In the next example, a controller is designed for the parallel structure using optimal control approach.

Direct MIMO Design

State space method is another popular method for designing controller, especially for plants with multiple inputs and multiple outputs. In this approach, a state feedback controller is first designed to achieve desired system dynamics assuming all states of the plant are available for feedback. This is followed by the design of an estimator that reconstructs the unavailable states of the plant.

Let the state space model of the dual-stage actuator be

$$\begin{cases} x_p(k+1) &= A_p x_p(k) + B_p u_p(k), \\ y_p(k) &= y_m(k) + y_v(k) = C_p x_p(k), \end{cases} \qquad (3.151)$$

where

$$A_p = \begin{bmatrix} A_m & 0 \\ 0 & A_v \end{bmatrix}, \quad B_p = \begin{bmatrix} B_m & 0 \\ 0 & B_v \end{bmatrix},$$

$$C_p = \begin{bmatrix} C_m & C_v \end{bmatrix}.$$

Here the vector $x_p = \begin{bmatrix} x_m^T & x_v^T \end{bmatrix}^T$ is the state vector and y_p is the displacement output of dual actuator, and $\{ A_m, \ B_m, \ C_m \}, \{ A_v, \ B_v, \ C_v \}$, x_m, x_v, y_m, y_v, u_m and u_v are the system matrices, states, displacement outputs and control inputs of micro-actuator model and VCM, respectively.

State feedback design does not include any integral action. However, we are interested to have integral control for the HDD servomechanism to overcome the effect of input bias. This can be achieved by augmenting the state space model by including two more states, i.e., the tracking errors of the secondary stage and the VCM [68],

$$\begin{align} x_{pi}(k+1) &= x_{pi}(k) + r(k) - y_p(k), & (3.152) \\ x_{vi}(k+1) &= x_{vi}(k) + r(k) - y_v(k). & (3.153) \end{align}$$

Then it follows that the generalized system is

$$\begin{cases} x(k+1) &= Ax(k) + Bu(k) + B_r r(k), \\ y(k) &= Cx(k). \end{cases} \qquad (3.154)$$

where $x = \begin{bmatrix} x_p^T & x_{pi}^T & x_{vi}^T \end{bmatrix}^T$, $C = \begin{bmatrix} C_p & 0 & 0 \end{bmatrix}$ and

$$A = \begin{bmatrix} A_p & 0 & 0 \\ -C_p & 1 & 0 \\ \begin{bmatrix} 0 & -C_v \end{bmatrix} & 0 & 1 \end{bmatrix},$$

$$B = \begin{bmatrix} B_p \\ 0 \\ 0 \end{bmatrix}, \quad B_r = \begin{bmatrix} 0 \\ 1 \\ 1 \end{bmatrix}.$$

Let us consider the following quadratic performance index,

$$J = \frac{1}{2} \sum_{k=0}^{\infty} \left(z(k)^T R_1 z(k) + u_k^T R_2 u(k) \right), \tag{3.155}$$

where $R_1 \geq 0$, $R_2 > 0$ are weights, and $z(k) = \begin{bmatrix} x_{pi} & x_{vi} \end{bmatrix}^T$. Then the design problem can be restated as: *find a proper controller for the generalized system (equation 3.154) such that the closed-loop system is stable, and the linear quadratic (LQ) performance J is minimized.*

To obtain the dual-stage controller, one can separate the design into the following two steps:

1. In the first step, we assume that all the states of the generalized system are available and design a static state feedback control law $u(k) = Kx(k)$, such that it solves the LQ optimal control problem. Since $\{\ A,\ B\ \}$ is stabilizable, if R_1 is selected such that $\left\{\ R_1^{\frac{1}{2}},\ A\ \right\}$ is detectable, then it follows that this problem of static state feedback LQ optimal control is solvable and the feedback gain can be obtained by:

$$K = -(R_2 + B^T PB)^{-1} B^T PA \tag{3.156}$$

where $P > 0$ is the unique stabilizing solution of the following Riccati equation:

$$A^T PA - P + R_1 - A^T PB(R_2 + B^T PB)^{-1} B^T PA = 0. \tag{3.157}$$

2. Since not all the states of the system in our problem are measurable, we need an observer to reconstruct the unmeasured states from the measurements of inputs and outputs. Since the states of the integrator, x_{pi} and x_{vi}, can be directly derived from equation 3.153, only the states for the model given by equation 3.151 need to be estimated. We can use the following state observer:

$$\begin{aligned} x_o(k+1) &= A_p x_o(k) + B_p u_p(k), \\ &+ L(C_v x_o(k) - v_o(k)). \end{aligned} \tag{3.158}$$

where x_o is the estimation of x_p. The observer gain L can be obtained by

$$L = -A_pQC_p^T(V_2 + C_pQC_p^T)^{-1} \qquad (3.159)$$

where V_1 and V_2 are some weighting matrices, and $Q > 0$ is the unique stabilizing solution of the following Riccati equation:

$$A_PQA_P^T - Q + V_1 - A_PQC_p^T(V_2 + C_pQC_p^T)^{-1}C_pQA_p^T = 0. \qquad (3.160)$$

Following equations 3.156 and 3.159, the desired controller can be obtained with the following state space description:

$$\begin{cases} x_k(k+1) &= A_kx_k(k) + B_ku(k), \\ &\quad + B_{kr}r(k) + B_{ky}y(k), \\ u(k) &= C_kx_k(k), \end{cases} \qquad (3.161)$$

where

$$A_k = \begin{bmatrix} A_p + LC_p & 0 & 0 \\ 0 & 1 & 0 \\ [\,0 & -C_v\,] & 0 & 1 \end{bmatrix},$$

$$B_k = \begin{bmatrix} B_p \\ 0 \\ 0 \end{bmatrix}, \quad B_{kr} = \begin{bmatrix} 0 \\ 1 \\ 1 \end{bmatrix}, \quad B_{ky} = \begin{bmatrix} -L \\ -1 \\ 0 \end{bmatrix}.$$

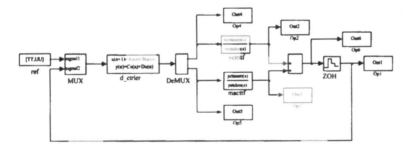

Figure 3.75: Dual-stage actuator system block diagram.

The simulation results presented here are obtained using the $MATLAB^{TM}$ SIMULINK. The SIMULINK model is shown in Figure 3.75.

Using the controller given above, the frequency responses of the closed-loop system are shown in Figure 3.76. It is obvious that the bandwidth of closed-loop system is about 2 kHz.

Figure 3.77 and Figure 3.78 show the simulation results of the closed-loop system with a reference input of $0.5\sin(200\pi t) + \sin(8000\pi t)$. As expected, the dual-stage system follows the reference input closely. The VCM mainly tracks the low frequency input, and the microactuator tracks the high frequency one.

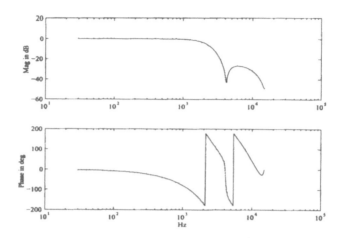

Figure 3.76: Frequency response of the closed-loop system.

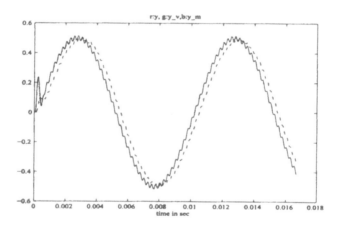

Figure 3.77: DSA output in response to reference input $r = 0.5 \sin(200\pi t) + \sin(8000\pi t)$. Solid line: y_p, dash-dot line: y_v, dotted line: y_m.

Simulation results for a 1 μm step response are shown in Figure 3.79 and Figure 3.80. It is observed that the micro-actuator moves very quickly to the setpoint first, and then slowly moves back to zero while VCM moves to the setpoint. This servo-mechanism significantly improves the dynamic performance of the system. It is also observed that neither the output of the micro-actuator nor that of the VCM exceeds its limit.

We can see from the above results that state feedback control design in conjunction with the state observer design works well for the dual-stage servo system in time domain. The design allows low frequency disturbance tracked by VCM and high frequency disturbance tracked by micro-actuator. However,

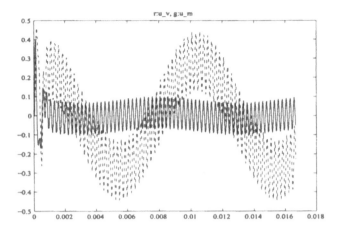

Figure 3.78: Control input of micro-actuator (solid line) and VCM (dashed line) corresponding to Figure 3.77.

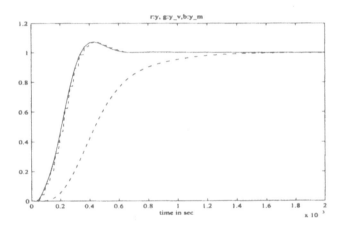

Figure 3.79: Response of dual-actuator model with 1 μm step input. Solid line: y_p, dash-dot line: y_v, dotted line: y_m.

the sensitivity transfer function and complementary sensitivity transfer function shown in Figure 3.81 reveal that, although the closed loop bandwidth is fairly high, the error rejection capability is not very good for the chosen controller and observer gains. The performance index of equation 3.155 does not explicitly include the frequency domain objectives. However, the frequency domain properties of the closed loop system can be manipulated indirectly by changing the weight matrices R_1, R_2, V_1 and V_2. These weight matrices should be reselected such that the frequency domain responses better suit the desired performance.

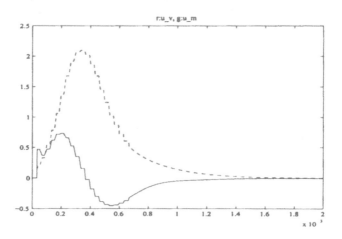

Figure 3.80: Control inputs of VCM (solid line) and microactuator (dashed line) corresponding to Figure 3.79.

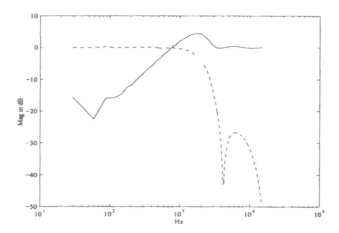

Figure 3.81: S (solid line) and T (dashed line).

3.7.3   PQ Method

The PID-type design of the parallel structure of dual-stage actuator discussed earlier in this section does not take care of the phase difference at the hand-off frequency between the two parallel paths. The PQ method, proposed in [170], includes the condition for proper hand-off into the design process. This method is a two-step design approach. In the first step, the design problem for the DISO plant is converted into a problem of single-input/single-output (SISO) design that satisfies the desired conditions for hands-off. Then, in the second step, a feedback controller is found for the modified SISO system ensuring stability of the feedback system and fulfilment of the specifications for performance and robustness. The control structure is shown in Figure 3.82.

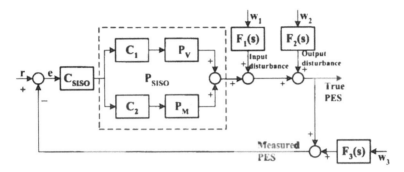

Figure 3.82: PQ method compensated system block diagram

In the first step of PQ design method, the controllers $C_1(s)$ and $C_2(s)$ are chosen simultaneously to address the issues of stable zeros, relative contribution of the two actuators to the combined output, and the interference between the two contributions. Let us define the ratio $R(s)$ between the transfer functions of the VCM path and the transfer function of the microactuator path,

$$\frac{P_V(s)C_1(s)}{P_M(s)C_2(s)} = R(s). \tag{3.162}$$

This ratio determines the relative share of contribution between the two actuators. The VCM actuator should have the dominant role for correcting errors in the low frequency range whereas the microactuator should contribute more for correcting high frequency errors. This division of task defines the desired shape of the ratio $R(s)$ which should have large magnitude ($\gg 1$) at low frequency and small magnitude ($\ll 1$) at high frequency. In the range of frequencies where $|R(s)|$ is close to 1, the outputs from VCM and microactuator have comparable magnitudes. Then the phase of $R(s)$ determines the relative phase between the outputs of $P_M(s)$ and $P_V(S)$, i.e., the amount of interference between the outputs of VCM and microactuator, and hence the overall magnitude of the output.

The transfer functions $C_1(s)$ and $C_2(s)$ are selected such that open loop transfer functions for the VCM ($C_1(s)P_V(s)$) and microactuator ($C_2(s)P_M(s)$), and an acceptable frequency response for $R(s)$ are obtained. After selecting these two compensators, the two parallel branches are combined to form a single-input single-output model P_{SISO}:

$$P_{SISO}(s) = C_1(s)P_V(s) + C_2(s)P_M(S). \tag{3.163}$$

In the second stage of the PQ method, a controller is designed for the SISO plant so that the overall system achieves the stability and performance of the design target. One can use any design method, such as lag-lead compensation, to deal with this SISO problem.

This approach is illustrated using the same design problem stated for the parallel structure. Since the VCM actuator has a phase lag of approximately 180° at frequencies below 1 kHz, we can choose a lag compensator with pole at $p_1 = -2/2\pi f_h$ as $C_2(s)$ to produce a phase delay of approximately 60° at f_h so that the phase difference between the outputs of PZT and VCM actuators is less than 120° at the hand-off frequency f_h. Accordingly we choose:

$$C_1(s) = 1, \tag{3.164}$$

$$C_2(s) = \left|\frac{k}{s^2}\Big/\frac{g_m}{p_1 s + 1}\right|_{s=j2\pi f_h} \frac{1}{p_1 s + 1}. \tag{3.165}$$

s.t. P_{SISO} has no non minimum phase zeros.

The new SISO plant model $P_{SISO}(s)$ that includes C_1 and C_2 has a -40 dB/dec slope at frequencies below f_h and about a -20 dB/dec slope for frequencies above f_h. One need to verify whether the zeros in P_{SISO} are stable or not.

To design the controller for the SISO plant P_{SISO}, we can elevate the low frequency gain and adjust the overall loop gain such that the compensated loop has a 0 dB crossover at f_m. A lag compensator with a pole at $p_c = -8/2\pi f_h$ and a zero at $z_c = -1.25/2\pi f_h$ can be used for this purpose. The compensator is selected as:

$$C_{SISO}(s) = k_c \frac{z_c s + 1}{p_c s + 1}, \tag{3.166}$$

with

$$k_c = \left|\frac{(p_c s + 1)}{(z_c s + 1)P_{SISO}}\right|_{s=j2\pi f_m}.$$

Results obtained for the design of servo controller for a dual-stage actuator using the PQ method are shown in Figures 3.83 to 3.89. For this design example, we chose $f_h = 400$ Hz, $f_m = 2000$ Hz, and achieve a open loop bandwidth of 2000 Hz. The phase margin and gain margin are overly optimistic because the actuator resonances are neglected in this example.

The PQ method provides a simple but effective way to allocate the control effort between two actuators of a dual-stage actuation system. One can use various optimal control methods instead of the simple lag compensator as shown above to design the controller for the compensated model P_{SISO}.

3.7.4 Decoupled Master-Slave Structure with Actuator Saturation

The decoupled master slave (DMS) configuration is shown in Figure 3.90. For this configuration, the open loop transfer function $L_{dms}(s)$, the closed-loop

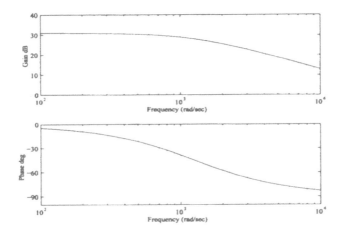

Figure 3.83: Frequency response of $C_2(s)$.

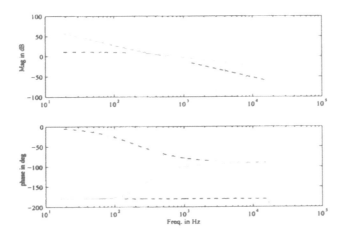

Figure 3.84: Compensated plant P_{SISO} (solid line). Dashed line: response of $C_1(s)P_V$; Dash-dot line: response of C_2P_M.

transfer function T_{dms} and the sensitivity transfer function S_{dms} are defined as

$$L_{dms} = (1 + C_M P_M)C_V P_V + C_M P_M,$$

$$T_{dms} = \frac{C_V P_V + C_M P_M + C_V P_V C_M P_M}{(1 + C_V P_V)(1 + C_M P_M)},$$

$$S_{dms} = \frac{1}{1 + C_V P_V}\frac{1}{1 + C_M P_M}, \qquad (3.167)$$

respectively if none of the the actuators is saturated. Furthermore, S_{dms} in equation 3.167 is equivalent to the product of the following two transfer func-

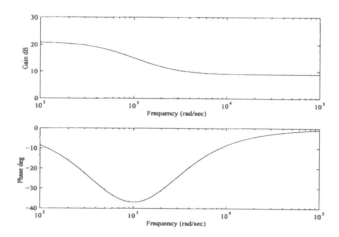

Figure 3.85: Frequency response of C_{SISO}.

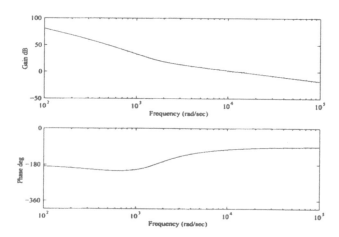

Figure 3.86: Frequency response of the compensated open loop $C_{SISO}P_{SISO}$.

tions:

$$S_{Vdms} = \frac{1}{1 + C_V P_V},$$ (3.168)

$$S_{\mu dms} = \frac{1}{1 + C_M P_M}.$$ (3.169)

It is easily understood that S_{Vdms} is equivalent to the sensitivity transfer function of a single stage VCM loop and $S_{\mu dms}$ is equivalent to the sensitivity transfer function of a single stage microactuator loop. The servo designer can select C_V and C_M separately to shape S_{Vdms} and $S_{\mu dms}$ individually and, as a result, the combined sensitivity transfer function S_{dms}.

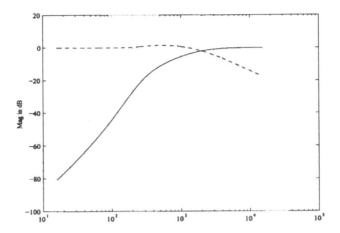

Figure 3.87: Sensitivity (solid line) and complementary sensitivity (dashed line) transfer functions.

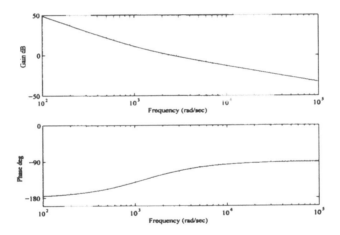

Figure 3.88: $R(s)$ frequency response. At below f_h, $|R(s)| > 1$. At above f_h, $|R(s)| < 1$

If we define

$$C_{VP} = C_V(1 + C_M P_m), \tag{3.170}$$

then the implementation of the DMS structure based controller is similar to that of the parallel structure.

Because of very small phase lag of the PZT actuator in the low frequency range, the simple lag-lead compensator (equation 3.149) for the VCM and a lag compensator for the PZT actuator (equation 3.150) also work for the DMS structure after some adjustments of the control parameters. For the plant in the previous example, we choose the lag-lead compensator such that the zero

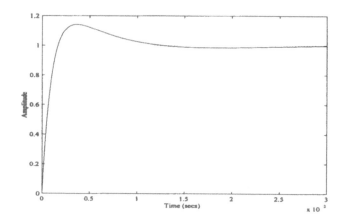

Figure 3.89: Closed-loop system step response.

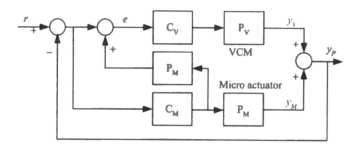

Figure 3.90: DMS configuration.

dB gain crossover occurs at 500 Hz for the VCM ($K_V P_V$) loop and at 2 kHz for the PZT loop. Figure 3.91 and Figure 3.92 show the results. We observe in Figure 3.91 a 76° phase lag at 2 kHz for the PZT path, whereas the gain and phase of the VCM path at 2 kHz are 0.65 dB and 177°, respectively. With the addition of the PZT loop in the compensation of the VCM, the gain becomes 9.42 dB and the phase delay becomes 170° at 500 Hz. Overall, the combined dual-loop has a gain of about 1.08 dB and a phase delay of 68° at 2 kHz. The phase margin is about 69° considering the gain drop. The overall bandwidth is slightly higher than the bandwidth of the microactuator path.

Actuator Saturation and Phase Margin

One important design consideration for the dual-stage actuator is to suitably allocate the control signals between the VCM and microactuator. It is, therefore, natural to examine the dependence of control system's performance on the saturation level of the microactuator. Let us limit the displacement of the

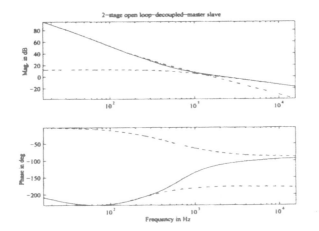

Figure 3.91: Open loop transfer function. Solid line: combined loop, dash-dot line: VCM path, dashed line: PZT path.

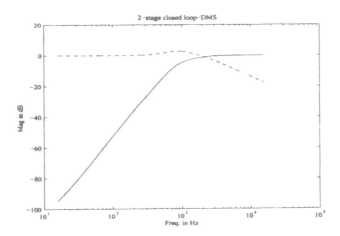

Figure 3.92: S (solid) and T (dashed) for DMS structure.

secondary stage actuator to ± 8 micro inches (μin) to investigate the possible problems due to the saturation in the secondary stage actuator. Figure 3.93 to 3.95 show the step responses of the dual-stage servo for step changes in reference by 8, 16, and 21 μin, respectively. It takes about 0.7 ms, 0.75 ms, and 1.7 ms, respectively, for the step responses to settle in these 3 cases, whereas, the overshoots of the combined output are approximately 44%, 87% and 100%, respectively.

It is interesting to note that in the linear control case shown in Figure 3.95, the gain margin (GM) of $C_V P_V$ was 9.5 dB. The maximum stable step change of 21 μin was about $20 \times \log_{10}(\frac{21}{8}) = 8.5$ dB. Assume that a 1 dB gain margin

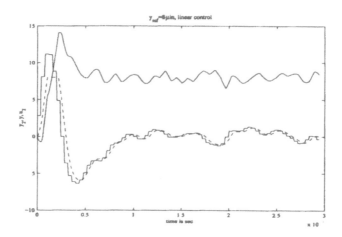

Figure 3.93: Response to 8 μin step input. Solid line: combined output, dashed line: microactuator output, dash-dot line: microactuator input.

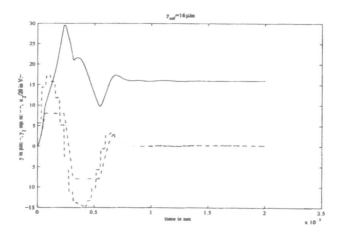

Figure 3.94: Response to 16 μin step input. Solid line: combined output, dashed line: microactuator output, dash-dot line: microactuator input.

is required to account for the numerical error due to sampling etc. We can see from the simulation results that the maximum allowable stable step change (MASSC) is close to $10^{(GM-1)/20} \times 8$. Table 3.8 summarizes the MASSC estimated using $10^{(GM-1)/20} \times 8$ versus the actual MASSC obtained from simulations. We see from this table that the MASSC can be predicted fairly well from the knowledge of the gain margin of $C_V P_V$.

Although increasing the gain margin of the VCM loop can increase the MASCC, this may require the VCM loop gain to be decreased and therefore, a relatively higher displacement range for the secondary stage actuator.

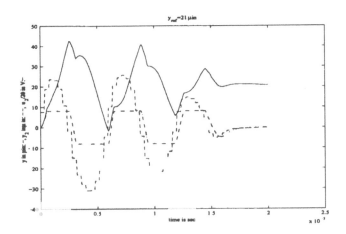

Figure 3.95: Response to 21 μin step input. Solid line: combined output, dashed line: microactuator output, dash-dot line: microactuator input.

Table 3.8: VCM Loop Gain Margin vs. MASSC

GM in dB	Estimated MASSC in μin	Actual MASSC in μin
14.6	42.9	40
12.9	31.7	33.5
11.6	27.1	**27**
10.4	23.6	23.5
9.5	21.3	21
8.6	19.2	19
7.8	18	18.1

Nonlinear Observer based DMS Control

In the previous designs, saturation in actuators was ignored and the controllers were designed and their performances were evaluated assuming non-saturating actuator. However, it is an usual practice to limit the magnitude of the input voltage applied to the microactuator to protect it from being damaged. That is,

$$u'_M = sat(u_M) = \begin{cases} \bar{u}_M & \text{if} \quad u_M \geq \bar{u}_M \\ u_M & \text{if} \quad \underline{u}_M < u_M < \bar{u}_M \\ \underline{u}_M & \text{if} \quad u_M \leq \underline{u}_M \end{cases} \qquad (3.171)$$

where $\underline{u}_M$ and $\bar{u}_M$ represent the minimum and maximum allowable control input to the microactuator.

Since the microactuator has a finite DC-gain, any limit on its input effectively puts a limit on the displacement effectuated by the microactuator. This limit on the microactuator's displacement can be incorporated in the design by modifying the DMS controller structure. Figure 3.96 shows the modified DMS structure which is similar to the DMS scheme (shown in Figure 3.90) except a block marked $sat(\cdot)$ included in the controller. To see how incorporating the microactuator model with saturation in the DMS controller helps to improve the stability of the control loop in presence of microactuator saturation, let us assume that $r = 0$, and the transfer function of a saturation section represented as $sat(\cdot)$.

Breaking the VCM loop open at the point e_V in Figure 3.96, we have the open loop transfer function of the VCM path as

$$O_v = C_V P_V (1 + C_M \hat{P}_M sat(\cdot)) \frac{1}{1 + C_M P_M sat(\cdot)}. \qquad (3.172)$$

Obviously, if $\hat{P}_M = P_M$, and assuming that the microactuator P_M and its model $\hat{P}_M$ have the same initial values, we have $\hat{y}_M(k) = y_M(k)$, resulting effectively in

$$(1 + C_M \hat{P}_M sat(\cdot)) \frac{1}{1 + C_M P_M sat(\cdot)} = 1. \qquad (3.173)$$

Consequently, the secondary stage actuator model does not appear in the VCM path, thus its saturation will not affect the VCM control loop's stability.

In contrast, in the usual linear control configuration shown in Figure 3.90, we have $|\hat{y}_M(k)| \gg |y_M(k)|$ if the microactuator is saturated for sufficiently long time. In this case, the effective open loop transfer function

$$O_v = C_V P_V (1 + C_M \hat{P}_M) \frac{1}{1 + C_M P_M sat(\cdot)} \qquad (3.174)$$

will have a higher effective gain than the case of equation (3.172), leading to potential instability.

Next, we assume that the microactuator P_M is represented by state space equations:

$$\begin{cases} x_M(k+1) &= F_M x_M(k) + \Gamma_M u'_M(k), \\ y_M &= H_M x(k), \end{cases} \qquad (3.175)$$

and the estimator $\hat{P}_M$ by

$$\begin{cases} \hat{x}_M(k+1) &= F_M \hat{x}_M(k) + \Gamma_M u'_M(k), \\ \hat{y}_M &= H_M \hat{x}(k). \end{cases} \qquad (3.176)$$

The error dynamics of the estimator is given by

$$\begin{aligned} \bar{x}(k+1) &= x(k+1) - \hat{x}(k+1), \\ &= F_M \bar{x}(k). \end{aligned} \qquad (3.177)$$

Since the microactuator model is assumed stable, so is the error dynamics (equation 3.177). Thus we have

$$\lim_{k \to \infty} y_M(k) - \hat{y}_M(k) = \lim_{k \to \infty} H_M \bar{x}(k) = 0. \tag{3.178}$$

Hence, we have that if P_M is stable and $\hat{P}_M = P_M$,

$$\lim_{k \to \infty} \hat{y}_M(k) - y_M(k) = 0. \tag{3.179}$$

It should be noted that for MEMS based microactuator with relative position error signal [125], y_M is available and thus can be used in the control scheme directly instead of $\hat{y}_M$. In case y_M is not available, its estimation $\hat{y}_M$ can be used as in Figure 3.96. In general, an open loop estimator does not guarantee asymptotic stability of the error dynamics [8].

Next, we prove using an example that even with a reduced order PZT actuated suspension model and parameter uncertainty, $\hat{y}_M$ is still a close estimate of y_M, and hence the modified algorithm is effective in dealing with the microactuator saturation.

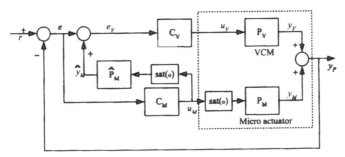

Figure 3.96: Dual-stage control with microactuator saturation taken into consideration.

3.7.5 Experimental Results

Although the design principle shown in the previous section applies to various types of dual-stage actuators, experiment using an actuated suspension is used here to show the effectiveness of the proposed nonlinear control scheme. against the original linear control scheme.

The nominal control loops, i.e., VCM path, microactuator path and overall dual-stage open-loop are shown in Figure 3.97 with

$$P_V(s) = \frac{2.04 \times 10^{21}}{s + 3.14 \times 10^4} \frac{1}{s^2 + 301.6s + 2.58 \times 10^5}$$
$$\frac{s^2 + 2073s + 4.3 \times 10^8}{s^2 + 1508s + 3.55 \times 10^8}$$

$$P_M(s) = \begin{cases} \dfrac{1}{s^2 + 1244s + 1.7 \times 10^9} \ \mu m/V, & (3.180) \\[2em] \dfrac{20 \times 5.45 \times 10^7}{s^2 + 2450s + 1.7 \times 10^9} \\[2em] \dfrac{s^2 + 4524s + 2.08 \times 10^9}{s^2 + 6032s + 3.64 \times 10^9} \ \mu m/V. & (3.181) \end{cases}$$

$$C_V(z) = \frac{1.6(z - 0.9875)^2}{(z - 0.2846)(z - 0.99)} \frac{z^2 - 1.416z + 0.9415}{z^2 - 1.297z + 0.92}$$
$$\frac{z^2 + 0.1525z + 0.95}{z^2 + 0.4155z + 0.0144} \ V/V, \qquad (3.182)$$

$$C_M(z) = \frac{1.227(z - 0.081)}{z - 0.9158} \frac{z^2 + 1.32z + 0.7856}{z^2 + 0.6232z - 0.277}$$
$$\frac{z^2 + 0.148z + 0.9}{z^2 + 0.439z + 0.84} \ V/V. \qquad (3.183)$$

The models of the plant is obtained from experimental frequency response measured using a laser Doppler Vibrometer (LDV) whose measurement scale is set at 2 μm/V. The gain of the microactuator driver is 20. The controller is implemented using a digital signal processor (DSP) in the commercially available DSP board **dSPACE 1103** with sampling frequency of 25 kHz. The VCM controller $C_V(z)$ includes two notch filters to cancel the resonances of the VCM actuator at around 3.0 kHz and 6.5 kHz. For the resonance-compensated VCM model, a lead-lag compensator is used to achieve desired gain margin and phase margin. As the microactuator's phase loss is less than that of the VCM, a lag compensator in series with two notch filters to cancel microactuator resonances at 6.5 kHz and 9.6 kHz is chosen as its controller $C_M(z)$ to achieve the target open-loop bandwidth. The sensitivity and the complementary sensitivity functions achieved with this design are shown in Figure 3.98.

Step Responses Considering Saturation

To show that the proposed method is effective in dealing with the problem of saturation in the secondary stage actuator, we limit the amplitude of the microactuator input to ±0.15 V which restricts the microactuator output to ± 0.12 μm, or ±4.8 μ at steady state. In reality, the microactuator is capable of producing much larger displacement. The rationale for studying this extremely small displacement range of microactuator includes,

1. this emulates the saturation of microactuator without reaching the hard limit, and makes a challenging test ground for the controller,

2. for very high performances, the microactutor is to be placed closer to the R/W head resulting in smaller mechanical amplification factor and hence limited range of displacement range, and

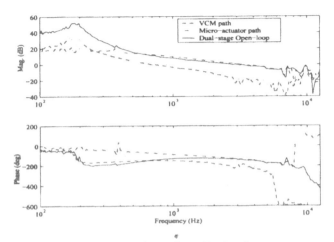

Figure 3.97: Open loop Bode plots.

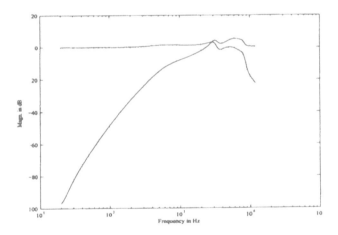

Figure 3.98: Sensitivity function (S) and Complementary sensitivity function (T).

3. if the controller is capable of utilizing a microactuator with smaller range of displacement without significant degradation of performance, the specifications for mechanical design can be relaxed and therefore the actuator can be optimized better.

Figure 3.99 shows a 4 μin step response of the dual-stage control system. The micro-actuator is not saturated so the linear controller and the nonlinear controller behave in similar manner. Figure 3.100 shows the linear controller response with a 10 μin step input. The position output is oscillatory, and the settling time is as long as 1.5 ms. Increasing the reference signal amplitude further makes the dual-stage control loop unstable. However, on the contrary, when the proposed nonlinear controller is used, the dual-stage loop remains stable even with a step input of 35 μin (Figure 3.101) or higher at the reference command.

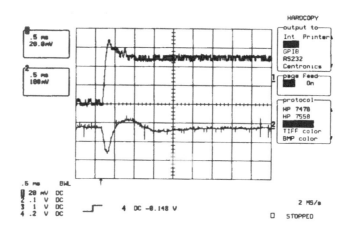

Figure 3.99: Response to 4 μin step input. Upper curve: combined output y_p, lower curve: microactuator control voltage u'_M.

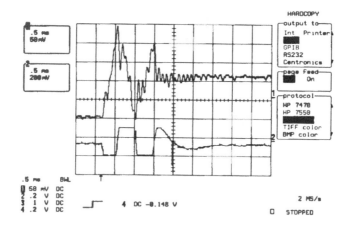

Figure 3.100: Response to 10 μin step input with linear control scheme. Upper curve: combined output y_p, lower curve: microactuator control voltage u'_M.

Effect of Saturation Level

In the studies of microactuator performance presented above, only the microactuator saturation caused by changes in external reference is considered.

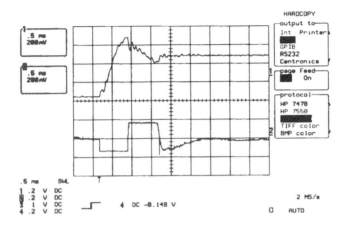

Figure 3.101: Response to 35 μin step input with nonlinear control scheme. Upper curve: combined output y_p, lower curve: microactuator control voltage u'_M.

The effect of secondary stage actuator stroke limitation on system performance is evaluated in the following experiments by analyzing 3σ (or the 3 times the standard deviation) of the dual-actuator combined output y_p. The experiment is repeated with the saturation range of the secondary stage set to different levels. The combined output y_p is measured, the data is acquired, and the standard deviation σ is computed for each of these cases. Figure 3.102 shows the comparison of 3σ for y_p with the secondary actuator saturation range set at ± 0.05, ± 0.1, ± 0.15, ± 0.2, ± 0.25, ± 0.3, ± 0.4, ± 0.5, and ± 0.7 V in this experiment.

As can be seen from the figure, the 3σ for the nonlinear controller degraded from 0.5 μin to 1.1 μin when the microactuator control voltage limit reduced from ± 0.7 V to ± 0.05 V. The linear controller, while achieving the almost same 3σ of 0.5 μin at a control voltage range of ± 0.7 V, could not maintain the closed-loop stable when the control voltage is reduced to slightly below ± 0.05 V.

Evaluation of Robustness

It should be noted that an accurate model of the microactuator is required for this nonlinear controller to make the estimate of the microactuator output a true representation of the actual output of the microactuator, i.e., $\hat{y}_M \rightarrow y_M$. Getting an accurate model of any mass-produced actuator is no doubt a ideal proposition. In reality, the properties of the microactator varies with environmental conditions and also from sample to sample. In such case, a natural doubt arises about the effectiveness of the controller - *is the proposed*

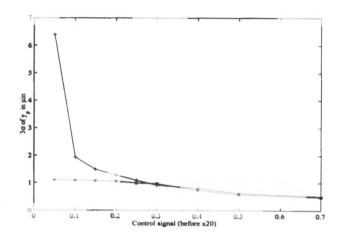

Figure 3.102: Combined output 3σ vs. saturation range from experimental results. '+': linear controller; '*': nonlinear controller.

controller robust enough in the face of variations in the actuator parameters? The robustness issue is addressed with the help of an experiment.

Figure 3.103 shows the responses of the system for different step inputs with $\hat{P}_M$ kept as constant and is equal to the DC gain of P_M. The microactuator control voltage is limited to ± 0.15 V. Figure 3.104 shows the corresponding $\hat{y}_M$ and y_M. It is clearly evident from these results that, even using just a gain to estimate the output of microactuator, the proposed scheme is able to retain a reasonably good stability when the microactuator is severely saturated.

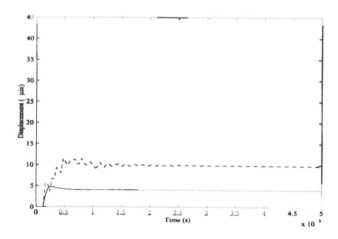

Figure 3.103: Combined output responses with reduced order microactuator models. $r=4$ μin (solid), 10 μin (dashed), and 35 μin (dotted).

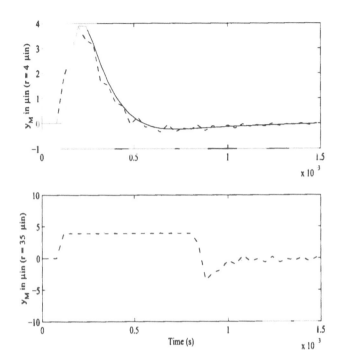

Figure 3.104: Microactuator and estimator responses with reduced order microactuator models for $r=4$ μin (y_M: solid line, $\hat{y}_M$: dashed line) and 35 μin (y_M: dotted line , $\hat{y}_M$: dash-dot line).

The robustness of the modified DMS scheme can be further investigated by evaluating the maximum MASSC. Table 3.9 summarizes the maximum MASSC obtained through simulation for three cases:

- **Case 1:** use equations 3.182 and 3.183 to control the plant described by the equations 3.180 - 3.181 according to the scheme presented in the block diagram of Figure 3.90,

- **Case 2:** use the proposed modified DMS scheme with a full order model of the microactuator $\hat{P}_M$ used to estimate the output of the secondary stage, and

- **Case 3:** use the modified DMS scheme but $\hat{P}_M$ is the DC gain of the P_M only.

Figure 3.105 plots the MASSC versus DC gain error expressed as a percentage for the linear control scheme. It shows that the MASSC decreases from 28 μin to 7.6 μin as the DC gain error varies from -50% to $+50\%$. For the modified nonlinear control scheme with full order and reduced order models

Table 3.9: MASSC for different control schemes

Control scheme	MASSC in μin
Case 1	10
Case 2	> 1000
Case 3	> 1000

for P_M, the MASSC can still be greater than 1000 μin when the DC gain of $\hat{P}_M$ fluctuates in between ±50% of its nominal value.

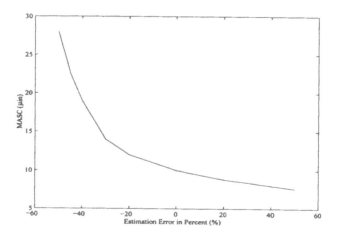

Figure 3.105: MASSC for linear control scheme with reduced order micro-actuator model.

It is observed from the above results that for the linear control scheme of Figure 3.105, a lower value for $\hat{P}_M$ increases the MASSC. However, lower value of $\hat{P}_M$ decreases the gain of the VCM path and affects the hand-off properties of the dual-stage loop even though the microactuator is not saturated. The proposed nonlinear modification, on the other hand, has a much larger MASSC with a saturating secondary stage. Furthermore, it retains all the properties of the original linear design with the microactuator not saturated.

In a dual-stage servo controller, the secondary stage actuator helps to achieve higher servo bandwidth but not necessarily lower sensitivity peak. MEMS based microactuators with frequency response similar to that of a double integrator $1/s^2$ will produce the same bode integral, and hence the water bed effect. Similar to the discussion on VCM, MEMS actuators having velocity and/or acceleration sensor can help to achieve a lower Bode integral value, and hence, more vibration rejection with the same servo bandwidth. The frequency responses of PZT actuated suspensions and sliders have constant gain for a wide range of frequencies. If resonant modes of such actuator are at very high frequencies, a simple gain can model the actuator fairly accurately [134]. In such case, the secondary stage actuator loop can bring down the peak of the sensitivity transfer function to practically less than 3 dB [46].

Control structures for dual-actuated HDD head positioning servomechanism and the basic concepts of the dual-stage actuation loop shaping are introduced in this section. Three different strategies, namely, the decoupled-master slave structure, the parallel structure and PQ method, for the design of controller for dual-stage servo are illustrated. A simple lag-lead compensators plus notch filters for high frequency resonances, if present, produces fairly satisfactory results for the dual-stage servomechanism if the actuators are not saturated. This fact is proven with the help of simulation of control loop. In reality, however, the microactuator saturation is unavoidable, and in such case, the response of the closed loop shows more oscillation when a large step is applied at the reference input. Stability of the loop is also deteriorated when there is saturation in the microactuator.

The performance and stability of the dual-stage servomechanism can be greatly improved even in the presence of microactuator saturation through a simple modification to the decoupled master-slave configuration. In this modification, a nonlinear model of the secondary stage actuator is used to estimate its output. Results presented here show that the proposed nonlinear controller retains all the properties of the original linear controller when the secondary stage actuator is not saturated and, when the secondary stage is saturated, the proposed design is found to show better stability of the dual actuator servo loop. The design is also robust in presence of uncertainties in the microactuator model. Additionally, the design method can be applied to other dual-actuator systems with a saturating secondary stage actuator such as power systems with flexible AC transmission devices [203]. A more generic anti-windup design approach is proposed in [78], however, the controller presented above is simple and yet effective. Simple and yet effective system is of great importance in the design and implementation of HDD servomechanism and many other practical servo systems. Dual-stage actuators coupled with multisensing and other advanced controls offer additional avenue for improvement of control performance. Interested readers may refer to the article by Pang *et al* [158] and the references therein for more information.

From the examples given so far illustrating different design approaches for ideal VCM model and PZT based microactuator model, it is easily comprehended that each method has its advantages as well as disadvantages. It depends on the designer to decide on the trade-off and choose the parameters accordingly so that the design goals are achieved.

Chapter 4

Spindle Motor Control

4.1 Magnetic Field Fundamentals

An electric machine is an electromagnetic (EM) system. It relies on the reactions of the magnetic field in the machine to realize the electromechanical energy conversion. Knowledge about the fundamentals of electromagnetic theory is, therefore, essential for the design, analysis and applications of electric machines.

In the operation of electric machine, the frequency of variation of the magnetic field is normally below the range of kHZ. Therefore, the displacement current, which must be taken into consideration for analyzing high frequency electromagnetic field, can be neglected. This makes it possible to use of some simplified EM models to describe and analyze electric machines.

In this section, the basic concepts of the magnetic field are introduced first, followed by the concepts of magnetic circuit that can be treated as a useful tool to simplify the magnetic field analysis in the machine.

4.1.1 Flux

Flux is a scalar, normally denoted by the symbol Φ. It is used to describe the total amount of magnetic field in a given region. Flux is depicted as lines (or flux lines) which can be visualized by a flow circulating out of the N-pole and into the S-pole of a magnet in a circulating path. Figure 4.1 shows an illustration of the flux lines generated by a permanent magnet. It should be noted that these lines do not cross one another. The SI unit of flux is *Weber*.

In the electric machine analysis, flux can be used as a quantity to describe the global performance of the magnetic system, or the global effect of the magnetic fields within a certain area, e.g., the flux linked with a winding.

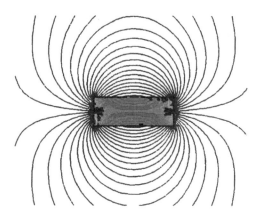

Figure 4.1: Flux lines generated by a permanent magnet

4.1.2 Flux Density (Magnetic Induction)

Flux density is a vector, and is normally denoted by the symbol $\bar{B}$. It describes the magnetic field distribution in the space and gives flux lines through a unit area. The SI unit of the flux density is **Tesla** (**T**), or **Weber per square meter** (**Weber/m²**). In the electrical engineering, Gauss is also used as the unit of flux density where 1 Gauss = 10^{-4} Tesla.

The relationship between the flux density and the flux can be described by

$$\Phi = \int_S \bar{B}.d\bar{S}. \tag{4.1}$$

In the equation, S is the area where the flux lines going through, and $\bar{B}$ is the flux density at the local area $d\bar{S}$.

In the electric machine analysis, the flux density is very import in analyzing the local field effects, e.g., the magnetic saturation at a point, and the leakage field in a local area.

4.1.3 Magnetic Field Strength, Permeability and Relative Permeability

Magnetic field strength is a vector, and is normally denoted by the symbol $\bar{H}$. It describes the intensity of magnetic field. The SI unit of magnetic field strength is **Ampere per meter** (A/m). **Oersted** (Oe) is also used in engineering as an alternative unit, and $1 \text{Oe} = 10^3/4\pi$ A/m.

Flux density and magnetic filed strength are closely related. In vacuum, $\bar{B}$ is proportional to $\bar{H}$,

$$\bar{B} = \mu_0 \bar{H} \tag{4.2}$$

where, the coefficient μ_0 is the permeability in vacuum whose value is $4\pi \times 10^{-7}$ in units of *Henrys per meter* (H/m).

Different materials show different behaviors when the magnetic field passes through them. But the concept of permeability can still be used to describe the relationship between B and H. For a given material, the relationship between the field strength and flux density can be described by

$$B = \mu H, \tag{4.3}$$

where μ is permeability of the material. Permeability of any material defines its ability to allow transmission of magnetic field through it. Permeability is also described as

$$\mu = \mu_r \mu_0 \tag{4.4}$$

where μ_r is known as the relative permeability of the material. For most of the materials, the value of μ_r is close to 1. If a material shows linear relationship between $\bar{B}$ and $\bar{H}$ when the magnetic field varies in a very wide range, then the material is called linear material in the context of electro-magnetic (EM) analysis.

4.1.4 Energy in Magnetic Field

The magnetic field contains energy, and its energy density is expressed as

$$w_m = \frac{1}{2}\bar{B} \cdot \bar{H}. \tag{4.5}$$

The SI unit of the energy density (w_m) is *Joule per cubic meter* (J/m^3). *MGOe* is also used as the unit where $1 MGOe = 10^5/4\pi \ J/m^3$. If the EM performance of the material is homogeneous and its B-H curve is monotonic, then the expression for energy density can be simplified by substituting equation 4.3 into equation 4.5,

$$w_m = \frac{1}{2}B.H = \frac{1}{2\mu}B^2 = \frac{\mu}{2}H^2. \tag{4.6}$$

Taking V as the volume of the whole system, the magnetic energy in the system is

$$W_m = \int_V \frac{B^2}{2\mu}dv = \int_V \frac{\mu H^2}{2}dv. \tag{4.7}$$

Operation of an electric motor depends on the conversion between the magnetic and mechanical energy. However, the EM structure of the motor is normally complicated and the magnetic material used in the motor normally operates in non-linear state. As a result, it is generally difficult to calculate the local magnetic energy density and the total magnetic energy of the motor. But it is important in the design and analysis of any electric machine.

4.1.5   B-H Curve

The relationship described by equation 4.3 can be expressed by the B-H curve. For linear materials mentioned before, the B-H curves are straight lines. However, there are materials whose B-H curves are not linear. In the design and analysis of electric machines, two types of materials are of concern as they play important roles in determining the EM performance of electric machines. These are soft-magnetic materials and hard-magnetic materials, whose B-H curves are very different from that of linear materials.

4.1.5.1. Soft-magnetic Material (Ferromagnetic Material)

The relative permeability of a soft-magnetic material is much higher than 1. These materials are made of iron, nickel, cobalt and manganese, or their compounds. These materials are also known as ferromagnetic materials. The magnetic performances of these materials also depend on the processes used to produce them such as heat and rolling treatments. Relative permeability of a soft-magnetic material is not constant - it varies when the magnetic intensity is changed.

It will be shown in section 4.1.6 that, for a given EM device, the field strength H is proportional to the current source that induces magnetic field in the device, even if the device structure is complicated and ferromagnetic materials are used. But the relationship between the flux density B and current source can be complex if ferromagnetic materials are used. The relationship between B and H can be expressed by the B-H curve of the material. Figure 4.2 shows a typical B-H curve of ferromagnetic material in the first quadrant of Cartesian coordinate system. The full B-H curve is located in the first and third quadrants, and is symmetrical about the origin. It is evident from the curve that the flux density varies slowly when H is small. Then for a wide range of H, B increases rapidly with increasing H and their relationship can be approximated as linear. For $H > H_t$, the flux density B is a non-linear function of H, but still monotonic. When the flux density is increased to a certain level (B_s), the material is saturated and further increase of the flux density is difficult. For practical purposes, the B-H curve of the the soft-magnetic materials can be considered linear in the range $(-H_t, +H_t)$ that is their permeability remains constant.

As their relative permeabilities are higher than 1 and their hysteresis loops (see section 4.1.5.2) are very narrow, the ferromagnetic materials are widely used in electric motors as stator core and rotor yoke (see section 4.3) to realize the required magnetic field path (or the magnetic circuit; see section 4.1.6). Magnetic performances of some ferromagnetic materials are shown in Table 4.1.

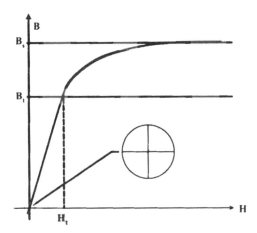

Figure 4.2: B-H curve of a ferromagnetic material

Table 4.1: Performance of some ferromagnetic materials

Material	Iron	4% Si steel	Monimax	Hypernic	Supermalloy
Maximum μ_x	5,000	7,000	35.000	70,000	800,000
Saturation flux density (Tesla)	2.15	2.0	1.5	1.6	0.8

4.1.5.2 Hard Magnetic Material (Permanent Magnet Material)

Hard magnetic materials (Permanent magnet materials) have the capability to produce magnetic field after they are fully magnetized. They exhibit a large hysteresis loop in the their B-H curves, as it is shown in Figure 4.3. It is easily observed that they are very different from the ferromagnetic materials. The hysteresis loop in the second quadrant is called demagnetization curve. In the design and analysis of permanent magnet electric machines, this part of curve is the most important in the entire hysteresis loop as the operation range of the magnet is normally in this quadrant. This is explained using an example in section 4.1.6.

There are three important points in the second quadrant of the B-H loop,

1. Point a: $H = 0$ and the magnetic flux density is equal to B_r which is known as the remanent magnetic flux density, or remanence. This value is a measure of the remaining magnetization when the magnetization field is removed. Therefore, this parameter can show the capability of the permanent magnet in generating the magnetic field.

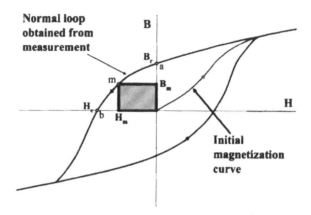

Figure 4.3: B-H curve of a permanent magnetic material

2. Point b: $B = 0$ and the magnetic field intensity is equal to H_c which is known as the coercive force, or coercitivity. H_c shows the value of the reverse field needed to remove the magnetization of the material after it is saturated magnetized. This parameter shows the capability of the magnet to counteract the influence of external field.

3. Point m: Maximum product resulting from B and H on the demagneti-zation curve, i.e., the largest rectangle which can be formed within the hysteresis curve in the second quadrant. This point indicates the maxi-mum energy that a magnetic material can supply to an external magnetic circuit (see section 4.1.6) when operating at this point, that is,

$$(BH)_{max} = Maximum(w_m) = H_m B_m. \tag{4.8}$$

In the design of electromagnetic devices, the operating point of the per-manent magnet material should be designed around the point M for utilizing the material effectively. A high $(BH)_{max}$ implies that the required magnetic flux can be obtained with a smaller volume of the material, making the device lighter and more compact.

In the spindle motor, the permanent magnet is made of bonded neumodium-iron-boron (NdFeB). This kind of material is made by binding rapid-quenching NdFeB powder. The powder is mixed with resin to form a magnet by com-pressing molding with epoxy or infecting molding with nylon. The magnet surface is coated with epoxy to prevent corrosion. This kind of material is fine in B_r, H_c and $(BH)_{max}$. Because the quenching powder is used, the magnetic property of the bonded NdFeB is isotropic. Therefore, it can be magnetized with multi-poles in radial direction. The magnet can also be made as a ring with a thin wall, and the cost of the ring is low in mass production.

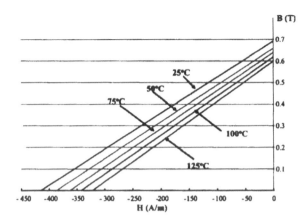

Figure 4.4: Demagnetization curves of a bonded NdFeB magnet

The performance of magnet is also related to the temperature of the environment. Increasing the temperature leads to weakening of the field generating capability, see Figure 4.4. In the application of permanent magnet materials, we must let the magnet to operate in the temperature lower than T_{max}, the maximum practical operating temperatures allowed for the magnet. The magnetic performance of the permanent magnet can be recovered when the temperature returns to its original value if the temperature is lower than T_{max}. For the bonded NdFeB magnet, T_{max} normally lies in the range from 120°C to 140°C.

4.1.6 Magnetic Circuit and Magnetomotive Force

Because of the complexity in the geometric and electromagnetic structures, analyzing the magnetic field in an electromagnetic device is normally a complicated procedure. Numerical methods, e.g., the finite element method (FEM) are often used to solve a boundary value problem defined by Maxwell equations and boundary conditions, and expressed by magnetic potential. After getting the discrete magnetic potential solutions, post-processing technology is employed to make clear the global electromagnetic performance of the device. Such complicated procedure relies on special software tools. In many cases, some simplifications can reduce the complexity and an engineering solution can be obtained easily. Magnetic circuit method is one such method which uses magnetic flux Φ, *reluctance R* and *magnetomotive force F* to describe the effect of magnetic field in the device. It can be used to analyze both linear and non-linear magnetic field problems, and is even applicable to the magnetic device with complicated electromagnetic (EM) structures [69],[70],[77]. The concept and application of magnetic circuit are explained next with the help of two examples.

Example-1: Magnetic circuit of an inductor

An electromagnetic device is shown in Figure 4.5. This is an inductor formed by a winding and a magnetic core made of steel. There is an airgap in the core. The core is of uniform cross section and the number of turns in the winding is N. Input current I into the winding, and the magnetic field is thus excited in the core as shown in Figure 4.5. To simplify the analysis, it is assumed that the permeability of the steel is infinite, and the magnetic field distributed evenly in the core and airgap, and there is no leakage field in the airgap.

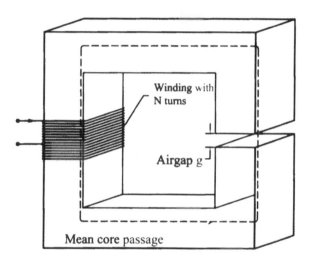

Figure 4.5: Magnetic circuit of an inductor

Using Ampere's law, the relationship between the magnetic flux and current can be described by the following close loop line integration

$$\oint_l \bar{H}.d\bar{l} = NI, \tag{4.9}$$

where N is the number of turns in the winding, and I is the current going through the winding.

Use the middle line of the core as the integration line l. Since the direction of the flux intensity $\bar{H}$ is the same as that of $d\bar{l}$, the integration can be simplified as

$$H_c L_c + H_g g = NI, \tag{4.10}$$

where,

$$
\begin{aligned}
H_c &\quad : \text{ flux intensity in the core,} \\
H_g &\quad : \text{ flux intensity in the airgap,} \\
l_c &\quad : \text{ length of the core, and} \\
g &\quad : \text{ length of the airgap.}
\end{aligned}
$$

Taking equation 4.3 into consideration, the relationship described by equation 4.10 can be rewritten as,

$$\frac{B_c L_c}{\mu_c} + \frac{B_g g}{\mu_0} = NI, \tag{4.11}$$

where,

$\quad B_c\quad$: flux density in the core,

$\quad B_g\quad$: flux density in the airgap,

$\quad \mu_c\quad$: permeability of the core, and

$\quad \mu_0\quad$: permeability of air.

Since the flux lines are continuous and the cross-section areas of the core and the airgap are same, it can be assumed that

$$B_c = B_g. \tag{4.12}$$

The permeability of steel is very large compared to that of airgap making the first term in the left hand side of equation 4.11 negligible. With that assumption, equations 4.11 and 4.12 can be combined into

$$B_c = B_g = NI\frac{\mu_0}{g}. \tag{4.13}$$

This equation can also be used to define the flux,

$$\phi_g = \phi_c = B_c A_c = NI\frac{\mu_0 A_c}{g} = \frac{NI}{\Re_g}, \tag{4.14}$$

in which, A_c is the area of cross section of the core, and $\Re_g$ is defined as the reluctance of the magnetic circuit,

$$\Re_g = \frac{g}{\mu_0 A_c}. \tag{4.15}$$

Defining the **Magnetromotive force** (MMF) as

$$F = NI, \tag{4.16}$$

the MMF consumed by the reluctance $\Re_g$ can be expressed as,

$$F_g = \phi \Re_g. \tag{4.17}$$

referring to the circuit shown in Figure 4.5, as there is only one significant reluctance $\Re_g$, the MMF consumed in the airgap reluctance is equal to the MMF generated by the winding current, that is

$$F = F_g. \tag{4.18}$$

It can be concluded from the above analysis that, for a given magnetic circuit, the magnitude of the flux can be calculated using the MMF consumed in the circuit and the reluctance of the circuit. If the MMF generated by the winding current is fixed, changing circuit reluctance by modifying the area of cross section or the length of airgap or both alters the induced flux.

While equation 4.17 may look simple, it includes all the factors affecting the magnetic field, e.g., structure of the core, materials used, input current, and number of turns in the winding. Using this magnetic circuit method, the quantity of the magnetic field obtained is flux, which is an integrated value, or global value, of the magnetic field. The flux density B and the field intensity H can be derived using equations 4.1, 4.3 and 4.9.

Structures of many EM systems are more complicated than the one shown in Figure 4.5. But the magnetic circuit model which may include multiple MMF sources and reluctances can still be used to describe these EM systems. The reluctances may be connected in series or in parallel to form a complicated network. Similarity between magnetic and electric circuits allows us to apply methods used in analyzing electric circuits to the analysis of a magnetic circuit network. Non-linearity of ferromagnetic materials can also be taken into consideration in the analysis [77].

It is evident from the above analysis that MMF can be linked directly to the flux density in the airgap

$$F_g = \left(\frac{g}{\mu_0}\right) B_g. \qquad (4.19)$$

This implies that the waveform of the MMF on the airgap has the same shape as the airgap flux density. This concept can be used to simplify the analysis of the effects of airgap fields in EM devices.

Example-2: Magnetic circuit of a permanent magnet motor

Let us consider the EM device shown in Figure 4.6, which is a simplified structure of motor. The stator core and rotor are made of steel, and their permeabilities are assumed infinite. The demagnetization curve of the permanent magnet materials is shown in Figure 4.7.

If the airgap length g is much smaller than the radius of the rotor, the surfaces of the stator core and the rotor on the two sides of the airgap can be considered equal,

$$A_g = \bar{r} l \theta_m, \qquad (4.20)$$

where, $\bar{r}$ is the average radius of the airgap, and l_m is the thickness of the motor core (Figure 4.6). Let us select the center line of the device (shown by broken line in Figure 4.6) as the line of integration. There is no current in this loop and the magnetic circuit is formed only by one magnet and two airgap.

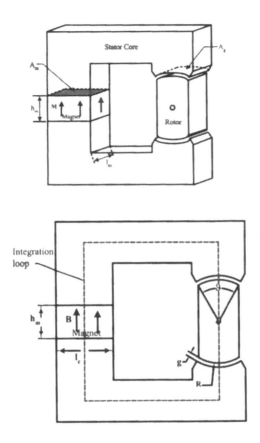

Figure 4.6: A magnetic circuit formed by permanent magnet, stator core and rotor

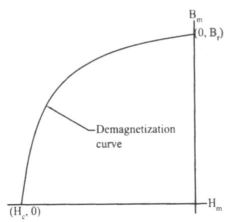

Figure 4.7: Demagnetization curve of the PM material

Therefore,

$$\Re_m \Phi_m + 2\Re_g \Phi_g = 0, \tag{4.21}$$

where, the subscripts m and g represent the magnet and airgap, respectively. As $\Phi_m = \Phi_g$, the equation 4.21 is reduced to,

$$\Re_m = -2\Re_g. \tag{4.22}$$

From Example-1 shown earlier, the airgap reluctance $\Re_g$ is positive. Therefore, the reluctance of the magnet $\Re_m$ is negative implying that the magnet has the capability of providing magnetic energy to external magnetic circuit. The following results can be deduced using the definition of reluctance introduced earlier,

$$\Re_g = \frac{g}{\mu_0 A_g} = \frac{g}{\mu_0 \bar{r} \theta l_m} \tag{4.23}$$

and

$$\Re_m = \frac{h_m}{\mu_m A_m} = \frac{h_m}{\mu_m l_c l_m}, \tag{4.24}$$

where, θ, A_m and μ_m are the angle of the rotor facing the stator core, the cross section area of the magnet, and the permeability of the magnet, respectively. From equations 4.22 - 4.24, it can be shown that

$$\mu_m = -\frac{\bar{r}\theta h_m}{2g l_c}\mu_0. \tag{4.25}$$

As the relationship between B_m and H_m can also be expressed as

$$B_m = \mu_m H_m, \tag{4.26}$$

we can use a straight line with slope μ_m to describe equation 4.26 on the demagnetization curve of the magnet, see Figure 4.8. This straight line is known as the *load line* in the PM device design. In the system operation, the magnetic performance of the magnet must meet both the relationships determined by the demagnetization curve of Figure 4.8 and the load line given by equation 4.26. The point of intersection between the load line and the demagnetization curve is the operating point from which the flux density B_m can be found. Consequently, the flux in the magnetic circuit can be calculated,

$$\Phi_m = \Phi_g = A_m B_m. \tag{4.27}$$

And the flux density in the airgap is,

$$b_g = \frac{\Phi_g}{A_g} = \frac{A_m}{A_g}B_m. \tag{4.28}$$

It can be easily concluded from equation 4.25 and Figure 4.8 that the parameters like airgap length g and magnet length l_m affect the capability of the permanent magnet to generate magnetic field. In the design of the EM system, these parameters should be adjusted so that the permanent magnet is utilized effectively, and the system possesses the required magnetic characteristics.

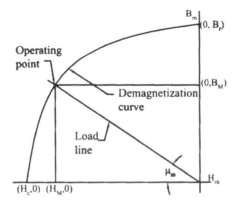

Figure 4.8: Demagnetization curve and load line

4.2 Electric Motor Fundamentals

The electric motor is used to convert electric power to mechanical power. Only electro-magnetic (EM) motors are considered in this book. For this kind of motors, the magnetic field is a necessary medium for realizing this conversion of energy. Electrical, magnetic and mechanical influences on the operation of the motor must be considered in the analysis of electric motors.

Based on the background knowledge in magnetic fields and its calculation described in section 4.1, some basic concepts used for analysis of electric machines and the application of the magnetic circuit method are discussed in this section.

4.2.1 MMF Generated by Distributed Winding

Let us consider a simple motor construction shown in Figure 4.9. This motor has a solid cylindrical stator inside a hollow cylindrical outer rotor. Several conductors are distributed on the surface of the stator. The space interval between adjacent conductors with the symbol A_+ is 15 degrees. It is the same for the conductors with symbol A_-. The distance between the centers of the conductors A_+ and A_- is 180 degrees. The stator core and rotor core are made of steel. The length of the motor is l_c, and the airgap length is g. Each conductor is formed by W_c wires and the current in each wire is I. To make the analysis simple and concise, both cores are assumed to have infinite permeability. This system is considerably more complicated than those introduced in section 4.1.6. It is of interest to know how the magnetic field is distributed in the airgap and whether this field can be determined using the magnetic circuit method.

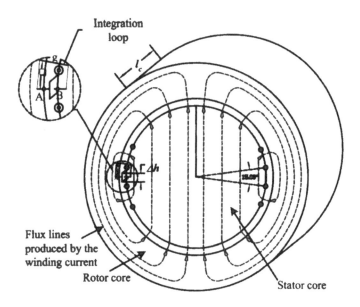

Figure 4.9: Magnetic field generated by the distributed winding with 1 pole-pair

Consider a magnetic circuit around the line AB going through the airgap, where the cross section of the circuit is $\Delta h l_c$, and the length is g as shown in Figure 4.9. Δh is selected to be so small that the magnetic field in this section of area $\Delta h l_c$ can be considered to be evenly distributed. From the symmetry of the electromagnetic structure, it can be said that no flux goes through the pass AB, which also means that the MMF acting on this part of airgap is zero, that is $F_{AB} = 0$.

In Figure 4.10, there are 4 closed loops indicated by broken lines in the left side of the motor. We can apply Ampere's law to each of these loops. As the permeabilities of the stator core and rotor core are infinite, the MMFs acting on the circuits in these cores are zero, and the loop circuit can be considered as formed only by two airgap reluctances, $\Re_{AB}$ and $\Re_g(\theta)$ shown in Figure 4.11. For such circuit, the MMF consumed in the airgap reluctances can be described as

$$F_{AB} + F_g(\theta) = \sum_j i_j, \qquad (4.29)$$

where, $\sum_j i_j$ is the net current surrounded by the loop. As F_{AB} is zero, the above equation can be rewritten as

$$F_g(\theta) = \sum_j i_j, \qquad (4.30)$$

where θ is the rotor position.

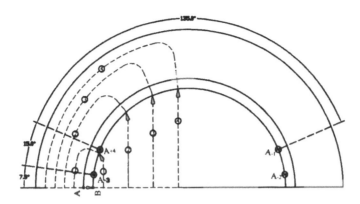

Figure 4.10: Cross-section of the motor showing the integration loops

It is easily observed from Figures 4.10 and 4.11 that, in the range $(0°, 7.5°)$, the right side of equation 4.30 is zero and therefore $F_g(\theta)$ is zero in the range $(0°, 7.5°)$. At $\theta = 7.5°$, $F_g(\theta)$ jumps to the value $W_c I$ and remains the same for the range $(7.5°, 22.5°)$ because no other conductors are included in the loop. Similarly, $F_g(\theta)$ jumps to $2W_c I$ at $\theta = 22.5°$ and remains the same for the entire range of $(22.5°, 157.5°)$. It can be easily comprehended that as θ continues to change from $0°$ to $360°$, the airgap MMF $F_g(\theta)$ shows sudden "jumps" at all positions where the conductors are located. So the MMF as a function of θ is shown by the waveform of Figure 4.12. For the electric machines, the conductors are installed in the stator slots. Therefore, it is not difficult to draw the MMF curve in the range $(0°, 360°)$, even for the motor with complicated distribution windings. As all the armature winding are installed in the stator slots, the MMF jumps can be assumed to occur at the position of slot centers. It was shown in equation 4.19 that the MMF $F_g(\theta)$ is proportional to the airgap flux density $B_g(\theta)$. Therefore, the waveform shown in Figure 4.12 also represents the variation of airgap flux density B_g as a function of rotor position θ. For the MMF curve, the positive area must be equal to its negative area as the flux outflow should be same as the flux inflow.

Using Fourier series, the waveform shown in Figure 4.12 can be expressed as

$$F_g(\theta) = \sum_{n=1}^{\infty} F_n sin(n\theta). \tag{4.31}$$

For a given motor, the amplitude of $F_g(\theta)$ changes with the variation of the armature current. Therefore, from equation 4.30, change in the input current changes the amplitude of the MMF $(F_g(\theta)$ and therefore the gap flux density $B_g(\theta)$. However, the shape of the MMF waveform is determined by the winding distribution, not the magnitude of current.

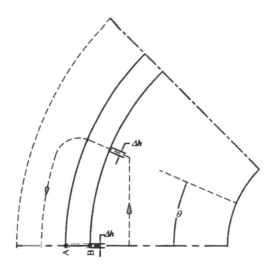

Figure 4.11: The reluctances in the loop circuit

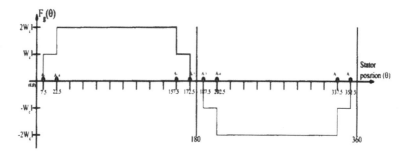

Figure 4.12: The MMF waveform of the distributed conductors

4.2.2   Rotating Magnetic Field, Pole-pair, Speed and Frequency

For the field distribution of motor shown in Figure 4.9, there is only 1 pole-pair of magnetic field produced by the winding is. The winding can be designed such that the number of pole-pairs produced is 2 as shown in Figure 4.13, or more. When the harmonic fields are considered, the relationship between the pole-pair of the airgap field and winding structure becomes complicated which will be discussed later in section 4.3.

It is clear that a rotating magnetic field can be generated in the airgap by rotating the magnets using mechanical force. However, for the design and analysis of electric machines, we are more concerned about the realization of rotating magnetic field using electrical means.

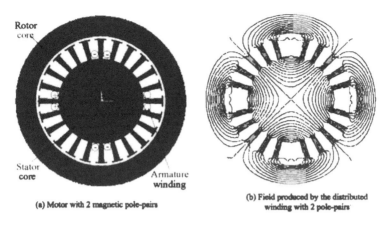

(a) Motor with 2 magnetic pole-pairs

(b) Field produced by the distributed
winding with 2 pole-pairs

Figure 4.13: The distributed winding with 2 magnetic pole-pairs

For the motor shown in Figure 4.9, the conductors A_+ and A_- are actually the two sides of a distributed winding. The winding is redrawn in Figure 4.14. The part of the coil in the area B is called *effective coil* and its length is about the same as the length of the stator core. The part of the coil in areas A and C are known as "end coil". For design and analysis of an electric motor, the effective coils are considered as the part of the coil structure that contribute to the EM torque for the electromechanical energy conversion to take place, whereas the end coils are the conductors required to form a closed electric circuit but do not contribute to the EM torque.

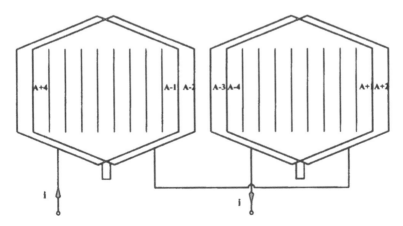

Figure 4.14: One-phase distributed winding

According to equation 4.31, the MMF generated by the windings can be considered the result of the sum of MMF harmonics, and its fundamental one is determined by the winding pole-pairs,

$$F_{a1}(\theta) = F_{p1}\sin(p\theta) = Ki_a\sin(p\theta) \qquad (4.32)$$

where, p is the pole-pair of the motor. The constant K is determined by the turns and structure of the winding, and it can be obtained through Fourier series analysis. If the winding current i varies with time as

$$i_a(t) = I_m\sin(\omega t) \qquad (4.33)$$

where, ω is the angular frequency of the current. Then the MMF waveform of equation 4.32 changes to

$$F_{a1}(\theta) = KI_m\sin(\omega t)\sin(p\theta) = \frac{KI_m}{2}[cos(\omega t - p\theta) - cos(\omega t + p\theta) \qquad (4.34)$$

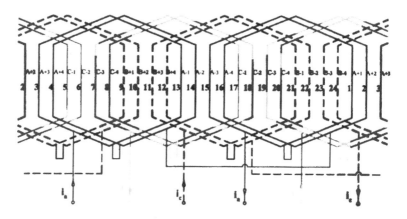

Figure 4.15: Three-phase distributed winding

Two other windings, B-phase winding and C-phase winding, can also be installed on the stator to form a set of 3-phase symmetric windings in the motor space. Figure 4.15 shows such 3-phase windings which is developed from the 1-phase winding shown in Figure 4.14. It is clear, the fundamental MMF harmonic of B-phase and C-phase windings can be expressed as

$$F_{b1}(\theta) = Ki_b\sin(p\theta - 120°),$$
$$F_{c1}(\theta) = Ki_c\sin(p\theta - 240°) \qquad (4.35)$$

Moreover, if the currents i_b and i_c vary with time,

$$i_b(t) = I_m\sin(\omega t - 120°), \qquad (4.36)$$
$$i_c(t) = I_m\sin(\omega t - 240°)$$

then the currents i_a, i_b and i_c form a 3-phase sinusoidal current set, which is symmetric in the time domain. In this case, the airgap MMF generated by the windings changes to

$$
\begin{aligned}
F_g(\theta, t) &= F_{a1}(\theta, t) + F_{b1}(\theta, t) + F_{c1}(\theta, t) \qquad (4.37) \\
&= \frac{3KI_m}{2} \cos(\omega t - p\theta) = F_{m1} \cos(\omega t - p\theta)
\end{aligned}
$$

Equation 4.37 is a typical expression of traveling wave, and its phase angle τ is determined by

$$
\tau = \omega t - p\theta. \qquad (4.38)
$$

Therefore, if an MMF value or a phase angle τ is given at certain time instant t, the position of the traveling wave θ at that phase angle can be determined using equation 4.38. It is clear that this position θ is a function of time.

For a given τ, differentiating both sides of equation 4.38 yields

$$
0 = \omega - p\frac{d\theta}{dt} \qquad (4.39)
$$

that is,

$$
\Omega = \frac{d\theta}{dt} = \frac{\omega}{p} = \frac{2\pi f}{p}, \qquad (4.40)
$$

where Ω is the angular speed of the traveling wave whose unit is radian per second, and f the frequency of the current in Hz (Hertz).

Conventional practice in engineering expresses rotational speed in units of *revolution per minute* or RPM and uses the symbol n to describe it. From 4.40, it can be shown that,

$$
n = \frac{60\Omega}{2\pi} = \frac{60f}{p}. \qquad (4.41)
$$

It shows that, the rotational speed of the MMF or of the airgap flux density is proportional to the frequency of the current, and inversely proportional to the pole-pair of the magnetic field. In other words, a rotating magnetic field can be generated by inputting the 3-phase time-symmetric alternating currents into the 3-phase space-symmetric windings. This is an electrical method to realize the rotational field. The field speed can be adjusted by changing the frequency of the current. Otherwise, for a given frequency of the current in the winding, the speed of the rotating field can be designed with a suitable magnetic pole-pair in the motor design stage.

We can also know from equation 4.31 that, besides the fundamental MMF, the winding-A, winding-B and winding-C also contain harmonic fields. It is not difficult to prove that, when the input currents are symmetric in the time domain, the mth order harmonic fields produced by the 3-phase winding can also form a rotating field. From equation 4.41, the speed of this harmonic field is $\frac{1}{m}$ of the speed of the fundamental field. Therefore, higher is the order of the harmonic field lower is the speed of the corresponding rotating field. It can also be shown that the summation of the triple MMF harmonics is zero for the 3-phase systems, and therefore, the effects of the triple MMF harmonics are normally not considered in the discussions of energy conversion.

4.2.3 Force and Torque Generated by Magnetic Field

Magnetic fields contain energy, and that can also appear as force/torque acting on the parts of the related electromagnetic system. The existence of the EM force/torque is the necessary condition in realizing the electromechanical energy conversion. The EM forces can be categorized into two types, the *Ampere's force* and the *reluctance force*.

4.2.3.1 Ampere's Force and Ampere's Torque

Let us consider a straight conductor of length $\bar{l}$ conducting a current I in an evenly distributed magnetic field. In this case, the conductor is acted on by the Ampere's force, $\bar{F}_A$. The force can be expressed as,

$$\begin{cases} \bar{F}_A &= \quad I\bar{l} \times \bar{B} \\ F_A &= \quad I \cdot l \cdot B \cdot \sin(\theta) \end{cases} \qquad (4.42)$$

where, θ is the angle between the conductor and the field (see Figure 4.16). The force is a vector and its direction is vertical to both the conductor and the magnetic field.

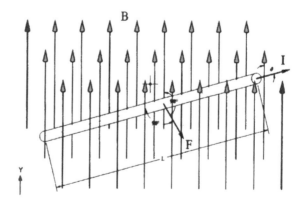

Figure 4.16: Ampere's force acting on a conductor

If the conductor is perpendicular to the magnetic field **B**, then the directions of the current, the magnetic field and the Ampere's force are given by the left-hand rule, as shown in Figure 4.17.

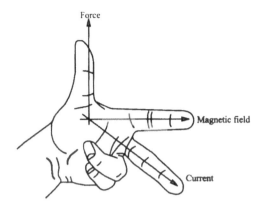

Figure 4.17: Ampere's force and left-hand rule.

If a coil is placed in a magnetic field, then the Ampere's force acts on both the effective sides of the coil. However, the direction of the force acting on one of the sides is opposite to that acting on the other side. These forces form a torque acting on the coil as illustrated graphically in Figure 4.18.

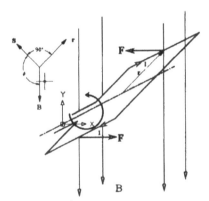

Figure 4.18: Electromagnetic torque acting on a coil.

In the figure, r is the distance between the conductors on the rotating shaft, S the direction of the surface formed by the coil, θ the angle between S and B. In the electric machine, this kind of magnetic field (B) is also called *exciting field*. If B is evenly distributed, then the torque T can be calculated as,

$$T = 2W_c lBIr \sin(\theta), \qquad (4.43)$$

where, W_c is the number of turns in the coil winding. It is obvious that the torque acting on the coil varies with the rotational angle of the coil, and the torque is maximum when the surface of the coil is vertical to the magnetic field (see Figure 4.18 and Figure 4.19). For a rotating coil, maximum torque occurs at the positions where the coil axis is vertical to the axis of the exciting magnetic field; the flux encircled by the coil is zero at these positions. On the other hand, the torque is zero at the position where the coil axis is aligned with or opposite to the axis of the exciting magnetic field, and the flux going through the coil (linked with coil) is maximum at these positions.

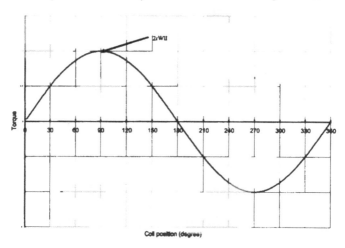

Figure 4.19: Torque acting on single coil as a function of the coil position, θ.

It is obvious from Figure 4.19 that, if there is no change in the current of the coil, the average torque of the coil over 360° is zero. That means the coil cannot sustain continuous rotation if the current is not changed in the coil. To make the coil rotate continuously, the direction of the current must be changed, or *commutated*, according to the position of the coil such that the average torque becomes positive in the direction of coil rotation. The commutation can be realized by the mechanical means, that is, using a commutator and brushes to change the current at the right positions as shown in Figure 4.20. The brush is made of conductive materials such as graphite. The commutation process for different rotor positions is illustrated in Figure 4.21.

Using such a commutation system, even if the input current is kept uni-directional, the current flowing in the coil can be made to alter at the right position, as illustrated in Figure 4.22. Correspondingly, the torque generated is always positive as shown in Figure 4.23.

It can be deduced from Figure 4.23 that, with the commutation system in use, the average torque is not zero and, therefore, the coil can rotate around its shaft continuously. However, the torque produced contains rich torque ripples. This mechanical commutation method, one with commutator and brush, is used in DC motors where the input current from the power supply is unidirectional but the current in the armature coils is alternating. The nature of the input current defines the nomenclature of *DC motor*.

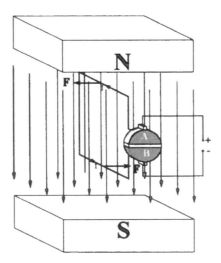

Figure 4.20: The single coil and commutator ($\theta = 90°$)

The torque ripple in the DC motor can be reduced significantly by increasing the number of coils. For the coil shown in Figure 4.19, if three coils AX, BY and CZ are used and if they are symmetrically distributed around the stator with intervals of 120°, then the torque produced by each coil is the same as that shown in Figure 4.23 but shifted from one another by 120°. The resultant torque is shown in Figure 4.24. Comparing the figures of Figure 4.19 and Figure 4.24, it is easily understood that use of multiple coils with proper arrangement can reduce the torque ripple.

Let us define a ratio r_{tam} as

$$r_{tam} = \frac{\text{Average Torque}}{\text{Maximum Torque}}. \tag{4.44}$$

For a multi-coil system equipped with commutation device, if the number of coil M is odd, then it can be proved that,

$$r_{tam} = \frac{2M}{\pi} \sin(\frac{\pi}{2M}). \tag{4.45}$$

And if M is even, then

$$r_{tam} = \frac{M}{\pi} \sin(\frac{\pi}{M}). \tag{4.46}$$

Therefore, when the number of coil, M, is increased to a certain level, r_{tam} tends to a value close to 1, i.e., the generated torque can be considered constant if the input current is constant.

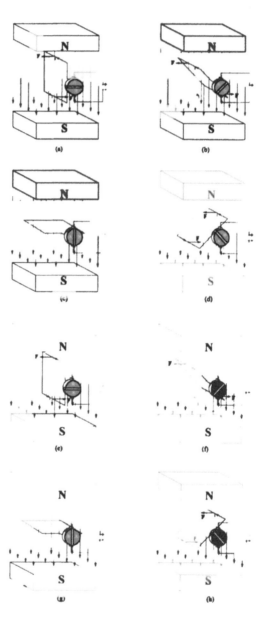

Figure 4.21: The single coil at different positions a: $\theta = 90°$, b: $\theta = 135°$, c: $\theta = 180°$, d: $\theta = 225°$, e: $\theta = 270°$, f: $\theta = 315°$, g: $\theta = 360°$ or $0°$, h: $\theta = 45°$.

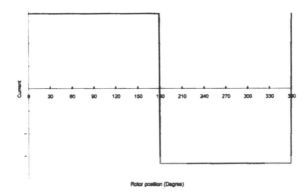

Figure 4.22: Current in a single coil when commutation is used.

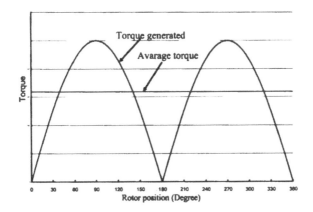

Figure 4.23: The torque waveform of a single coil with mechanical commutation in use.

If the exciting field is constant, the torque of each coil is proportional to the input current at a given position implying that the global torque generated by the multi coils is also proportional to the input current as it is the sum of torque produced by all the coils. Therefore, if the number of coils is big, then the relationship between the global torque and input current can be described by the following equation,

$$T_{em} = K_t I. \qquad (4.47)$$

where K_t is a constant, and generally called as the *Torque Constant*. This constant is determined by the motor structure, number of coils, coil turns and the magnetic field produced by the field exciting system. In a DC motor, these coils can be connected together to form electrical paths between the brushes connected to positive and negative pole of power supply [80]. Figure 4.25 shows the connection of three coils which forms one loop of windings.

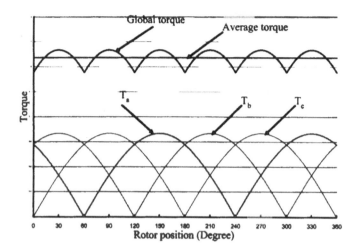

Figure 4.24: Torque produced by the DC EM system with 3 phase coils T_a: torque produced by coil-AX, T_b: torque produced by coil-BY, T_c: torque produced by coil-CZ

The mechanical commutation system has been successfully used in DC motors for many years. But it has some inherent problems which are of serious concern in many applications. Such problems include,

1. Spikes: Commutation process results in sudden jumps in the current waveform (Figure 4.22). These jumps causes sparks between the surfaces of commutator and brushes inducing electromagnetic interference (EMI) as the magnetic field energy stored in the inductance of the coils must be released in a very short time. The sparks can damage the contact surfaces of the brushes and commutator, and reduce the lifespan of the commutation system.

2. Speed: The commutator is formed by many segments as shown in Figure 4.25. Due to centrifugal force, such mechanical structure cannot withstand high speed rotation. In addition, the presence of brushes also puts limit on the maximum surface speed of the commutator as the brush material cannot stand the intensive wear and tear caused by high speed.

3. Vibration and noise: The brushes must contact the commutator surface with certain pressure. In high speed operation, such mechanical contact can induce severe acoustic noise and vibration.

4. Maintenance: The brush must be in contact with the commutator surface during operation of motor, making wear and tear of brushes and commutator inevitable. Consequently, regular maintenance is necessary which may not be feasible in many applications, e.g., HDD.

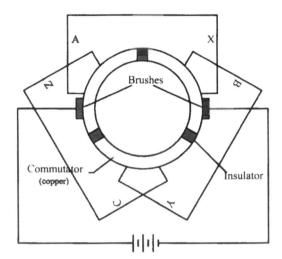

Figure 4.25: Connection of three symmetric coils

All the problems mentioned above can be avoided by using electronic commutation system, where no brush or commutator is used. Following the fast developments in power electronics and drive technology, the electronic commutation technology has been progressing rapidly since 90's. The motor that uses electronic system to detect the rotor position and to commutate armature current is called brushless DC (BLDC) motor.

In DC and BLDC motor systems, a rotating part moves with respect to the other part that remains stationary. The components producing the excitation field should be installed on one side and armature winding that carries current is placed on the other. In a DC motor, the excitation filed producing components are on the stator while armature winding is on the rotor. However, this arrangement is reversed in the BLDC motor where the armature windings are installed on the stator side and the components generating the excitation field are mounted on the rotor. In both DC motor and BLDC motor, the excitation field can be produced by permanent magnets (PM). Such motors are called as PM DC motor and PM BLDC motor, respectively.

When mechanical commutation system is used, detection of rotor position and commutation of armature current are realized together using the same device. However, in BLDC motor, the rotor position is detected by some sensing mechanism, and the coil current is commutated using power electronic system. Figure 4.26 shows the basic structure of a 3-phase BLDC motor, and Figure 4.27 shows the electronic commutation circuit (H bridge). In DC motor, commutation is determined by the armature coil position, whereas in BLDC motor, commutation is determined by the position of the excitation field produced by the permanent magnets on the rotor.

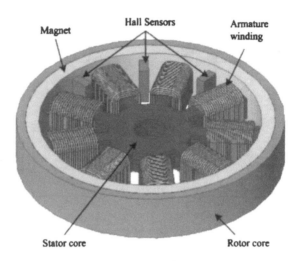

Figure 4.26: The structure of PM BLDC motor

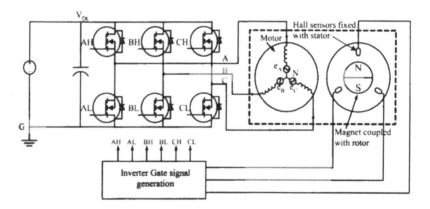

Figure 4.27: The 3 phase coils of PM BLDC motor driven by 3-phase full bridge (with sensor)

It will be shown in section 4.3 that the rotor position can be detected using additional sensors such as *Hall-sensor* and *encoder*, or form the signals produced in the motor operation, such as *back-EMF signals*.

Similar to the DC motor, increasing the phase number of the BLDC windings can reduce the torque ripple of the motor. It will also be shown in section 4.4 that, the linear relationship of $T(I)$ described by equation 4.47 can also be used in the analysis of BLDC motor.

The current in armature windings of any electric machines also induces magnetic field, that is, both the fields produced by the magnets and armature windings exist in the PM DC motors and PM BLDC motors. Therefore, the Ampere's torque can also be considered as the result of the reaction between the stator field and rotor field.

4.2.3.2 Reluctance Torque

Besides the Ampere's torque, another type of torque known as *reluctance torque* may also be induced during the operation of an electric machine. The mechanism behind the generation of the reluctance torque is quite complicated. However, it can be explained using the concepts of the global electromagnetic energy and virtual work. Let us use the EM system shown in Figure 4.6 to explain this kind of torque. To simplify the description, it is assumed that the magnet has an ideal demagnetization curve with infinite H_c, as shown in Figure 4.28.

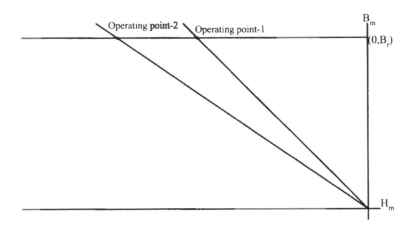

Figure 4.28: The ideal demagnetization curve of the permanent magnet

The motor shown in Figure 4.6 is formed by a stator and a rotor. The stator consists of permanent magnet and the stator core. The magnet is used to produce the excitation field. The rotor is simply a core capable of rotating about its shaft, but there is no armature winding and current on the rotor.

It can be stated, referring to the permanent magnet with the characteristic shown in Figure 4.28, the flux density going through the magnet is constant at B_r. Therefore, the flux ϕ_m produced by the magnet is constant. Neglecting the leakage field, the airgap flux ϕ_g is same as ϕ_m, as explained in section 4.1.6. Therefore, ϕ_g is constant and independent of the rotor position,

$$\phi_g = \phi_m = B_r A_m = B_r l_m l_c, \tag{4.48}$$

where, l_m and l_c and A_m are the thickness, the width, and the cross-section area of the magnet, respectively; see Figure 4.6. Suppose the rotor is rotated by an angle of $\theta°$ where $|\theta| < \frac{\delta}{2}$ as illustrated in Figure 4.29.

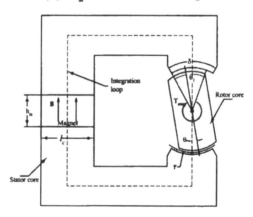

Figure 4.29: Moving the rotor by a small angle

The effective angle coupling the surfaces of the stator and the rotor is

$$\theta_1 = \delta - |\theta|. \tag{4.49}$$

Therefore, the reluctance of this magnetic circuit, $\Re_m(\theta)$, is formed by the reluctances of the two airgaps, that is,

$$\Re_m(\theta) = 2\Re_g(\theta), \tag{4.50}$$

and

$$\Re_g(\theta) = \frac{g}{\mu_0 \bar{r} \cdot (\delta - |\theta|) \cdot l_m}, \tag{4.51}$$

where, $\bar{r}$ is the average radius of the airgap. Using the discussions presented in section 4.1.4, the magnetic energy of whole system is

$$\begin{aligned} W_g &= 2w_g V_g = \frac{1}{\mu_0} B_g^2 V_g = \frac{1}{\mu_0} \left(\frac{\phi_g}{A_g}\right)^2 V_g \\ &= \frac{1}{\mu_0} \left(\frac{B_r l_c l_m}{\bar{r}(\delta - |\theta|) l_m}\right)^2 [\bar{r}(\delta - |\theta|) l_m g] \\ &= \frac{g l_m}{\mu_0} \frac{(B_r l_c)^2}{\bar{r}(\delta - |\theta|)} = \frac{1}{2}(B_r l_c)^2 l_m^2 \Re_g(\theta). \end{aligned} \tag{4.52}$$

If the center of the rotor is not aligned with the center of the stator, i.e., $\theta \neq 0$, then the magnetic field in the airgap generates an EM torque to make these centers aligned. This torque can be calculated by using virtual work method [69]. The principle of this method is based on the fact that the EM energy of the motor varies with rotor position and this variation can induce an EM torque $T_r(\theta)$ acting on the rotor. This torque can be expressed as a function of θ,

$$T_r(\theta) = -\frac{dW_g}{d\theta} = -\frac{1}{2}(B_r l_c)^2 l_m^2 \frac{d\mathcal{R}_g(\theta)}{d\theta} = \begin{cases} -\frac{gl_m}{\mu_0}\frac{(B_r l_c)^2}{\bar{r}(\delta-\theta)^2}, & 0 < \theta < \frac{\delta}{2} \\ \frac{gl_m}{\mu_0}\frac{(B_r l_c)^2}{\bar{r}(\delta+\theta)^2}, & -\frac{\delta}{2} < \theta < 0. \end{cases}$$

$$(4.53)$$

Such a torque cannot be explained directly from the concept of the Ampere's force as there is no current-carrying conductor in the motor. From equation 4.53, it is the variation in reluctance that can be considered as the cause of this kind of torque in an electric motor. Hence, this torque is normally called the *reluctance torque*.

The torque as a function of θ, which is obtained using the equation above, is shown in Figure 4.30. Using virtual work and extremum methods, it is not difficult to prove that the torque at equals zero. If the rotor is rotated a little from the position $\theta = 0°$ by external disturbance, it can be deduced from equation 4.53 that the rotor will return to the original position when the disturbance disappears. Therefore, the position $\theta = 0°$ is a stable point for the motor shown in Figure 4.29.

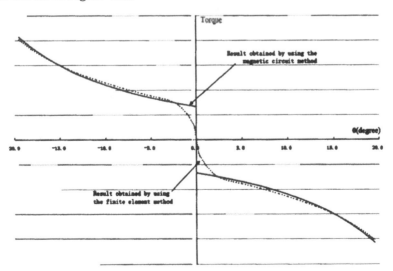

Figure 4.30: Reluctance torque

Figure 4.30 also shows the same torque calculated using the finite element method, which is an accurate numerical method widely used in electromagnetic system design and analysis. It shows that, in comparison to the finite element method, the calculation of reluctance torque using the magnetic circuit method produces fairly accurate result for a wide range of rotor position. However, for a narrow range of rotor position near $\theta = 0$, the magnetic circuit method gives erroneous result. However, the magnetic circuit method gives an analytical expression of the reluctance torque, which is very helpful in the analysis of the torque performance of a device, and can simplify the computation.

The virtual work principle is used in the analysis described above. If the magnetic field distribution is known, this method is an effective tool for calculation of the electromagnetic torque/force, which include both the Ampere's and reluctance torque/force.

4.2.4   Cogging Torque and Unbalanced Magnetic Pull

Putting slots on the stator core makes the installation of winding easy and reliable. However, the slots also induce the cogging torque problem in the PM motors. This torque is related to the shape of the slots, number of slots, number of magnetic pole-pairs, and the magnetization of the magnets [109], [70].

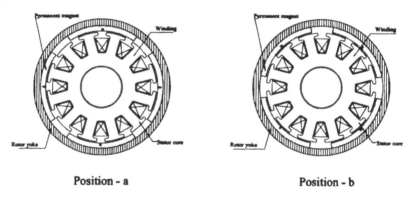

Position - a   Position - b

Figure 4.31: Cogging produced in a PM motor with 12-slots and 2-pole-pairs

Generation of cogging torque is explained here using the PM motor shown in Figure 4.31. In this motor, the magnets are mounted on the inner surface of the rotor. When the rotor rotates from position a to position b, the total teeth surface facing the magnet is increased. Following the discussions in section 4.2.3.2 on generation of reluctance torque, the increased surface area of the teeth facing magnet means decrease in reluctance of the main magnetic circuit when the rotor is rotated from position a to position b, and reluctance torque is induced. It can be shown that, if the ratio between the slot number and the magnetic pole is not an integer, the variation in reluctance is not very significant in the rotation of motor, although it still exists to some extent [77]. In hybrid stepping motors [108], cogging torque can be used to produce holding torque. However, such torque is not desired in the operation of HDD spindle motor, which will be explained later in section 4.3.

Besides the electromagnetic torque, which is tangential to the rotor surface, the magnetic field in the airgap of the motor also induce radial force. The local radial force in the airgap of the motor is usually much higher than the local tangent force that produces the electromagnetic torque. When the geometric center of the rotor is aligned with the center of the electromagnetic field, the local radial forces balance each other and the resultant force caused by these radial forces is zero. However, if the two centers are not aligned, the resultant radial force is non-zero, which is known as the *Unbalanced Magnetic Pull* (UMP).

The UMP may also be introduced due to problems with component quality. For example, when the magnet ring is not symmetrically magnetized or when the stator core dimensions are not correct, UMP is generated. The unbalance can also be induced if the motor is not correctly assembled. The UMP caused by these factors is called *extrinsic UMP* [16]. On the other hand, the UMP may produced by the electromagnetic structure itself. Figure 4.32 shows the magnetic field in a spindle motor with 9 stator slots and 4 magnetic pole-pairs. This structure shows an unbalance between the left and right sides of the EM field, inducing the UMP. As the local field changes with different position of the rotor, the UMP also changes as shown in Figure 4.33. This UMP is not caused by the qualities of the components and production, but by the motor structure itself, and it is thus known as the *intrinsic UMP*.

Figure 4.32: Magnetic field in a 9-slot/4-pole-pair spindle motor (obtained using FEM)

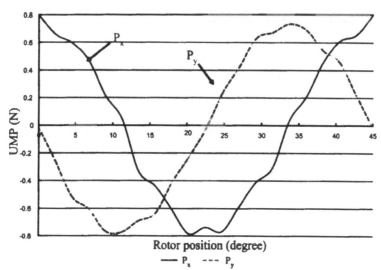

Rotor position (degree)
—— P_x --- P_y

P_x: UMP component in x-direction, P_y: UMP component in y-direction

Figure 4.33: UMP generated in the motor operation: P_x - UMP component in X direction and P_y: UMP component in Y direction

Normally, UMP is detrimental for the application of spindle motor as it produces acoustic noise and vibration. The quality control of components and production is critical for reducing UMP. Selecting the suitable match between the slot number and magnetic poles can avoid the intrinsic UMP. In [16], it is proven that, if the slot number is even, the intrinsic UMP can be eliminated. The orders of UMP harmonics are even.

4.2.5 Generation of back-EMF

When a coil rotates in a magnetic field, electromotive force (EMF) is induced in the coil, and the induced EMF can be expressed as

$$e = -N\frac{d\phi}{dt} \tag{4.54}$$

where, ϕ is the flux going through the area surrounded by the coil, and N is the number of turns of the coil.

When the coil is in an evenly distributed magnetic field and rotates at a constant angular speed Ω shown in Figure 4.34, the flux linked with the coil is a function of time,

$$\phi = BS\cos(\theta) = BS\cos(\Omega t + \alpha). \tag{4.55}$$

where,

   α : the initial position of the coil,

   S : the area formed by the coil, and

   θ : the angle between the magnetic field and the area formed by the coil.

Therefore, the EMF induced in the coil is

$$e = \Omega N B S \sin(\Omega t + \alpha). \tag{4.56}$$

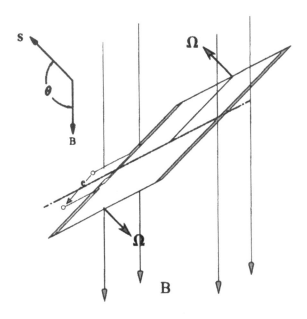

Figure 4.34: Induced EMF in a coil.

When given coil, placed in a fixed magnetic field, rotates at a constant speed, EMF is induced in the coil. The induced EMF varies sinusoidally in the time domain (Figure 4.35), and its amplitude is proportional to the rotational speed of the coil.

From Figure 4.34 and equations 4.54 to 4.56, the peak values of the back-EMF appear at the positions where the coil axis is vertical to the magnet axis, and the zero crossing positions of the back-EMF happen at the locations where the coil axis is along with or opposite to the magnet axis. These are similar to the Ampere's torque discussed in section 4.2.3.

Using the commutation system shown in Figure 4.20, the alternating EMF can be rectified to one shown in Figure 4.36. It contains DC component as well as harmonics.

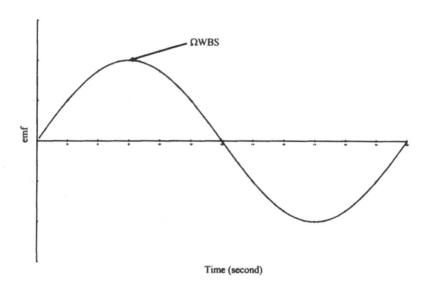

Figure 4.35: EMF induced in a single conductor rotating at constant speed.

For a DC motor, similar to the case of torque generation, increasing the coil number and distributing the coil symmetrically in the space can reduce the harmonics of the EMF output from the brush. Figure 4.37 shows the EMF output of Δ-connected 3-phase symmetric coils.

In normal operation of an electric motors, the EMF induced in the winding always opposes the variation in the input current. Therefore, in electric machine analysis, the EMF induced is also called the *back-EMF*.

Similar to EM torque generation in DC motor, increasing the number of winding (or phase number) makes the back-EMF approximately constant. Equations 4.45 and 4.46 can still be used to describe the relationship between the peak and average values of EMF. When the motor rotates, the EMF of each coil varies linearly with the rotational speed. So the EMF from the commutation system can be described by

$$E = K_e \Omega, \tag{4.57}$$

where, K_e is constant for a given motor and is known as the back-EMF constant. Its value depends on the motor and the winding structure. For a multi-coil DC motor, equation 4.57 is a fairly accurate description of the relationship between back-EMF appearing on the brush-pair and motor speed. For BLDC motor with multi-phase coils, the relationship described by equation 4.57 is still valid if high order harmonics are neglected. Introduction of back-EMF constant enables us to analyze DC motors and BLDC motors from the point of view of electric circuits, where K_e describes motor's EM structure and Ω describes motor speed. The equivalent circuit of motor shown in Figure 4.38 can be used to analyze electrical performance of a DC motor.

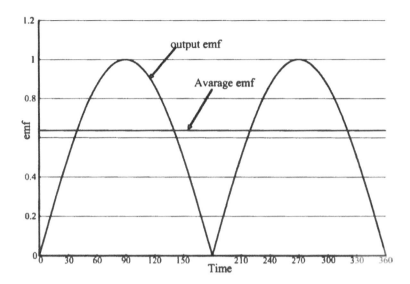

Figure 4.36: The output EMF from the brushes of commutation system (for a single coil).

In this circuit, R_a and L_a are the resistance and inductance of the armature winding, respectively. The external voltage applied on the terminals of the winding is V, while the back-EMF produced in the motor operation is E. The dynamics of this equivalent circuit model of DC motor can be expressed as

$$V = i \cdot R_a + L_a \frac{di}{dt} + E. \tag{4.58}$$

In the steady state, the input current is constant and the above equation reduces to

$$V = i \cdot R_a + E. \tag{4.59}$$

The conversion between the electro-magnetic power P_{em} and the mechanical power P_{me} is possible because of the existence of the back-EMF. The relationship between these two powers can be expressed as

$$P_{em} = P_{me} \quad \Leftrightarrow \quad E \cdot I = T_{em} \cdot \Omega. \tag{4.60}$$

Further explanation on conversion between P_{em} and P_{me} is given in section 4.3.10.

One can easily deduce from equations 4.47 and 4.60 that,

$$\begin{cases} T_{em} = K_t \cdot I, \\ E = K_e \cdot \Omega. \end{cases} \tag{4.61}$$

This implies that the torque constant (K_t) and back-EMF constant (K_e) of a motor have the same magnitude when SI units are used.

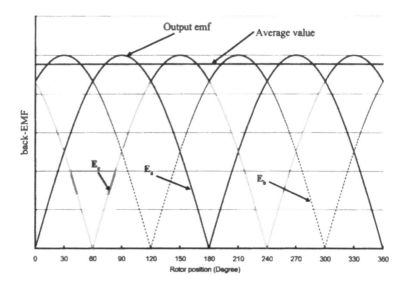

Figure 4.37: The EMF output from the brushes of commutation system.

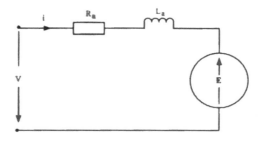

Figure 4.38: The equivalent circuit of a DC motor.

4.2.6   Electrical Degree and Mechanical Degree

In the analysis of an electric motor, both electrical and mechanical performances must be taken into consideration. In motor operation, these performances vary cyclically. For motors with one magnetic pole-pair, i.e., $p = 1$, the electrical cycle matches with the mechanical cycle, or in other words, the electrical signal varies one cycle in one revolution of the motor, as shown in Figure 4.39. For such a motor, analysis in the mechanical domain is equivalent to the analysis in electrical domain. However, if the magnetic pole-pair p is more than one, then the variation of electrical signal will complete p cycles in one mechanical revolution. Figure 4.40 shows the variation in back-EMF of a motor with 3 magnetic pole-pairs ($p = 3$) in one revolution. In order to relate the results of analysis in the electrical domain to those in the mechanical domain, the terminology of *electrical degree* and *mechanical degree* are jointly used in the analysis of electrical machines. The former is used to describe the position in electrical domain and is denoted by θ_e, and the second one defines the physical position of the rotor in space and is denoted by θ_m. It is obvious that rotating the rotor by a mechanical degree θ_m leads to the variation in electrical signal by an angle $p\theta_m$. That is, the relationship between θ_e and θ_m is,

$$\theta_e = p\theta_m. \tag{4.62}$$

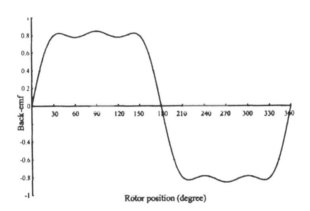

Figure 4.39: Back-EMF induced in the winding (p=1).

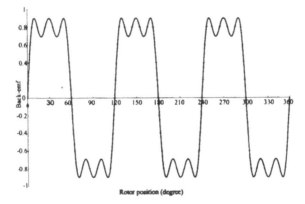

Figure 4.40: Back-EMF induced in the winding (p=3).

4.2.7 Armature reaction

In the operation of a motor, besides the excitation field, there exists another magnetic field induced by the current in the armature winding which affects the distribution of the airgap field. Assuming this process is linear, the effects of the excitation field and the armature current field can be considered separately, and the total airgap field can be obtained by summing up the two fields according to the principle of superposition. However, if the stator core or the rotor core or both operates in the saturation state, the total airgap field will be smaller than the sum of the two individual fields. The difference increases with increasing magnitude of armature current. From the point view of motor control, this effect is equivalent to a reduction in the excitation field with increase in drive current. This phenomenon is known as *armature reaction*.

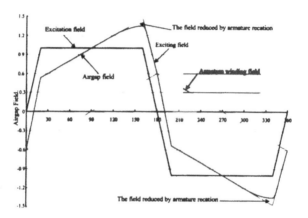

Figure 4.41: Airgap field formed by exciting field and armature winding field.

The conditions for generating armature reaction are,

1. the saturation of stator core, or rotor core, or both is serious, and

2. the field produced by the armature winding current can affect the airgap field obviously.

It was mentioned in sections 4.2.3 and 4.2.3 that the EM torque and back-EMF of a DC or a BLDC motor can be described as

$$T_{em} = K_t I, \tag{4.63}$$
$$E = K_e \Omega.$$

If the armature reaction of the motor is severe, the torque constant K_t and back-EMF constant K_e cannot be taken as constants. They are reduced when the input current is increased. The nonlinearity of the torque constant and the back-EMF constant makes the precision control of the motor complicated.

4.2.8 Conditions of Magnetic Field for Producing Ampere's Torque

As explained in section 4.2.2, the rotating field can be generated by either mechanically rotating the rotor or electrically changing the current between pole-pairs. Moreover, it is explained in section 4.2.3 that the EM torque in an electric motor is produced by the interaction between the rotor field and the stator field. Can all the stator and rotor fields interact to generate the EM torque? What are the necessary conditions for these fields to generate the torque?

Both the stator and rotor can produce magnetic field in the airgap. The total airgap field is

$$B_g(\theta, t) = B_r(\theta, t, \delta_r) + B_s(\theta, t, \delta_s) = \sum_m r_m(\theta, t, \delta_r) + \sum_n s_n(\theta, t, \delta_s), \quad (4.64)$$

where, the r_m and s_n are the mth order rotor harmonic and the nth order stator harmonic fields, respectively. The reference positions of the rotor and stator are δ_r and δ_s, respectively.

If these harmonic fields rotate in space at speeds of Ω_{rm} and Ω_{sn} respectively, then these harmonics can be expressed as

$$r_m(\theta, t, \delta_r) = R_m \sin[m(\Omega_{rm}t - \theta + \delta_{rm}) + \delta_r], \quad (4.65)$$
$$s_n(\theta, t, \delta_s) = S_n \sin[n(\Omega_{sn}t - \theta + \delta_{sn}) + \delta_r], \quad (4.66)$$

where, δ_{rm} and δ_{sn} are the reference positions of the harmonics r_m and s_n, respectively.

The energy of the magnetic field in the airgap is

$$W_g = \frac{gl_c\bar{r}}{2\mu_0} \int_0^{2\pi} B_g^2 d\theta = \frac{gl_c\bar{r}\nu_0}{2} \int_0^{2\pi} B_g^2 d\theta, \quad (4.67)$$

where,

l_c : effective length of the motor (see Figure 4.9),
$\bar{r}$: average radius of the airgap,
μ_0 : permeability of air,
g : width of the airgap, and
ν_0 : the inverse value of μ_0, known as the reluctivity of air.

$$(4.68)$$

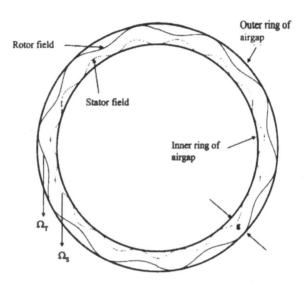

Figure 4.42: Stator field and rotor field in the airgap

We can apply the principle of virtual work to calculate the electromagnetic torque produced by such airgap field (refer to section 4.2.3 and [69]. Assuming a small angle of rotation of the rotor, i.e., δ_r, the following expression for the torque calculation can be obtained,

$$
\begin{aligned}
T_{EM} &= -\frac{dW_g}{d\delta_r} = -\nu_0 g l_c \bar{r} \int_0^{2\pi} B_g \frac{dB_g}{d\delta_r}\, d\theta, \qquad (4.69) \\
&= -\nu_0 g l_c \bar{r} \int_0^{2\pi} \left[\sum_m r_m(\theta,t,\delta_r) \sum_n s_n(\theta,t,\delta_s) \right] \\
&\qquad \sum_m R_m \cos[m(\Omega_r t - \theta + \delta_m) + \delta_r]\, d\theta \\
&= \sum_{m,n} T_{mn}(t).
\end{aligned}
$$

In equation 4.69, $T_{mn}(t)$ is the torque harmonic produced by the interaction between the rotor harmonic field r_m and the stator harmonic field s_n. Using the orthogonality of the triangular function, it can be derived from equation 4.69 that,

$$
T_{mn}(t) = \begin{cases} 0, & m \neq n \\ T_{kk}(t) = \\ -\pi \nu_0 g l_c \bar{r} R_k S_k \sin[k(\Omega_{rk} - \Omega_{sk})t + (\delta_r - \delta_s)], & m = n = k. \end{cases}
$$

$$(4.70)$$

According to equation 4.70, the interaction between the mth order rotor harmonic field and the nth order stator harmonic field has the following characteristics:

1. If the stator field has different pole-pairs with the rotor field, no torque is produced by these two fields,

2. If the stator field has same pole-pairs with the rotor field, and their relative speed is not zero, the torque produced varies with time, and the average torque in the time domain is zero, and

3. If the stator field has the same pole-pairs with the rotor field, and their relative speed is zero, the torque produced by the interaction is constant.

From the explanations given above, we can find the effect of each harmonic of the stator field and the rotor field in generating the EM torque. Only the interaction between the stator harmonic and the rotor harmonic with the same magnetic pole-pair can generate the EM torque. This is the necessary condition for producing the Ampere's torque. We can tell, knowing the relative speeds of the rotor field harmonic and the stator field harmonic in the airgap, whether the interaction can generate effective torque that makes the rotor rotate continuously, or just produce torque ripple that induces vibration.

4.3 Spindle Motors used in HDD

Varieties of motors suitable for different applications are available in the market, but not all can be used in HDD. For instance, the EMI (Electro-magnetic Interference) in conventional DC motor is quite high and the EMI sensitivity of HDD electronics makes these motors inappropriate for use in HDD. Furthermore, particles generated by the brushes of DC motor, and the necessity of regular maintenance of DC motor are additional reasons for not using them in HDD. The major factors that must be considered before selecting a motor for HDD spindle are explained below. Issues arising from these factors are addressed. The BLDC motor with surface mounted permanent magnets on the rotor is a suitable candidate for application in HDD.

4.3.1 Special Requirements for HDD Spindle Motor

The EMI affecting the electronics system of HDD is a key factor to be considered while selecting the spindle motor. Particles released inside the drive enclosure and the range of speed are two other important points to consider. Moreover there are some special requirements on the spindle motor, which are listed below.

1. The ratio of EM (electro-magnetic) power to the volume of motor must be high. Hard disk drive is a compact electromechanical system and, therefore, very limited space is available for mounting any component including the spindle motor. But, on the other hand, the motor must have the capability of driving the disks to rotate at the required speed, and must generate torque high enough to drive that load. This requirement is crucial especially for small form factor HDD's.

2. The HDD spindle motor operates at high speed for long time. Higher spindle speed is always in demand in order to reduce the latency in reading/writing of data. The industry have experienced this trend of continuously increasing trend of spindle speed over past years, and some products have already reached rotational speed of 15,000 rpm or higher. Moreover, an HDD is required to operate continuously for years. Many motors cannot stand such high speed operation for such long time due to the limitations in the structures and components used in those motors.

3. Accurate speed is another important requirement for HDD spindle motor. Any variation in the spindle speed causes jitter between the readback information and the read clock. With increasing linear density of recording, small jitter in synchronization can cause erroneous detection of binary bits. Therefore, the accuracy of the rotational speed must be maintained and the spindle speed must be controlled robustly.

4. Reliability of the motor must be good. The HDD is a key component for mass data storage in PCs and other computing systems, and it may operate continuously for years. Although the HDD may spend the most of its running time in idle mode, i.e., no reading or writing of data being performed, the spindle motor operates continuously while other components remain inactive.

5. Leakage magnetic field must be low. HDD utilizes the magnetic field to read/write data onto the media. Hence, the process of data access is very sensitive to external magnetic field, which includes the leakage magnetic field from the spindle motor. This issue will be especially crucial when the industry will adopt perpendicular recording technology.

6. Runout is the horizonatal and vertical movement of the spindle and disks with respect to the center of rotation, and it may be contributed by many factors such as misalignment of stator and rotor, defects in bearing, misalignment between spindle and disks etc. Runout must be kept low as it influences severely the recording density in HDDs (see section 4.3.7). Therefore, reducing the runout, especially the non-repeatable runout (NRRO), of the spindle motor is very important for the HDD products. This requirement has propelled the development of special bearings like the fluid dynamic bearing, which are now used widely in the HDD products. The issue of runout is elaborated again in section 4.3.7.

7. Acoustic noise and vibration must be kept at low level. HDDs are now used in many consumer electronic products, e.g., high-end video recording and PDA phone. Any component used in these products must have low acoustic noise and vibration.

8. Low Contamination: Modern HDD products are using Magnetoresistive (MR), Giant Magnetoresistive (GMR), Tunnelling Magnetoresistive (TMR), and Current Perpendicular to Planes (CPP) materials as the read head sensor. The head performance is thus very sensitive to the contamination and particles from the components used. The major contamination from the spindle motor is from the leakage of bearing lubrication oil. How to prevent the leakage effectively is an important consideration in the HDD industry.

9. For obvious reasons, efficiency of spindle motor must be high. The efficiency refers to not only low power consumption but also low heat production. This is important for guaranteeing the life and reliability of the HDD.

10. Low cost: This is self explanatory. The trends of HDD products are driven by the market. Therefore, low cost is an essential requirement for such a widely used product.

Only a few motor structures can meet the above mentioned requirements. From the 80s, this role has always been played by the permanent magnet BLDC motor. The analysis in the following sections explains in detail the importance held by the PM BLDC motor as the choice for spindle motor in HDD.

4.3.2 Back-EMF in Spindle Motor

Permanent magnet AC motors (PMACM) are available with two kinds of back-EMF waveforms, (i) sinusoidal back-EMF and (ii) trapezoidal back-EMF with 120° flattop as shown in Figure 4.43. A PMACM with sinusoidal back-EMF is traditionally known as permanent magnet synchronous motors (PMSM), and that with trapezoidal back-EMF is called permanent magnet brushless DC (BLDC) motors. These two kinds of PMACMs should be driven with different drive modes. These drive modes are explained in section 4.4.

It is also discussed later that the spindle motor must use very limited stator slots to realize multiple magnetic pole-pairs. It is difficult to realize trapezoidal back-EMF using such EM structure. As a result, sinusoidal back-EMF is more commonly used in most of the spindle motors. These motors are normally required to be low in their back-EMF harmonics to simplify the quality control procedure, which is very important in mass production. The following discussion is restricted to the analysis of spindle motors with sinusoidal back-EMF.

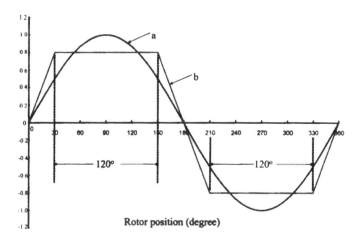

Figure 4.43: Two typical back-EMF waveforms of PMACM a: sinusoidal, b: trapezoidal.

The number of cycles in the back-EMF waveform for one revolution is equal to p, the number of pole-pairs in the motor. For an ideal spindle motor operating at constant speed, and when the rotor position is referred to using mechanical degree, its back-EMF can show the following performance,

$$e(\theta) = e\left(\theta + \frac{m360°}{p}\right), \tag{4.71}$$

where m is an integer number.

From equation 4.71, the interval between two successive zero-crossing positions (ZCP) of back-EMF should be $180°/p$ (as illustrated in Figure 4.44). However, like any other component, the motor is neither perfect in its mechanical dimensions nor magnetically homogenous. So the intervals between the ZCPs are not exactly $180°/p$, which is illustrated in Figure 4.44. These errors are called *pole-jitter* in the nomenclature used for analysis of motor.

The zero-crossings of the back-EMF waveform are widely used to detect the rotor position and to measure the speed of the spindle motor in HDD products. This issue is explained in details in section 4.4. Any jitter in the locations of zero-crossings introduces error in determining the rotor position as well as in sensing the speed of the motor. In the application of HDD, the accuracy and precision of the spindle motor's angular speed is very important. Therefore, the existence of pole-jitters is not a desirable feature for the spindle motors.

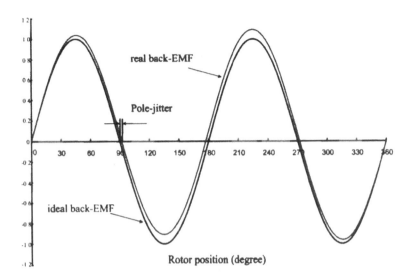

Figure 4.44: Pole jitter of back-EMF waveforms.

4.3.3 Load of Spindle Motor

In HDDs, the disks for storing information are mounted on the shaft of the spindle motor, and they rotate together with the rotor of the motor. Depending on the designed capacity of the HDD, it can come with one disk or more disks. When more than one disks are used, they are separated from one another using a *spacer* put between them. All the disks are rigidly fixed to the shaft of the spindle using *clamps*. A spindle shaft with two disks is illustrated in Figure 4.45. Increasing the number of disks mounted on the shaft has a direct consequence on the capacity of the HDD. Therefore, many HDD products use 3, or even more, disks.

The load of the spindle motor consists of the friction between the disk surface and the fast flowing air inside the enclosure (also known as the *windage friction*) and the friction between the sliders and disk surfaces. The former is the main contributor to the load of spindle motor. The relationship between the windage friction and the speed of rotation is very complicated. The load due to the windage friction can be described by the equation,

$$p_f = C_f D_f^{4.6} n^{2.8}, \tag{4.72}$$

where, C_f is a coefficient whose value depends on the structure of HDD structure and number of disks used, D_f the diameter of the disk, and n is the rotational speed of the motor [60]. Therefore, the load of the spindle motor is affected significantly with changes in rotational speed and diameter of the disks.

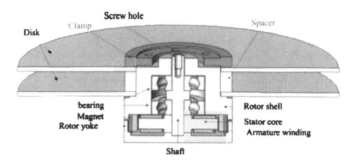

Figure 4.45: The spindle motor with disks

As a consequence of the rapid growth in areal density in magnetic recording, many manufacturers nowadays produce hard disk drives with only one disk or even only one surface of a disk to get the required capacity. For disks with same dimensions, reduction in number of disks can reduce the load of the spindle motor as both windage loss and loss due to friction between slider and disk are reduced. Use of fewer disk also lowers the cost of an HDD by reducing costs of recording media, magnetic heads, and related electronic components.

Though higher speed of spindle rotation increases the loss in the spindle system, high RPM is a desirable feature as it reduces the latency time of data access and hence increases the rate of data transfer between the HDD and the host computer [187].

4.3.4 Motor Configuration

The basic structure of the spindle motor is shown in Figure 4.45. The rotating parts of the motor include the bearing cover, the rotor shell, the outer rings and related parts of the bearing, the rotor yoke and the magnet. The inner ring of the bearing and stator core are rigidly fixed to the base of the motor.

Rotor yoke, made of steel, is necessary for the PMACM as it increases the airgap field produced by the permanent magnet. Yoke has another function in motors with the outer rotor structure shown in Figure 4.46. The rated speed of the spindle motor is very high which creates a strong centrifugal force acting on the permanent magnet. The permanent magnet ring is made from the bonded NdFeB material (see section 4.3.5) which is very weak in mechanical strength. The rotor yoke protects the magnetic ring from the strong centrifugal force during high speed operation.

Depending on the location of the electromagnetic components in the motor, the spindle motors are classified into two types: *Underslung motors* and *In-hub motors*. The one shown in Figure 4.45 is an underslung spindle motor, where the EM part of the motor is located under the disks. Key components of the motor are shown in Figure 4.46.

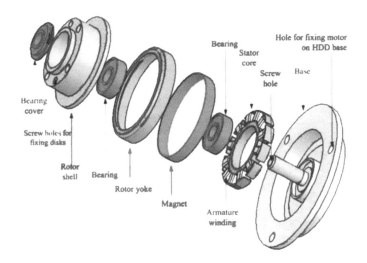

Figure 4.46: Key components of underslung spindle motor.

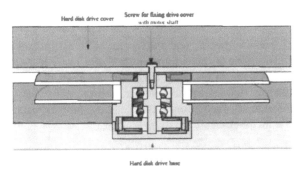

Figure 4.47: The spindle motor with disks installed inside the drive enclosure.

For this type of motor, one end of the shaft is fixed on the motor base which itself is fixed on the base-plate of HDD. The other end of the shaft can be screwed onto the HDD cover through a hole on the shaft (Figure 4.47). This ensures high stiffness for the entire spindle system.

Leakage of lubrication oil from the bearing can contribute to the contamination inside the drive enclosure which is not good for the long life of the HDD. Contamination from lubrication oil of the bearing is prevented in some motors using ferro-fluid sealing as shown in Figure 4.48. The ferro-fluid is a special kind of liquid whose distribution can be changed by applying a magnetic field. In this type of sealing, the airgap between the rotating parts and the stationary parts is filled with ferro-liquid, and a permanent magnet is used to produce the magnetic field to increase viscosity of the fluid. This sealing prevents leakage of lubrication oil without affecting the spinning of rotor.

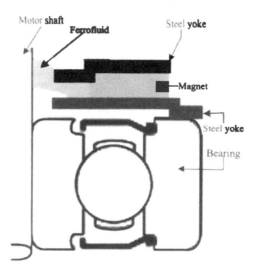

Figure 4.48: Prevention of oil leakage using ferro-fluid sealing.

In an in-hub spindle motor, the electromagnetic components of the motor is located inside the rotor, as shown in Figure 4.49. The EM parts of the motor is in the middle of the bearings. The length of the shaft of this kind of spindle motor is longer than that of the underslung motor, and it is able to produce enough EM torque for the motor operation. These motors are used in the disk drives that use multiple disks.

Hard disk drives with low rated speed of spindle use inner rotor as shown in Figure 4.50. These are used especially for small form factor HDD products.

4.3.5   Magnetic Ring

Spindle motors use permanent magnets to produce the excitation magnetic field, and to realize the conversion from the electrical energy to mechanical energy. Therefore, the performance of the spindle motor is linked closely to the performance of the magnet.

The permanent magnet is manufactured as a ring and installed on the surface of the motor (see Figure 4.45 and Figure 4.46). Its pole-pair determines the magnetic pole-pair of the motor. Bonded-NdFeB magnets are used in the spindle motors. It is isotropic, and this characteristic is important to the spindle motor because the magnetic poles must be magnetized in radial direction. This material can be easily made into a ring at low cost. To protect from rusting, the surface of the magnet is coated with anti-corrosion film. The B-H demagnetization curve of a bonded NdFeB magnet is shown in Figure 4.4. Its performance also depends on the environment.

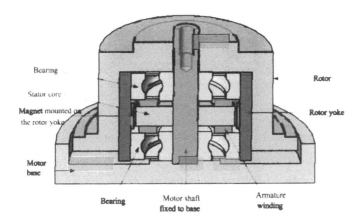

Figure 4.49: Spindle motor with in-hub structure.

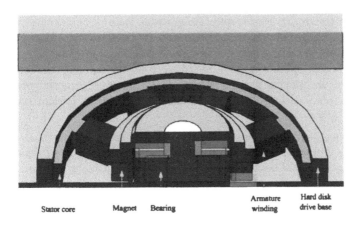

Figure 4.50: Spindle motor with inner rotor in a hard disk drive.

a: Stator core for outer rotor spindle motor   b: Stator core for inner rotor spindle motor

Figure 4.51: Laminated stator core of spindle motor: (a) stator core for outer rotor spindle motor and (b) stator core for inner rotor spindle motor.

Since the bonded NdFeB magnet can produce strong magnetic field, and it is located in the gap between the stator core and rotor core, the armature current does not have significant effect on the magnetic field in the airgap. Therefore, the dominant magnetic field in the airgap is produced by the permanent magnet ring, even when the motor is in operation and the drive current flows. With this kind of EM structure, the armature reaction is weak, and the inductances, including the self and mutual inductances, of the armature windings are small.

4.3.6   Stator Core

The stator core is used to form the required magnetic field pass, or magnetic circuit, on the motor stator, and is also the fixture for the armature windings. Figure 4.51 shows two typical stator cores. Compared to large AC motors, there are few slots in stator core of the spindle motor. Since the dimensions are small for the spindle motors, use of too many slots makes the teeth thinner. This results in poor mechanical strength of the core. Moreover, production of the stator core becomes difficult. Number of slots in typical HDD spindle motors is 6, 9 or 12.

The magnetic pole-pair of spindle motor is usually more than 2 (see section 4.2.4. Multiple magnetic pole-pair structure makes the frequency of the stator magnetic field be high when the motor is rotating at its rated speed, and eddy current is easily induced in the motor core. To reduce the eddy current, the stator should be laminated with silicon steel sheets [69], [77].

4.3.7   Spindle Motor Bearings

Similar to the large motors, the ball bearing is a logical choice for the spindle motor thanks to its low friction and high reliability. The structure of the ball bearing used in HDD spindle is shown in Figure 4.52.

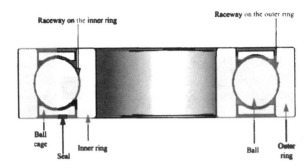

Figure 4.52: Structure of the ball bearing used in spindle motor

These bearings are small in size and, therefore, sensitive to the ingress of foreign particles. It is important to take measures to prevent external particles from entering the inside of bearing races. Rolling balls made of steel are usually used in ball bearings. Many manufacturers use ceramic ball bearings in order to obtain improved performance of the motors operating at high speed. Rolling balls of these bearings are made of ceramic materials like silicon nitride and zirconia, which possess properties such as good heat resistance and low thermal expansion, and have longer lifespan.

4.3.7.1 Spindle Motor Runout

The runout induced during the motor operation is an important issue in the selection of bearing used in the spindle motors. The concept of runout was explained in chapter 2 and in section 4.3.1 of this chapter. Runout is the lateral and vertical movement of the spindle away from its nominal rotating motion. Some of these motions are periodic and repeat themselves at regular intervals; these are called repeatable runout (RRO). There are other types of eccentricity in the rotation of spindle which do not repeat, and are known as non-repeatable runout (NRRO).

Factors affecting the RRO of the spindle motor are mostly related to the defects found in motor components such as bending of rotor shaft, misalignment of the rotor etc. It may also be caused by the problems in bearing, e.g., misalignment of bearing components. Factors contributing to NRRO, on the other hand, are complex and difficult to be attributed to any specific feature of the spindle components. In the ball bearing operation, all the balls run in the raceways between the inner and outer rings, and also rotate about their own axes. The speeds of rotation are different for different balls and they vary with time. The direction of rotational axis of a ball changes randomly with time (see Figure 4.53). When the inner ring of the bearing is fixed and outer ring rotates from position 1 to 2 and 3, all the balls rotate with asynchronous speed, and the axis of each ball may rotate in different directions.

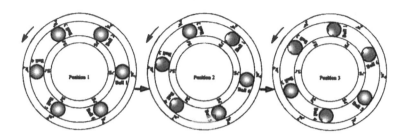

Figure 4.53: Movements of balls in the ball bearing.

As there is clearance between the cage, the rolling balls and raceway rings and the waviness on the component surfaces, the position and state of each ball are uncertain in the motor operation, even at constant rotational speed. These uncertainties make the center of the rotating ring move irregularly as a function of time causing NRRO.

Increase in both track density and linear density is necessary to meet the requirements for rapidly growing demand for higher areal density. The runout of spindle motor becomes an impediment to the realization of high density of data tracks. However, the RRO is repeatable and therefore can be modelled. Its influence on the tracking error of head positioning servomechanism can be mitigated using precision actuator and advanced servo control algorithms. Such methods are explained in details in chapter 3. However, the motion caused by NRRO is uncertain and, therefore, cannot be modelled. Its spectrum spans a significantly wide range of frequencies making it difficult to be compensated by the feedback control. Even though the movement due to NRRO of spindle motor contributes to displacement of quite small magnitude (in the scale of of 10^{-8} m), the NRRO has become the major factor limiting the achievable track density. This issue therefore drew attention of many researchers over past years. Besides compensating in the head positioning servo control loop, killing the source, i.e., reducing effectively the NRRO of the spindle motor itself has been an important research topic for years.

In ball-bearing spindle motors, all balls are in contact with the raceways and each of these contacts occurs on a very small area. This makes the spindle very sensitive to mechanical shock. This problem is particularly severe for hard disk drives used in mobile applications. Ball-bearings are also the major contributors to the acoustic noise in the hard disk drives. This fact was experimentally verified by researchers [15]. The noise of the ball-baring is generated by the mechanical contacts between two solid parts with one moving relative to the other, e.g., balls to rings, balls to cage, and cage to rings. The regular and irregular movements of the rolling elements make the spectrum of the acoustic noise rich in frequencies. Spectrum of the acoustic noise generated by a spindle motor running at 5,400 RPM is shown in Figure 4.54.

Acoustic noise of HDD should be as low as possible, especially for applications in consumer electronics. Industrial and academic research over the past years has been achieving continuous improvement in the NRRO and acoustic performances of spindle motors. However, such improvements have reached near saturation and these achievements are not enough to meet the requirements demanded by the rapid development of high density data recording.

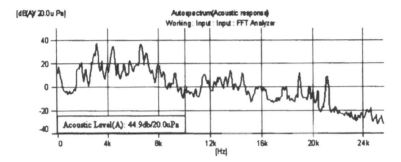

Figure 4.54: Acoustic noise spectrum of a spindle motor with ball bearing (5,400 rpm, PWM drive).

4.3.7.2 Fluid Dynamic Bearing Motors

Because of its excellent performances in reducing NRRO and acoustic noise, fluid dynamic bearing (FDB) is gradually replacing ball bearings used in the HDD spindles. In FDB, the function of the balls of the ball-bearing is realized using an oil film filled in the gap between the sleeve and the shaft. Grooves are created on the shaft or sleeve as shown in Figure 4.55. When the shaft rotates, pumping effect produced by the groves enhance the *squeeze film action* [75] generating hydrodynamic pressure wave in the oil film and makes the rotating surface float on the oil film. Because of this effect of levitation, the sleeve and the shaft are not in contact. This helps to avoid local overheating. As there is no solid connection between rotating and stationary parts, the problems of horizontal and vertical movements caused by the contact defects or deformation due to shock etc found in the rolling balls are eliminated.

FDB motors used in HDD spindle can be classified into two types according to the operational state of the motor shaft - moving shaft and fixed shaft - depending on whether the shaft rotates or not. Figure 4.55 shows the basic structure of a moving shaft FDB. The shaft of this kind of FDB is fixed with the rotor (Figure 4.56). For a spindle motor with fixed shaft, the sleeve is fixed with the rotor, shown in Figure 4.57. As one end of the shaft is fixed on the motor base and the other screwed onto HDD cover, the fixed shaft FDB motor has better mechanical performance than the moving shaft FDB. However, the construction is simpler for moving shaft FDB motor making it less costly.

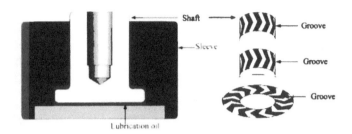

Figure 4.55: Basic structure of an FDB spindle.

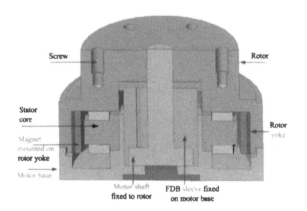

Figure 4.56: Spindle motor with moving shaft FDB.

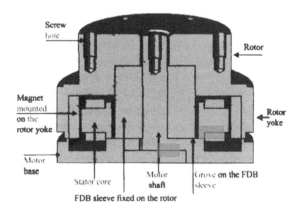

Figure 4.57: Spindle motor with fixed shaft FDB.

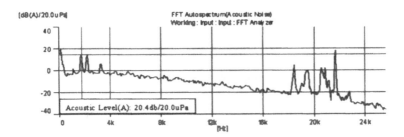

Figure 4.58: Spectrum of acoustic noise for spindle motor with fixed shaft FDB.

The NRRO and acoustics performances of the FDB spindles are much better than those of the ball bearing motors. The spectrum of the acoustic noise in an FDB spindle motor is shown in Figure 4.58. Improvement is clearly evident when this spectrum is compared with the spectrum of the acoustic noise for a ball-bearing spindle shown in Figure 4.54. The data for these spectra are collected experimentally from two spindles of same geometry, one with a ball-bearing motor while the other with a FDB motor.

It is easily concluded from these results that the acoustic noise of the spindle motor is reduced significantly if FDB spindle is used. The NRRO is also reduced significantly. These two are the major benefits of using FDB spindle. The friction between the oil film and metal parts can still induce certain level of acoustic noise and oil whirl when the motor is in motion, and irregular whirl can induce NRRO. However, the amplitudes of such noise and NRRO are much smaller than those caused by the rolling contacts in ball-bearing.

In an FDB spindle, the airgap between the sleeve and the shaft, and the depth of the groove are only several micro meters. The tolerance of the geometrical dimensions must be controlled very strictly during the poduction of these FDB spindles. This requirement makes production of FDB components and assembling of bearing difficult which, in effect, increases cost. Effective sealing of the lubricant oil is another issue to be addressed while selecting FDB motors for any application. Nevertheless, the developments in FDB design, precision machining and precision assembling have successfully enabled the application of FDB in increasing number of spindle motors.

For the FDB spindle motor, as there is more conforming surface through the lubricant to the ball bearings, these spindles have better shock resistance. However, if the motor operates at very high speed, the centrifugal force and high temperature caused by the friction make the sealing of oil difficult. Therefore, FDB spindle motor is not a good choice for HDDs with rotational speed higher than 20,000 RPM.

4.3.7.3 Aerodynamic Bearing

The structure of the aerodynamic bearing (ADB) is similar to that of the FDB, but the oil film between the sleeve and shaft is replaced by air pad. The problems caused by the movement of fluid is eliminated. The NRRO and acoustic noise performance of ADB spindle motor are better than those of FDB spindle motor. Since the linkage between the stationary part and rotating part of an ADB motor is an air pad, this kind of motor can operate in very high speed, usually in the range greater than 20,000 RPM.

In order to realize the levitation effect using air pad, the airgap in the ADB must be made narrower than that of the FDB. Tolerances for dimensions of the ADB motors are stricter. All these are challenges to the spindle motor manufacturers.

4.3.8 Winding Structure and the Airgap Field Produced by the Winding

Current flowing through the armature windings produces the magnetic field required for operation of motor. Interation between the field produced by armature current and the motor excitation field produces electromagnetic torque which acts on the rotor and reacts on the stator (see section 4.2.3).

Distributed windings are normally used in many electric machines, especially the big ones, so that the MMF waveform produced by the armature current is close to sinusoidal (see section 4.2.1). However, limited number of slots and multi magnetic pole EM structure forbid the usage of distributed winding in HDD spindle motors. The big space taken by the winding end parts is another problem for the distributed windings (Figure 4.59). For utilizing the limited stator slots to realize multi magnetic pole-pair and compact structure, concentrated windings are used in the spindle motors.

Figure 4.60 shows a typical concentrated winding used in the spindle motor, where one coil is wound around one tooth of the stator core. As there is no overlap between the end coils of adjacent windings, the space taken by the winding ends is very much reduced, see Figure 4.61.

To realize multiple magnetic pole with limited slots, each winding is put between the windings of the other two phases. As a result three consecutive slots form one winding cycle. Figure 4.60 shows an example using 9 slots to realizing three winding cycles. The symbols A and X, B and Y, and C and Z in this figure represent the two sides of the windings of A-phase, B-phase and C-phase, respectively. The distance between the centers of neighboring coils is 120 electrical degrees. Connecting the windings of the same phase in series obtains high torque constant. The three phase windings can be either Y-connected or Δ-connected. We can obtain the MMF for A-winding using the method mentioned in section 4.1.6; this MMF is shown in Figure 4.62.

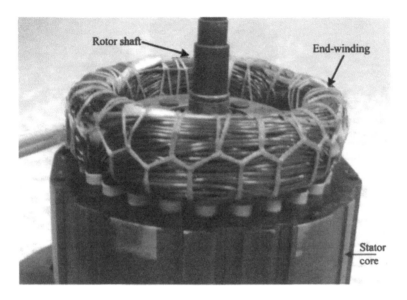

Figure 4.59: Winding ending of a 3-phase AC motor.

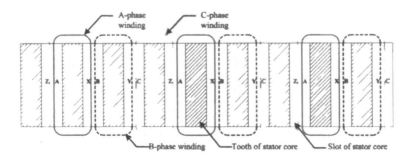

Figure 4.60: Concentrated winding used in spindle motor (9 slots).

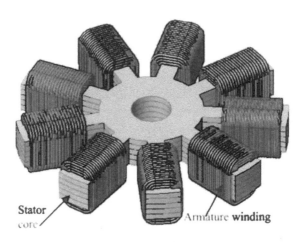

Figure 4.61: The concentrated winding of a spindle motor.

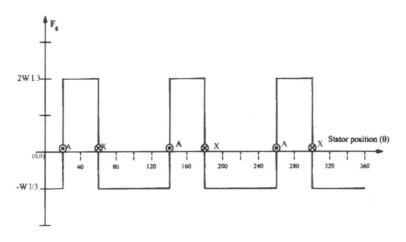

Figure 4.62: MMF generated by A-phase winding shown in Figure 4.60.

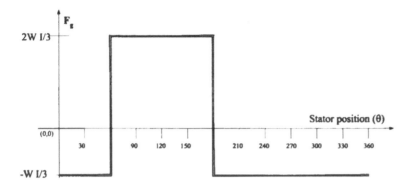

Figure 4.63: MMF expressed in electric degree

Figure 4.63 shows one electrical cycle of the MMF waveform. We can use this waveform for the Fourier analysis of the MMF waveform in the entire motor airgap. This waveform can also be used to analyze the distribution of airgap field as they are similar, according to the analysis given in section 4.2.1.

The result of the Fourier analysis of the MMF generated by one phase winding in the 9-slot motor is presented in Table 4.2. The spectrum of the airgap MMF is shown in Figure 4.64.

Table 4.2: MMF Harmonics produced by the one phase winding in the 9-slot spindle motor with normal winding format

Order of Harmonic	1	2	4	5	6	7
Harmonic Value	$\frac{\sqrt{3}}{\pi}$	$\frac{\sqrt{3}}{2\pi}$	$\frac{\sqrt{3}}{4\pi}$	$\frac{\sqrt{3}}{5\pi}$	$\frac{\sqrt{3}}{7\pi}$	$\frac{\sqrt{3}}{8\pi}$

The results presented in Table 4.2 and Figure 4.64 show that the airgap field generated by the concentrated winding is rich in harmonics. According to the conditions for torque generation introduced earlier in section 4.2.8, the pole-pair of the magnet installed on the rotor determines the motor pole-pair, and it must match one of the harmonic fields generated by the airgap MMF. It is logical to select the fundamental harmonic of the MMF to match the motor pole-pair as it is the strongest among all the harmonics of the airgap fields. The pole-pair of the fundamental harmonic is equal to the cycle number of the winding on the stator core. Therefore, for a spindle motor with 9 slots, the magnetic ring can be made with 3 pole-pairs. In a similar way, the number of motor pole-pairs can be obtained for spindle motor with different numbers of stator slots. For examples, if there are 6 stator slots, then the 3-phase winding can form 2 cycles, and the motor pole-pair should be chosen as 2. If the number of slots is 12, the winding cycle is 4, and the motor pole-pair can be 4.

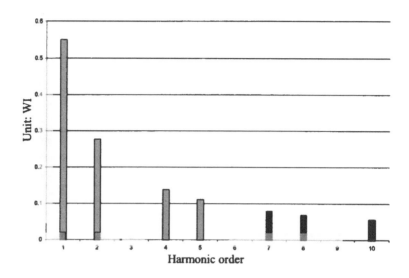

Figure 4.64: Spectrum of the MMF shown in Figure 4.62.

The number of magnetic pole-pairs of the motor is increased if the 2nd order harmonic field can be selected to match the pole-pair of the PM ring. Therefore, the motor with 9 slots can have 6 pole-pairs. In the same way, the spindle motors with 6 slots and 12 slots can have 4 and 8 pole-pairs, respectively. The 3rd harmonic does not contribute to energy conversion (see section 4.2.2). The 4th and higher harmonics contribute to energy conversion, buts their amplitudes are too small and they should not be taken into consideration. Otherwise, it results in very poor power density and efficiency of the motor. Only the fundamental and the second harmonic are used to determine the number of pole-pairs while designing the spindle motor. So far it explained that every three slots produce one, or two, magnetic pole-pairs. However, for a stator core with 9 slots, a special winding can realize 4 magnetic pole-pairs, as shown in Figure 4.63. The intervals between the centers of these three phase windings are 120° (mechanical). Therefore, they still form a symmetrical winding set in the range of 360° mechanical degree. The MMF waveform generated by one phase winding for this structure is shown in Figure 4.66, and the spectrum of the MMF is shown in Figure 4.67.

It can be concluded from the spectrum that the winding shown in Figure 4.65 is a special case as its fundamental harmonic is very low, but the fourth harmonic is the strongest. Therefore, it is reasonable to use the fourth order harmonic as the motor pole-pair. In this way, 4 pole-pairs can be realized with the 9-slot structure. A significant advantage of this structure of motor lies in the fact that the cogging torque is very small [32], [92]. Unfortunately, however, this structure introduces unbalanced magnetic pull problem [14], [17].

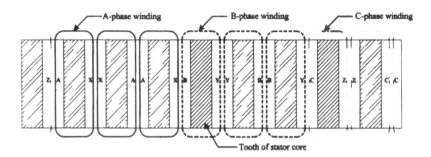

Figure 4.65: Concentrated winding for 9-slots/4-pole-pair spindle motor.

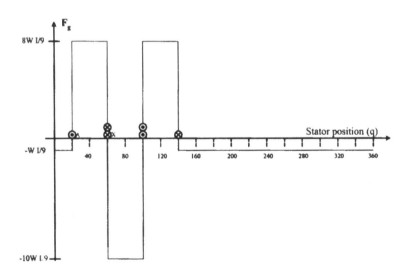

Figure 4.66: MMF waveform generated by the winding shown in Figure 4.65.

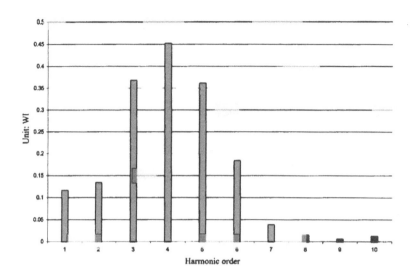

Figure 4.67: The spectrum of the MMF generated by the single phase winding shown in Figure 4.65.

4.3.9   Cogging Torque, UMP and Armature Reaction

The spindle motor in HDD uses concentrated windings, multiple magnetic pole-pair and surface-mounted magnet. During the operation of these motors, cogging torque and UMP are generated which is explained in sections 4.2.3 and 4.2.4.

Bonded NdFeB magnet, surface mounted on the rotor, is used in the HDD spindle motors. As the relative permeability of the bonded NdFeB magnet is close to 1, the surface mounted magnet makes the gap length between the stator core and the rotor core big (see Figure 4.68).

Since the motor core is made of steel, the permeability of each core is much higher than that of air. The main reluctance in the magnetic circuit of the motor comes from the airgap between the stator core and the rotor core. This makes it difficult for the current in the armature to produce strong radial magnetic field in the airgap. Considering that the NdFeB magnet is powerful in producing the magnetic field, the dominant radial airgap field is governed by the magnet. Figure 4.69 shows an example, where, it is found that even with an armature current 5 times the rated current, the effect on the airgap radial field is insignificant.

(1)   *Cogging torque generated in the spindle motor operation*
As the concentrated windings are used and installed in the slots of stator core, the cogging torque is unavoidable in the spindle motor. Figure 4.70 shows the cogging torque produced in a spindle motor with 9 slots and 6 pole-pairs.

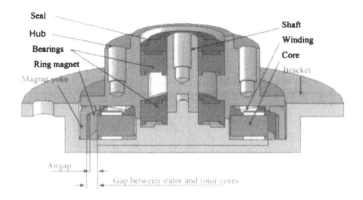

Figure 4.68: The gap between the stator core and rotor core

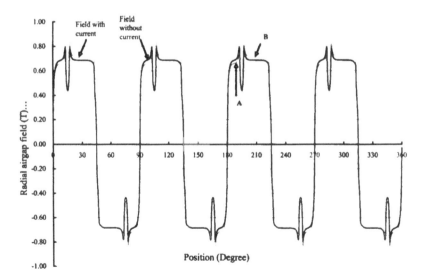

Figure 4.69: The radial direction airgap field in a spindle motor, A: The state without armature current and B: The state with 5 times of rated armature current.

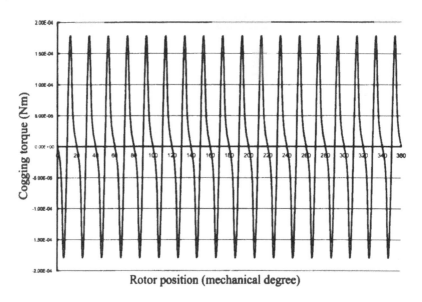

Figure 4.70: Cogging torque produced in a spindle motor with 9-slots and 3-pole-pairs (obtained by using finite element method).

As the frequency of the cogging torque is quite high in the motor operation (refer to the example shown in Figure 4.70), and the sensorless method can only detect few rotor position in one revolution, it is difficult to compensate for the cogging torque using drive current when the sensorless drive mode is used. The cogging torque is not desired in the spindle motor operation as it causes vibration and acoustic noise. Minimizing the cogging torque is an important task to accomplish in the design of spindle motor. Effective ways of reducing the cogging torque in the spindle motor include use of suitable match between the slot number and magnetic poles [70], optimizing the slot shape, and magnetizing the magnetic ring with reasonable field distribution [135], [32].

(2)   *UMP induced in the spindle motor operation*

Generation of UMP in a motor has been explained and the factors contributing to the UMP are described in section 4.2.4. As the airgap field is almost independent to the drive current, both the intrinsic and extrinsic UMPs of the spindle motor can be considered independent of the motor drive mode. The characteristic of the UMP is determined by the EM structure of the spindle motor during the stages of design and production. The UMP produces acoustic noise and vibration in the motor operation. The orders of the UMP harmonics are even [17]. A good spindle motor must show low UMP. The UMP is related to the number of slots in the stator core; using even number of slots can avoid the intrinsic UMP [17]. The extrinsic UMP can be reduced by adopting strict quality control of the motor components and the motor assembly process.

(3)   Armature reaction produced in the spindle motor operation

The conditions for generating the armature reaction are explained in section 4.2.7. Though the stator core and rotor core are usually saturated in the spindle motor, the armature reaction in the spindle motor is very weak as the airgap field is almost independent of the armature current. Therefore, the torque constant K_t and back-EMF constant K_e (refer to sections 4.2.3 and 4.2.3) can be considered constant in spindle motor operation.

4.3.10   Electromagnetic Power, Electromagnetic Torque and Motor Losses

Power loss is inevitable in operation of any motor. Losses of power in both electrical components and mechanical components contribute to the overall power loss. The electrical losses include *copper loss* and *iron loss*. The copper loss is the loss of power taking place in the resitances of armature windings, and can be calculated as

$$p_{cu} = \sum_{j=a,b,c} i_j^2(t) R_j, \tag{4.73}$$

where, i_j and R_j are the current and resistance of the jth phase winding. The influence of the copper loss in a motor is taken into consideration by including the circuit resistance in the equivalent circuit of Figure 4.71.

The electrical conductivity and magnetic hysteresis loop of the soft-magnetic materials are major factors contributing to iron loss. As the magnetic field rotates in the stator of the spindle motor, the local fields alternate and induce EMF in the stator cores. Since the conductivity of the core material is not zero, *eddy current* is produced in the core and results in eddy current loss. This loss can be reduced by using stator core which is laminated with silicon steel sheets or some other soft-magnetic materials, as shown in Figure 4.51. The magnetic pole-pair of spindle motor is typically not smaller than three and rotational speed is usually few thousand RPM. So the frequency of the local fields can be quite high (equation 4.41). Therefore, the lamination sheets must be very thin, typically 0.35 mm or thinner. For a given stator core, the eddy current loss is directly proportional to the frequency and the amplitude of the magnetic field.

Soft-magnetic materials exhibit hysteresis characteristics, i.e., there is hysteresis loop in the B-H curve of these materials. The hysteresis loop is much smaller for soft-magnetic materials compared to that of permanent magnet materials (see section 4.1.5). When the magnetic field varies and alters between

opposite polarities, the hysteresis loop induces loss in the magnetic material [77], [133]. Such loss is also related to the frequency and amplitude of the
magnetic field.

In a spindle motor, the bonded NdFeB ring is surface mounted on the rotor
yoke and the magnets produce strong magnetic field. So the motor field is
determined almost by the rotor magnet. The local fields on the magnet and
rotor core can be considered invariable, and no iron loss takes place in the rotor
during the operation of spindle motor. When the motor speed is constant, both
the frequency and the maximum value of the field passing through the stator
core are almost constant, and the iron loss can thus be considered independent
of motor load or drive current. In the rated operation, the iron loss of spindle
is normally much smaller than the copper loss.

EM power and EM torque are two important concepts used in the analysis
of electric machines. The former refers to the electrical power that is transferred from the stator, through the airgap, to the rotor with the help of the
magnetic field. This power is equivalent to the product of the EM torque and
the speed of the motor. The equivalent circuit shown in Figure 4.61 can also
be used to describe the generation of the EM power where the current passing
through the back-EMF generates the electromagnetic power, and

$$\begin{cases} P_{em} = E \cdot I = K_e \cdot \omega \cdot I \\ P_{em} = T_{em}\omega = K_t \cdot I \cdot \omega. \end{cases} \tag{4.74}$$

During the operation of spindle motor, only part of input electrical power
is converted into electromagnetic power, or transferred to the rotor. Part of
the input electrical power is wasted as copper loss. Since the dominant field is
generated by the permanent magnet on the rotor, the iron loss takes place only
when the rotor rotates. So the iron loss can be assumed to take place after
the electrical power has been transferred to the rotor. The equivalent circuit
shown in Figure 4.55 cannot describe any phenomenon that takes place after
the EM power has been created, e.g., the effects of iron losses.

There are also mechanical losses, e.g., friction loss in the bearings and
windage loss on inner and outer surfaces of rotor, when the motor is in operation. These losses depend on the mechanical structure of the motor, but are
independent of the drive current and load condition if the speed is fixed. As
these losses are caused by mechanical factors, it is difficult to describe them
using electric and magnetic circuits. Therefore, the equivalent circuit of Figure 4.38 describes only the conversion process from input electrical power to
electromagnetic power.

According to the analyses presented in this section, we can get the power
flow in the spindle motor as shown in Figure 4.71. If no disk is mounted on
the rotor, the output power of the motor is zero. When the disks are mounted,
the input power must be increased to compensate for the mechanical losses
caused by the inclusion of disks. This power increment is the output power of
the motor which is mechanical power.

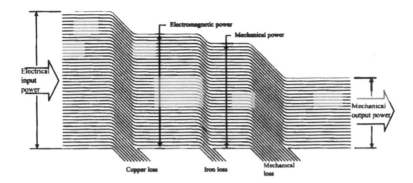

Figure 4.71: Power flow in spindle motor.

4.4 Spindle Motor Drive System

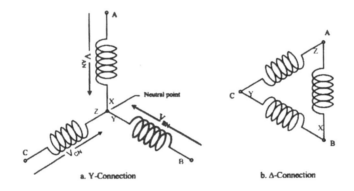

Figure 4.72: Three phase winding connections.

As mentioned in section 4.3, spindle motors used in HDD are generally PM AC motors with surface-mounted permanent magnets. There are usually three phases with the phase windings are arranged either Y-connected, or Δ-connected (Figure 4.72). The Y-connected winding is the most prevailing choice because of its low cost. In the Y-connected winding, besides the winding terminals A, B and C, the neutral point may also be used. The back-EMFs induced in the armature windings of the motor are normally symmetrically sinusoidal. Unlike the conventional method of driving PM AC motors with sinusoidal back-EMF, the spindle motor is usually driven as a BLDC (*Brushless DC*) motor. An introduction to the BLDC drive system is given, and the performance of a motor is explained for different BLDC drive modes. Since the motor drive is linked directly to electrical quantities, to make the explanation concise, the degree mentioned in the following is electrical degree (defined in section 4.2.6), unless stated otherwise.

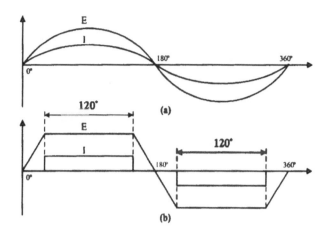

Figure 4.73: PMSM drive mode (a) and BLDC drive mode (b).

4.4.1   What is the BLDC mode?

It has been mentioned in section 4.2.5 that PMACMs are categorized into two groups according to the shape of their back-EMF waveforms. The PMACMs whose back-EMF is trapezoidal with 120° flattop are usually called brushless DC (BLDC) motors [153], and the PMACMs with sinusoidal back-EMF are usually named as permanent magnet synchronous motor (PMSM) [37]. These back-EMF waveforms are illustrated in Figure 4.73. It is a widely accepted practice to drive a PMSM using sinusoidal drive current and a BLDC motor using 120° square current (Figure 4.73). For the sinusoidal drive current (PMSM drive mode), a complex motor drive system is usually required as the current must be adjusted instantaneously in small magnitude according to the variation of the rotor position. Besides the PMSM drive mode requires continuous and accurate detection of rotor position, which adds to the difficulty to the realization of PMSM drive mode. On the contrary, the BLDC drive mode is easy to implement as only a few rotor positions are needed, and the commutation of current is required only at these limited rotor positions. The three phase back-EMFs and the desirable drive currents of the BLDC motor are illustrated in Figure 4.74.

A BLDC motor is usually driven by an inverter shown in Figure 4.75. A position sensor is needed to identify the required rotor position for current commutation. From Figure 4.74, we can see that at any time instant, only two phases are energized with drive current. The switching sequence of the energized two phases are (AH, BL) (AH, CL) (BH, CL) (BH, AL) (CH, AL) (CH, BL) according to the symbols shown in Figure 4.75. The procedure is same as the one introduced in section 4.2.3. This drive mode with only two phases energized at any time instant is called the BLDC drive mode, or in short, the BLDC mode.

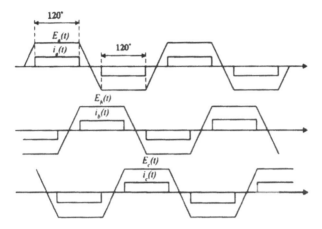

Figure 4.74: Ideal back-EMF and current waveform in the BLDC motor.

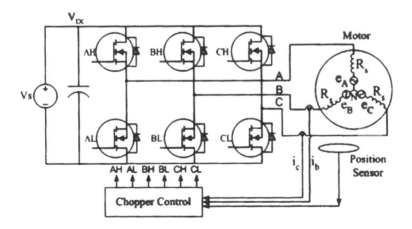

Figure 4.75: Schemes for motor drive system

4.4.2   Detection of Rotor Position in Sensorless Drive

It has been mentioned earlier that knowing the rotor position is an essential requirement for commutation of current and generation of effective torque in PMAC motors (both BLDC motors and PMSMs). The electronic commutation means generating proper commutation sequence at the correct position of the rotor to switch the power electronic devices of the inverter bridge. Physical sensors such as Hall-effect sensors, resolvers and position encoders can be used to obtain the rotor position. However, it was explained earlier in section 4.3.1 that the use of physical sensor is not a suitable choice for the HDD application due to limitation of space, cost restriction, and system reliability. As a consequence, the PMAC motor drive methods that do not need accurate physical position sensors are receiving wide attention. These methods are commonly referred to as position sensorless control method, or self sensing method.

The sensorless control methods for PMAC motors presented in different published articles can be broadly categorized into two groups following the classification of PMAC motors: (1) the sensorless methods for BLDC motors and (2) the sensorless methods for PMSMs. The difference between these two types of methods lies in the way the phases of a motor are energized. In the sensorless BLDC control methods, only two phases are energized at any time instant [12], and only a few rotor position signals are required. On the contrary, all three phases conduct at the same time in the sensorless PMSM control methods, and a continuous position signals is required [49]. Different methods can be used to find the rotor position of PMAC motors when a drive control scheme without a position sensor is to be used [104]. A brief summary of these methods is presented next.

4.4.2.1 Review of Methods for Sensorless Detection of Rotor Position

Some of the sensorless methods used to detect rotor position are,

1. based on voltage and current measurement [49],

2. hypothetical d-q model [139],

3. stochastic filtering [20],

4. self observers [180],

5. detecting variations in inductance [117], and

6. nonconventional methods such as use of neural network or fuzzy logic or both [177].

These methods are generally complex and computation intensive, demanding powerful digital signal processors (DSPs) for the motor control system.

Therefore, though these methods have been used successfully in other applications, they have never been found appropriate for HDD.

Some of the sensorless methods used for detecting rotor position in BLDC are

1. direct back-EMF detection [93],

2. back-EMF integration [12] [131],

3. integration of back-EMF third harmonic [146],

4. conducting state of the free-wheeling diodes of the inverter [153], and

5. phase current sensing [129].

These sensorless methods for detecting rotor position of the BLDC motor are less computationally intensive and the software complexity can be greatly reduced using suitable hardware.

According to the categorization of the PM AC motors, the spindle motors used in HDDs belong to the PMSMs because they have sinusoidal back-EMF. However, as the concerns like cost and reliability, BLDC drive mode is still used in HDDs to drive the spindle motors. This is not an optimal way to drive the spindle motor from the points of EM torque generation and motor efficiency (it will be discussed in section 4.4.4), but it can be implemented with less cost and less dependence on high-power DSPs and yet with satisfactory performance.

From above introduction, two states are included in the BLDC drive mode: energized state and silent state as shown in Figure 4.76. The drive current goes through the winding in the energized state to produce EM torque. In the silent state, there is no current in the winding, and the phase back-EMF can thus be obtained by measuring the phase voltage in the silent state.

4.4.2.2 Different Modes of Operation for BLDC Drives

Three BLDC drive modes are widely used: constant current BLDC mode (CC-BLDC mode), constant voltage BLDC mode (CV-BLDC mode) and pulse-width modulation BLDC mode (PWM-BLDC mode). In the CV-BLDC mode, the line voltage between the two terminals of the motor is kept constant in the 120° energizing state. If the drive voltage needs to be adjusted, a DC regulator is used to change the DC link voltage while keeping the two transistors in the "full-on" status [102]. Figure 4.77 illustrates the inverter gate signals in the CV-BLDC mode.

Similarly, in the CC-BLDC mode, the current in one phase winding is kept constant in its 120° energizing state [100]. Current references for different loads are used to compare with the measured current and a hysteresis current controller is used to adjust the drive current to meet the reference value.

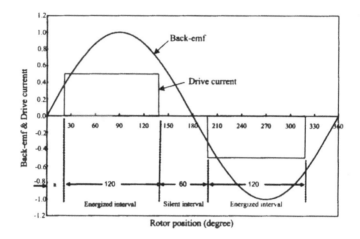

Figure 4.76: The states of operation in the constant current BLDC drive mode.

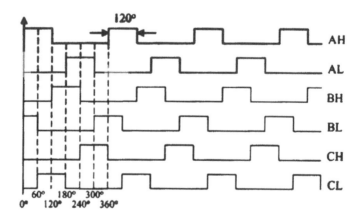

Figure 4.77: Gate signals for the three-phase inverter in the CV-BLDC mode.

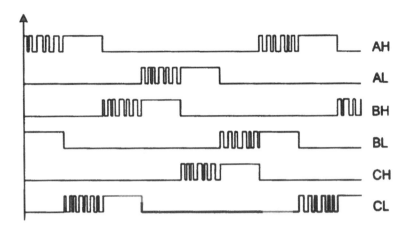

Figure 4.78: Gate signals for the three-phase inverter in the CC-BLDC mode.

In the PWM-BLDC mode, one of the transistors in the "on" state is turned on and off at the PWM frequency so that either the drive voltage is controlled by a fixed PWM duty cycle or the current is adjusted by means of an integrated current control circuit [174]. The PWM signal can be applied only on one transistor while keeping another transistor "full-on" during the complete step, or both the high side and the low side transistors are switched off together during each PWM off period. In this way, the voltage regulator can be saved which is needed in the CV-BLDC circuit, and in normally the efficiency of the drive system can be improved. If the PWM BLDC mode works at 100% duty cycle, it will be the CV-BLDC mode.

The gate signals in the CC-BLDC mode are actually PWM signals with varying duty cycles, which are different from those with fixed duty cycles in the PWM BLDC mode. Simulated gate signals for these two modes are shown in Figure 4.78and Figure 4.79, respectively.

Using CC-BLDC mode or the PWM BLDC introduces significant noises superimposed on the voltages, as shown in Figure 4.80, which is not desirable in some applications.

4.4.2.3 Detection of Rotor Position using back-EMF Signal

In using the BLDC mode, only the rotor positions signals at the commutation points are needed. It is clear that detecting the rotor position accurately at these points are critical to this drive mode. In the spindle motor, the sensorless method based on direct back-EMF detection is normally used to detect the rotor position for realizing the BLDC drive.

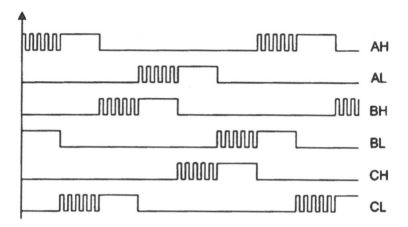

Figure 4.79: Gate signals for the three-phase inverter in the PWM BLDC mode.

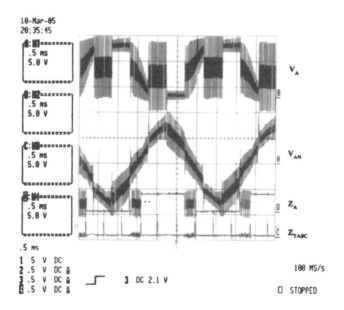

Figure 4.80: Terminal and phase voltages in PWM BLDC mode.

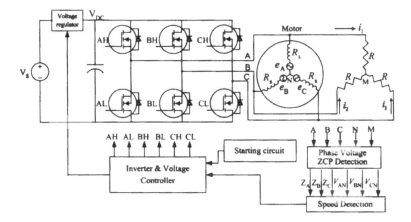

Figure 4.81: Schematic diagram of the drive system for spindle motor operated in BLDC mode.

When the back-EMF of one phase is at its zero crossing position (ZCP), the terminal voltage of this phase equals to the neural point voltage; see Figure 4.1.a. From section 2.5, the ZCPs take place at positions where the coil direction is aligned with, or opposite to, the magnet axis on the rotor. For the surface-mounted spindle motors, the armature reaction is usually quite weak, which means the back-EMF waveforms are not influenced by the currents in the motor operation, and the ZCPs of the phase back-EMF are not affected either. Therefore, the ZCPs in the spindle motor are reliable in detecting the rotor position and speed of the motor.

In the motor operation, as there are three phase windings, the number of ZCPs in one revolution is 6 times the number of pole-pairs. Therefore, the more pole-pairs, the more information of the rotor can be obtained, and the more accurate speed could be realized. This is one of the reasons why the HDDs prefer to use the spindle motor with multiple magnetic pole-pairs, and 4 and 6 pole-pairs are the most prevailing choices.

Detection of ZCP in CV-BLDC mode

The phase voltage means the voltage difference between the phase winding terminal and neutral point; see VAN, VBN and VCN in Figure 4.72. However, the actual neutral point of the spindle motor used in HDDs may not be available. Therefore, as shown in Figure 4.81, a Y-connected resistance circuit is used to create a virtual neutral point of the three-phase armature windings.

A virtual neutral point M is created by the resistance circuit. Since the back-EMFs in the armature windings of the spindle motor are symmetrically and sinusoidal, the virtual neutral point M has the same voltage potential as the real neutral point N of the motor (shown in Figure 4.81). The proof is given in the following.

From Figure 4.10, the stator voltages for the three phases can be written as:

$$
\begin{cases}
V_{AN} = e_A + R_A i_A + L_{AA}\frac{di_A}{dt} + M_{AB}\frac{di_B}{dt} + M_{AC}\frac{di_C}{dt} \\
V_{BN} = e_B + R_B i_B + L_{BB}\frac{di_B}{dt} + M_{BA}\frac{di_A}{dt} + M_{BC}\frac{di_C}{dt} \\
V_{CN} = e_C + R_C i_C + L_{CC}\frac{di_C}{dt} + M_{CA}\frac{di_A}{dt} + M_{CB}\frac{di_B}{dt}
\end{cases}
\tag{4.75}
$$

or,

$$
[V] = [e] + [R][i] + [L_p][\frac{di}{dt}],
\tag{4.76}
$$

where,

$e_A,\ e_B$ and e_C :   phase back-EMF of winding A, B and C, respectively,

$R_A,\ R_B$ and R_C :   resistance of winding A, B and C, respectively,

$L_{AA},\ L_{BB}$ and L_{CC} :   self-inductance of winding A, B and C, respectively,

$M_{AB}, M_{AC}, M_{BA}, M_{BC}, M_{CA}, M_{CB}$:   mutual inductance between windings,

$$[V] = [V_{AN}\quad V_{BN}\quad V_{CN}]^T,$$

$$[e] = [e_A\quad e_B\quad e_C]^T,$$

$$[i] = [i_A\quad i_B\quad i_C]^T,$$

$$[\frac{di}{dt}] = [di_A/dt\quad di_B/dt\quad di_C/dt]^T,$$

$$
[R] =
\begin{bmatrix}
R_A & 0 & 0 \\
0 & R_B & 0 \\
0 & 0 & R_C
\end{bmatrix}, \text{ and}
$$

$$
[L_p =
\begin{bmatrix}
L_{AA} & M_{AB} & M_{AC} \\
M_{AB} & L_{BB} & M_{BC} \\
M_{AC} & M_{BC} & L_{CC}
\end{bmatrix}.
$$

As mentioned in the previous chapter, the micro PMSM used in HDDs has many special characteristics, one of which is that the Y-connected three phases are symmetrical and the variation of the self-inductances and mutual-inductances are very small. So, the analysis can be simplified using the assumptions,

$R_A = R_B = R_C = R,$

$$L_{AA} = L_{BB} = L_{CC} = L,$$
$$M_{AB} = M_{AC} = M_{BA} = M_{BC} = M_{CA} = M_{CB} = M \text{ and}$$
$$i_A + i_B + i_C = 0.$$

Therefore,

$$[V] = [e] + [R][i] + [L][di/dt], \tag{4.77}$$

where,

$$[L] = \begin{bmatrix} L & M & M \\ M & L & M \\ M & M & L \end{bmatrix}.$$

Moreover, the back-EMF in the armature windings of a micro PMSM used in HDD are symmetrically sinusoidal, that is,

$$\begin{cases} e_A &=& K_e \omega_r \sin(\omega_r t) \\ e_B &=& K_e \omega_r \sin\left(\omega_r t - \frac{2\pi}{3}\right) \\ e_C &=& K_e \omega_r \sin\left(\omega_r t + \frac{2\pi}{3}\right) \end{cases} \tag{4.78}$$

Therefore,

$$V_{AN} + V_{BN} + V_{CN} = 0. \tag{4.79}$$

For the resistance circuit shown in Figure 4.81, the sum of the three currents flowing into the circuits is zero,

$$i_1 + i_2 + i_3 = 0, \tag{4.80}$$

which means

$$V_{AM} + V_{BM} + V_{CM} = 0. \tag{4.81}$$

Now looking back to Figure 4.81, one can easily derive the voltage equations,

$$\begin{cases} V_{AN} + V_{NM} + V_{MA} &=& 0 \\ V_{BN} + V_{NM} + V_{MB} &=& 0 \\ V_{CN} + V_{NM} + V_{MC} &=& 0 \end{cases} \tag{4.82}$$

By adding these three equation, one can get,

$$(V_{AN} + V_{BN} + V_{CN}) + 3V_{NM} - (V_{AM} + V_{BM} + V_{CM}) = 0. \tag{4.83}$$

Substituting equations 4.79 and 4.81 into equation 4.83,

$$V_{NM} = 0. \tag{4.84}$$

Therefore, the virtual neutral point M is at the same voltage as the real neutral point N and can be used to obtain the phase winding voltages,

$$\begin{cases} V_{AN} &= V_A - V_M \\ V_{BN} &= V_B - V_M \\ V_{CN} &= V_C - V_M \end{cases} \tag{4.85}$$

As discussed in section 4.2.1, the phase back-EMF can be detected in the silent interval of the phase winding. In this interval, there is no current in the winding and the phase back-EMF can be written as

$$\begin{cases} e_A(t) &= V_{AN}(t) = V_A(t) - V_M \\ e_B(t) &= V_{BN}(t) = V_B(t) - V_M \\ e_C(t) &= V_{CN}(t) = V_C(t) - V_M \end{cases} \tag{4.86}$$

Therefore, the zero-crossing points of a phase back-EMF can be obtained using a zero-crossing detection circuit by observing the difference between the voltages of winding terminal and the virtual neutral point in the silent state.

In the practical drive circuit, the three terminal voltages (V_A, V_B and V_C), neutral point voltage V_N (if available), and virtual neutral point voltage V_M, all with respect to the ground, are connected to the phase voltage ZCP detection circuit shown in Figure 4.81. Three phase voltages can be restored using equation 4.86 if the neutral point N is not provided. Otherwise, if V_N is available, we can simply replace V_M with V_N in equation 4.86. The obtained phase voltages of the spindle motor in Y connection are illustrated in Figure 4.82.

Finding zero-crossings in the back-EMF signal using the circuit above may introduce errors due to the influence of commutation of current. However, such errors can be detected by exploiting our knowledge of the logical relationships between the drive voltages and the ZCPs [102]. After eliminating the false ZCPs, we are left with the pulses corresponding to the correct ZCPs of the back-EMF of all three phases, and the interval between any two consecutive ZCPs is equal to 60°. Figure 4.83 is an example showing the phase voltages and their correct ZCPs.

Detection of ZCP in CC-BLDC mode and PWM-BLDC mode

Detection of ZCP in the CC-BLDC and PWM BLDC mode is more complicated than that in the CV-BLDC mode because of superposition of high frequency PWM signal on V_N for these modes. Two possible ways of detecting the ZCPs in these modes are

1. Detection of ZCP during the ON state of PWM, and

2. Detection of ZCP at the end of the OFF state of PWM,

implementations of which are shown in Figure 4.84 and 4.85, respectively.

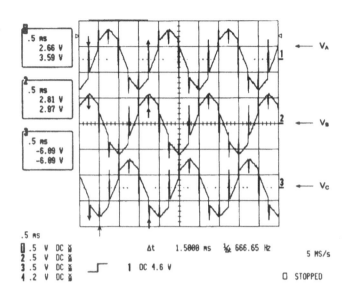

Figure 4.82: Three phase voltages of the spindle motor in BLDC mode.

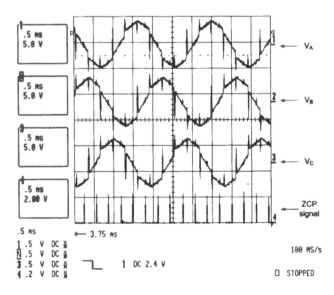

Figure 4.83: Three phase voltages and the corresponding true ZCP pulses.

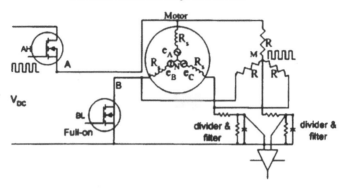

Figure 4.84: Detection of back-EMF zero-crossing points during PWM ON state.

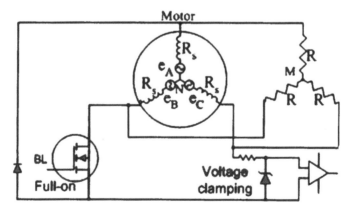

Figure 4.85: Detection of back-EMF zero-crossing points at the end of PWM OFF state.

4.4.2.4 Determining the Speed in Sensorless BLDC Drive

Detection of speed in BLDC mode can be implemented using numerical difference method and can be realized using a counter or timer. The counter is incremented by clock pulse from a precise crystal oscillator. If the change in counter value between two consecutive ZCPs is N, then the mechanical speed (Ω) of the rotor can be calculated as

$$\Omega = \frac{60 f_e}{p} = \frac{60 f_c}{6Np} = \frac{10 f_c}{Np}, \tag{4.87}$$

where, p is the number of pole-pair in the motor, f_c is the frequency of the crystal oscillator, and f_e is the electrical frequency. If a timer is used instead of the counter to measure the time difference Δt between two consecutive ZCP events then the mechanical speed is

$$\Omega = \frac{60 f_e}{p} = \frac{60}{6\Delta t p} = \frac{10}{\Delta t p}, \tag{4.88}$$

These methods of speed detection assumes the absence of voltage noise causing error in ZCPs and the effects of pole jitter (discussed in section 4.3.2). Unfortunately, pole jitter is unavoidable as neither the magnetic field is perfectly symmetric nor the dimensions of the components and quality of the assembling process can be controlled perfectly. While realizing the closed loop speed control, an average of several values of the measured speed can be taken into consideration to reduce the fluctuations caused by errors.

Once the rotating speed of the motor is obtained, the drive voltage can be controlled to maintain precise and accurate speed of the motor under closed loop control. In the CV-BLDC mode, the voltage regulation is realized using an adjustable DC regulator to modify the DC link voltage. In the PWM-BLDC mode, the drive voltage can be controlled by adjusting the PWM duty cycle. For the CC-BLDC mode, the current reference is set for a certain speed and the drive current is controlled to meet the reference value.

4.4.3 Starting of Spindle Motor

There exists a problem in using the back-EMF detection method for finding the rotor position. The back-EMF is either zero or very low at standstill or during low speed operation, and the rotor position cannot be detected. For the PMSM driven by the sensorless method mentioned above, the motor must be first started from standstill using some special procedures and brought up to a speed suitable for sensing back-EMF. Therefore, in the application of sensorless BLDC drive mode, how to start the motor effectively without a sensor detecting rotor position is an important concern.

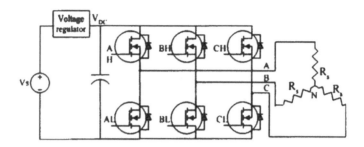

Figure 4.86: Simplified drive circuit at low speed.

4.4.3.1 Review of Different Methods for Sensorless Starting

The spindle motor can be started from standstill using one of the following two methods:

1. Detection of inductance variation and

2. Frequency-skewed or voltage-skewed drive signal.

As the permanent magnet is mounted on the surface of the rotor of spindle motors used in HDD, the motor inductance and variation of inductance are very small. Therefore the 1st method mentioned above is difficult to realize in HDD spindle motor, and the second one is the most commonly used method in the HDD industry. In this approach, a six-step signal is applied open-loop with gradual increase in frequency to spin up the motor from standstill.

4.4.3.2 Starting of Spindle Motor using Skew Frequency

Since the starting procedure works only at low-speed range, the effect of the inductance can be neglected. So an inverter connected to a star-connected 3-phase resistive circuit can be used to illustrate the starting sequence, as shown in Figure 4.86.

The six-step starting method are widely used for starting of the PMSM. With this method, the inverter gate is switched every 60°. The gate signals of the MOSFETs are illustrated in Figure 4.87, in which each gate signal remains HIGH for 120°. With this pattern of gate signal, waveforms of the phase voltages (V_{AN}, V_{BN} and V_{CN} of Figure 4.86) are as shown in Figure 4.88 with their amplitudes equal to $|V_{DC}/2|$. These two pictures are used just to show the basic principles of this method; the phenomenon of frequency skew or voltage skew is not shown here. It is clear that the phase voltages have sudden jumps, which will cause jumps in the drive currents as well. As a consequence, oscillation is induced in the rotor by the jumps in the current waveform, making it difficult to start the motor smoothly.

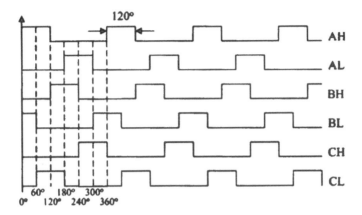

Figure 4.87: Gate signals for the three-phase inverter in six-step starting.

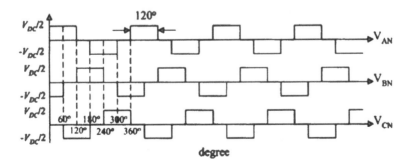

Figure 4.88: Three phase voltages in six-step starting.

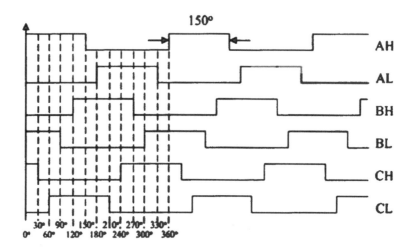

Figure 4.89: Gate signals for the three-phase inverter in twelve-step starting.

An alternative to the above mentioned method, the *twelve-step starting* [131], improves the starting performance. In the twelve-step pattern, the inverter gate is switched over every 30°, and hence the power electronics devices in the converter circuit conduct for 150°, instead of 120° in the six-step starting. The corresponding gate signals are shown in Figure 4.89 and the phase voltages with step amplitudes $V_{DC}/3$, $V_{DC}/2$ and $2V_{DC}/3$ are illustrated in Figure 4.90.

To realize the twelve-step pattern, an additional switching state is inserted between every two successive states of the six-step sequence, making it twelve states in the starting sequence. The switching sequences of both the six-step and twelve-step are listed in Table 4.4.

Table 4.4: Switching sequence for six-step and twelve-step starting method

Starting Method	Switching Sequence
Six-step	(AH, BL) (AH, CL) (BH, CL) (BH, AL) (CH, AL) (CH, BL)
Twelve-step	(AH, BL) (AH, BL, CL) (AH, CL) (AH, BH, CL) (BH, CL) (AL, BH, CL) (BH, AL) (AL, BH, CH) (CH, AL) (AL, BL, CH) (CH, BL) (AH, BL, CH)

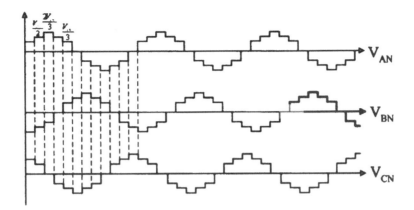

Figure 4.90: Three phase voltages in twelve-step starting.

The changes in current vector between two successive switching states in six-step and twelve-step are illustrated in Figure 4.91. Let's assume that the current flows from A to B during a state and the corresponding current vector is $\bar{I}_1$. In the next state of the six-step method, the current flows from A to C with corresponding vector $\bar{I}_{6-step}$; the variation in the vector is $\Delta\bar{I}_{6-step}$. On the contrary, current flows from both A to C and A to B in the next stage of twelve-step starting, and the current vector is illustrated as $\bar{I}_{12-step}$, current vector variation as $\Delta\bar{I}_{12-step}$. We can see that both the magnitude and angle of the vector variation are smaller in the twelve-step method than in the six-step method. So the magnetic field produced by the armature windings varies smoother in the twelve-step method, making it less prone to exciting resonance of the rotor during the starting process. The twelve-step method makes the starting easier and smoother.

Both these schemes can be easily realized using electronics. Details are not included in this book, but the interested readers may refer to [131] for elaborate analysis on the spin-up capabilities of these two methods. The results introduced in the literature show that, the twelve-step starting is more robust.

4.4.4 Spindle Motor Driven in Sensorless BLDC Mode

The performance of a motor is dependent on its drive mode. What is the most suitable drive mode for the spindle motor discussed in section 4.3.1? The spindle motor used in HDD is compact in EM structure and mechanical structure, weak in armature reaction, and has sinusoidal back-EMF in the constant speed. What is a suitable drive mode for such motor so that the requirements mentioned in section 4.3.1 can be met? It is shown using the analysis presented below that the sensorless BLDC drive mode is a good choice for HDD spindle motor.

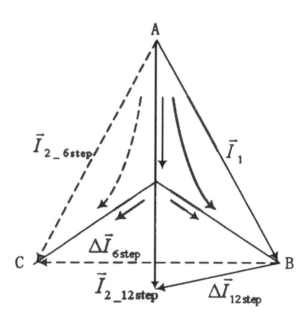

Figure 4.91: Current vector in the 6-step and 12-step.

4.4.4.1 The Optimal Commutation Angle of Spindle Motor

The back-EMF of a spindle motor can be defined by the expression in equation
4.89. Back-EMF of one of the phases is shown in Figure 4.92.

$$
\begin{cases}
e_A &=& E_m \sin(\omega_e t) \\
e_B &=& E_m \sin\left(\omega_e t - \frac{2\pi}{3}\right) \\
e_C &=& E_m \sin\left(\omega_e t + \frac{2\pi}{3}\right)
\end{cases}
\tag{4.89}
$$

When the BLDC drive mode is used, the exciting state spans over $120°$ in
the space domain, and the silent state spans over $60°$. However, the commutation angle α could be set at different value in the motor operation as shown
in Figure 4.92.

Varying the commutation angle α affects the performance of the motor. It is
important to find the commutation angle that results in optimal performance of
the motor. Optimal performance means the use of minimum effective current
to produce the required electromagnetic torque. Under this condition, the
copper loss is minimum while generating the required torque. In the analysis
presented below, the speed of the motor is assumed constant and the effects
of the spindle motor inductance are assumed negligibly small. Moreover, for
simplicity and conciseness of analysis, the magnetic pole-pair is set to one.

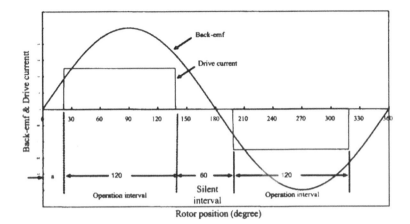

Figure 4.92: Operation states of three-phase BLDC spindle motor.

Therefore, the equation 4.89 is changed to

$$\begin{cases} e_A &= & E_m \sin\left(p\omega_e t\right) = E_m \sin\left(\theta\right) \\ e_B &= & E_m \sin\left(\theta - 120°\right) \\ e_C &= & E_m \sin\left(\theta - 240°\right) \end{cases} \qquad (4.90)$$

The angular frequency is same as the angular speed of the motor, i.e., $\omega = \Omega$. It is also assumed that the three phase windings are symmetric, the phase resistances of the windings are same and equal to R_a. The three phase windings are Y-connected.

When the CC-BLDC mode is used, the drive current is kept constant in the energized state as shown in the waveform of Figure 4.92. For the CV-BLDC mode, the power transistors keep the line voltages of the motor terminals constant during the energized state. The waveform representing one of the terminal voltage is shown in Figure 4.93.

4.4.4.2 Optimal Commutation Angle for CC-BLDC Mode

For a three-phase spindle motor driven using constant current mode, the following equations describe the phase currents:

$$i_A(\theta) = \begin{cases} I_m & , & \alpha \le \theta < \alpha + 2\pi/3 \\ -I_m & , & \alpha + \pi \le \theta < \alpha + 5\pi/3 \\ 0 & , & others \end{cases} \qquad (4.91)$$

$$i_B(\theta) = i_A\left(\theta - \frac{2\pi}{3}\right) = \begin{cases} I_m & , & \alpha - 2\pi/3 \le \theta < \alpha \\ -I_m & , & \alpha + \pi/3 \le \theta < \alpha + \pi \\ 0 & , & others \end{cases} \qquad (4.92)$$

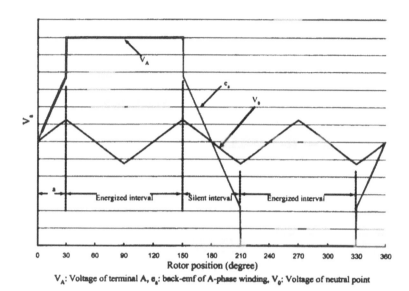

V$_A$: Voltage of terminal A, e$_a$: back-emf of A-phase winding, V$_0$: Voltage of neutral point

Figure 4.93: Line voltage waveforms of CV-BLDC mode.

$$i_C(\theta) = i_A\left(\theta - \frac{4\pi}{3}\right) = \begin{cases} I_m & , & \alpha - 4\pi \le \theta < \alpha - 2\pi/3 \\ -I_m & , & \alpha - \pi/3 \le \theta < \alpha + \pi/3 \\ 0 & , & others \end{cases} \quad (4.93)$$

In these equations, I_m is the magnitude of the constant current delivered by the power supply during the energized period. During the silent interval, the phase current is set to zero.

The EM torque generated by the motor can be expressed by the equation,

$$T_{em}(\theta, \alpha) = \frac{P_m}{\Omega} = \frac{e_A(\theta)i_A(\theta, \alpha) + e_B(\theta)i_B(\theta, \alpha) + e_C(\theta)i_C(\theta, \alpha)}{\Omega}. \quad (4.94)$$

Since the current waveform is related to the commutation angle, both the average torque and the torque ripple get affected when different commutation angles are used. Figure 4.94 shows the EM-torques produced for two different commutation angles, $\alpha = 20°$ and $\alpha = 20°$.

The average EM torque produced by a spindle motor driven in the CC-BLDC mode is

$$\begin{aligned} \bar{T}_{em} &= \frac{1}{2\pi}\int_0^{2\pi} T_{em}(\theta, \alpha)d\theta = \frac{3}{2\pi\Omega}\int_0^{2\pi} i_A(\theta, \alpha)e_A(\theta)d\theta \\ &= \frac{6}{2\pi\Omega}\int_0^{\alpha+2\pi/3} E_m I_m \sin(\theta)d\theta = \frac{3\sqrt{3}E_m I_m}{\pi\Omega}\sin\left(\alpha + \frac{\pi}{3}\right) \end{aligned} \quad (4.95)$$

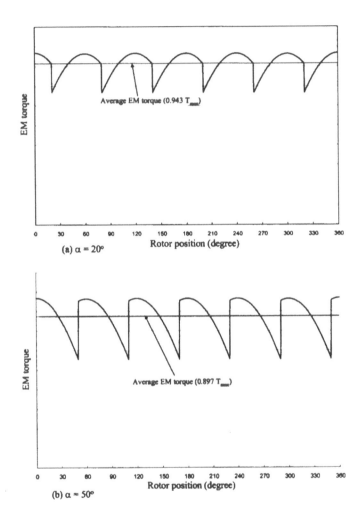

Figure 4.94: EM torque produced by different commutation angle in using constant current drive mode.

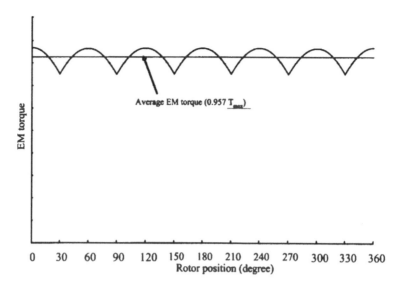

Figure 4.95: EM torque produced by optimal commutation angle in CC-BLDC drive mode.

It is obvious from the above equation that the EM torque $\bar{T}_{em}$ is a function of the commutation angle α. This equation also shows that,

$$\max[T(\alpha)] = \bar{T}_{em}\left(\frac{\pi}{6}\right). \tag{4.96}$$

Therefore, the optimal commutation angle for the CC-BLDC drive mode is $\alpha_{opt} = \pi/6$ as this angle can produce the maximum effective EM torque for a given current. The average EM torque generated with this optimum commutation angle is

$$\max[T(\alpha)] = \bar{T}_{em}\left(\frac{\pi}{6}\right) = \frac{3\sqrt{3}E_m I_m}{\pi\Omega}. \tag{4.97}$$

The EM torque generated by the optimal commutation angle is shown in Figure 4.95. The idea of using constant current in the energized state comes from the drive method for trapezoidal back-EMF motors. However, the spindle motors used in HDD nowadays have sinusoidal back-EMF. If a constant current is used for these motors, the generated torque will contain ripples even for optimal commutation angle. These ripples are caused by the drive mode itself and will be called as intrinsic torque ripple for the CC-BLDC drive mode.

Let us define the following objective function to assess the amount of ripple in the produced torque,

$$O(\alpha) = \frac{1}{\bar{T}_{em}^2(\alpha)} \int_0^{2\pi} [T_{em}(\theta,\alpha) - \bar{T}_{em}(\alpha)]^2 d\theta. \tag{4.98}$$

Then the value of α that minimizes this objective function is the optimal commutation angle. It can be easily shown that the minimum occurs at $\alpha = \pi/6$ and,

$$\min O(\alpha) = O\left(\frac{\pi}{6}\right) = \frac{\pi(72 - 72\sqrt{3} + 3\sqrt{3}\pi + 4\pi^2)}{108}. \tag{4.99}$$

In other words, while using CC-BLDC mode, $\alpha = \pi/6$ is also the optimal commutation angle from the point of minimizing the torque ripple. Moreover, the equation 4.48 supports the notion introduced earlier in section 4.2.3 that torque is linearly proportional to the drive current. Therefore, the EM torque of the motor can be controlled by changing the drive current.

4.4.4.3 Optimal Commutation Angle for CV-BLDC Mode

In the CV-BLDC drive mode, the voltage difference between motor terminals and ground can be expressed as,

$$u_A(\theta) = \begin{cases} U_{dc} & , \quad \alpha \leq \theta < \alpha + 2\pi/3 \\ 0 & , \quad \alpha + \pi \leq \theta < \alpha + 5\pi/3 \\ e_a(\theta) + V_N & , \quad others \end{cases} \tag{4.100}$$

$$u_B(\theta) = u_A\left(\theta - \frac{2\pi}{3}\right) = \begin{cases} U_{dc} & , \quad \alpha - 2\pi/3 \leq \theta < \alpha \\ 0 & , \quad \alpha + \pi/3 \leq \theta < \alpha + \pi \\ e_b(\theta) + V_N & , \quad others \end{cases} \tag{4.101}$$

$$u_C(\theta) = u_A\left(\theta - \frac{4\pi}{3}\right) = \begin{cases} U_{dc} & , \quad \alpha - 4\pi \leq \theta < \alpha - 2\pi/3 \\ 0 & , \quad \alpha - \pi/3 \leq \theta < \alpha + pi/3 \\ e_c(\theta) + V_N & , \quad others \end{cases} \tag{4.102}$$

The currents in the phase winding can be expressed as

$$\begin{cases} i_A &= [u_A(\theta) - e_A(\theta) - V_N]/R_a \\ i_B &= [u_B(\theta) - e_B(\theta) - V_N]/R_a \\ i_C &= [u_C(\theta) - e_C(\theta) - V_N]/R_a \end{cases} \tag{4.103}$$

It is obvious from these equations that the commutation angle α certainly affects the drive current, and therefore, affects the EM torque generated in the motor, which is illustrated by an example in Figure 4.96 showing the torque produced for two different commutation angle, $\alpha = 20°$ and $\alpha = 50°$.

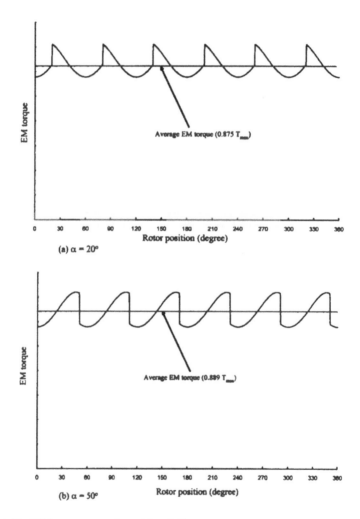

Figure 4.96: EM torque produced by different commutation angle in CV-BLDC drive mode.

In the CV-BLDC mode, the average EM torque is

$$\bar{T}_{em}(\alpha) = \int_0^{2\pi} T_{em}(\theta)d\theta$$

$$= \frac{6E_m}{2\pi\Omega R_a} \int_\alpha^{\alpha+\pi/3} \left[\sin(\theta) - \sin\left(\frac{2\pi}{3}\right)\right] \left[V_{dc} - E_m \sin(\theta) + E_m \sin\left(\theta - \frac{2\pi}{3}\right)\right] d\theta$$

$$= \frac{3E_m}{8\pi\Omega R_a} 4V_{dc}[3\cos(\alpha) + \sqrt{3}\sin(\alpha)] - E_m[4\pi + 3\sqrt{3}\cos(2\alpha) + 9\sin(2\alpha)]$$

$$\tag{4.104}$$

In CV-BLDC mode, any variation in the commutation angle α changes both the average torque and the armature current. Therefore, equation 4.104 cannot be used for determining analytically the optimal commutation angle. It is different from the case of CC-BLDC mode introduced in section 4.4.2. We must use a different method to find the optimal angle.

In the CV-BLDC mode, the drive voltage is constant during the energized state. Therefore, the average copper loss of the winding is related to the commutation angle, and it can be calculated as

$$P_{cu} = \frac{R_a}{2\pi} \int_0^{2\pi} [i_A^2(\theta) + i_B^2(\theta) + i_C^2(\theta)]d\theta \tag{4.105}$$

$$= \frac{3R_a}{\pi} \int_0^{\alpha+\pi/3} \left[V_{dc} - E_m \sin(\theta) + E_m \sin\left(\theta - \frac{2\pi}{3}\right)\right]^2 d\theta$$

$$= \frac{3R_a}{\pi} \frac{\pi V_{dc}^2}{3} - 3E_m V_{dc} \cos(\alpha) - \sqrt{3}E_m V_{dc} \sin(\alpha) +$$

$$\frac{E_m^2}{8}[4\pi + 3\sqrt{3}\cos(2\alpha) + 9\sin(2\alpha)].$$

Now we can define the ratio between the average EM torque and the copper loss as the objective function, i.e.,

$$O(\alpha) = \frac{\bar{T}_{em}(\alpha)}{P_{cu}(\alpha)}. \tag{4.106}$$

The value of α that maximizes this objective function is the optimal commutation angle. It can be found by equating the derivative of $O(\alpha)$ to zero and then solving for α.

$$O'(\alpha) = \frac{dO(\alpha)}{d\alpha} = 0. \tag{4.107}$$

By solving this equation, we get,

$$O'\left(\frac{\pi}{6}\right) = 0. \tag{4.108}$$

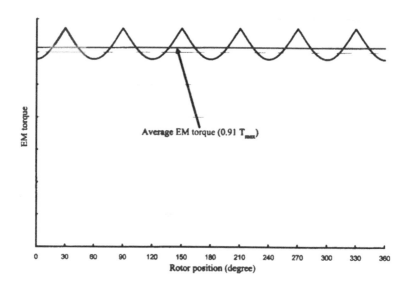

Figure 4.97: EM torque produced by optimal commutation angle with constant voltage drive mode.

And therefore,
$$O\left(\frac{\pi}{6}\right) = \max\left[O(\alpha)\right] =$$

$$\frac{\pi[-18(3E_m - 4V_{dc})^2 + 4\pi^2(3E_m^2 + 4V_{dc}^2) + \sqrt{3}\pi(9E_m^2 - 80E_m V_{dc} + 24V_{dc}^2)]}{18E_m^2[2\pi E_m + sqrt3(3E_m - 4V_{dc})]^2}.$$
$$(4.109)$$

Comparing this result with that of equation 4.96, we conclude that the optimal commutation angle is same for both the CV-BLDC mode and the CC-BLDC mode. The average torque with this optimal commutation angle is,

$$\bar{T}_{em}\left(\frac{\pi}{6}\right) = \frac{3E_m}{8\pi\Omega R_a}[8\sqrt{3}V_{dc} - E_m(6\sqrt{3} + 4\pi)].\qquad(4.110)$$

It is observed from the equations 4.104 and 4.110 that the EM torque of the motor driven by CV-BLDC mode is linearly proportional to the drive voltage. This performance is similar to the DC motors, i.e., changing the drive voltage can change the motor torque and the motor speed [52].

Typical optimal torque and current waveforms obtained using optimal commutation angle with CV-BLDC drive mode are shown in Figure 4.97 and Figure 4.98, respectively. It can be proven that the torque ripple of CV-BLDC is minimum at the optimal angle. However, the torque ripples still exist even if the optimal commutation angle is used. These ripples will be called as intrinsic torque ripple of the CV-BLDC drive mode.

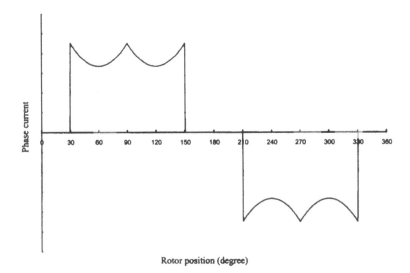

Figure 4.98: The A-phase current waveform produced by optimal commutation angle with constant voltage drive mode.

4.4.5   Acoustic Noise in Spindle Motor Driven in BLDC Mode

Many factors can induce acoustic noise in spindle motor operation. The bearing is the major source of acoustic noise in the spindle motors using ball-bearing. If fluid dynamic bearings are used, this source of noise is eliminated and the acoustic performance of the spindle and, therefore, of the HDD is greatly reduced. This issue was discussed in section 4.3. Since the noise contributed by bearings is negligible in FDB spindles, the noise caused by the electromagnetic sources becomes evident. The magnetic sources of the acoustic noise can be categorized as

1. the deformation of motor parts caused by the radial field [231],

2. unbalanced magnetic pull (UMP) [16], [17], and

3. torque ripples [120], [228].

For PMSM with surfaced PM ring, the airgap magnetic field is almost independent of the drive current (refer to section 4.3.9). Therefore, dependence of the acoustic noise caused by the deformation of motor parts on the drive current is insignificant. It is not difficult to distinguish this kind of noise during experiments. The acoustic noise caused by the deformation of spindle motor parts is very weak [131].

As explained earlier in section 4.3.9, both intrinsic and extrinsic UMPs of the spindle motor depend on the EM structure of the motor, but they are not related with drive mode used. In other words, the acoustic noise of the spindle motor caused by the UMP is independent to the drive mode.

Cogging torque cannot produce effective drive torque, but causes torque ripples in the motor operation. It was discussed in section 4.2.4 that cogging torque is caused by the EM structure of the spindle motor. As the order of the cogging torque is much higher than the fundamental torque, and the position signals obtained with sensorless method are limited, it is difficult to use drive current to create an opposite torque to compensate for the cogging torque in the sensorless BLDC driving. Therefore, the cogging torque can only be minimized by proper design of the motor. This is an important requirement for the design of spindle motors for application in HDD.

As the drive mode can affect the EM torque generated in the motor operation, it can certainly affect the acoustic noise generated in the motor operation.

4.4.5.1 Effects of EM Torque Ripple on Acoustic Noise

It was explained in section 4.4.4 that the operation torque in BLDC mode consists of intrinsic torque ripples which is caused by the BLDC mode itself. In using drive circuit to drive the motor, current commutations induce 'jumps' in the current waveform (illustrated in Figure 4.99). As a result, the current waveform becomes rich in harmonic components. These harmonics generates torque ripples and contribute to acoustic noise in the motor operation. This kind of torque ripples is known as extrinsic BLDC torque ripples. Now the question is, caused in the EM sources causing the acoustic noise, what are the roles of the intrinsic and extrinsic torque ripples?

A special CV-BLDC drive mode, presented in [130], can eliminate the intrinsic torque ripple. However, the measurements of acoustic noise in spindle motors show that, eliminating the intrinsic torque ripple has very little effect on the level of acoustic noise, as shown in Figure 4.100 and Figure 4.101.

When the spindle motor is driven using sinusoidal current, i.e., the PMSM drive mode, both the intrinsic and the extrinsic torque ripples are eliminated. Acoustic measurements on spindle motors using this drive mode show significant reduction in noise [131].

Figure 4.102 and Figure 4.103 show the experimental results from the test on acoustic noise in a FDB spindle motor driven by normal CV-BLDC drive and PMSM drive, respectively. It is obvious from these results that the PMSM drive mode can reduce the acoustic noise level at different speeds of operation. Since the PMSM drive mode induces neither the intrinsic torque ripple nor the extrinsic torque ripple, and since the intrinsic torque ripple contributes very little to the acoustic noise, it can be concluded that the extrinsic torque ripple is the major EM source of acoustic noise. Therefore, in order to reduce acoustic noise, the extrinsic torque ripples caused by sudden jumps in current waveform must be kept low, i.e., the current jumps in commutation must be small.

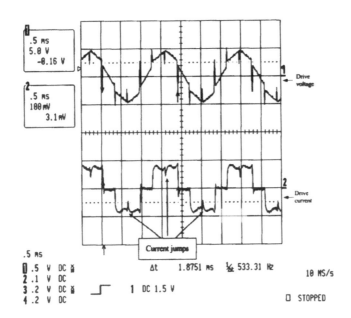

Figure 4.99: EM torque produced by optimal commutation angle with constant voltage drive mode.

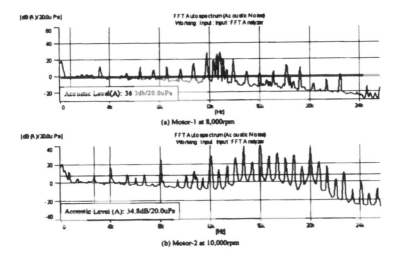

Figure 4.100: Spectra of acoustic noise of two FDB spindle motors driven by CV-BLDC mode with intrinsic torque ripple.

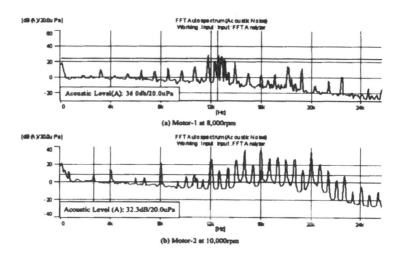

(a) Motor-1 at 8,000rpm

(b) Motor-2 at 10,000rpm

Figure 4.101: Spectra of acoustic noise of two FDB spindle motors driven by CV-BLDC mode without intrinsic torque ripple.

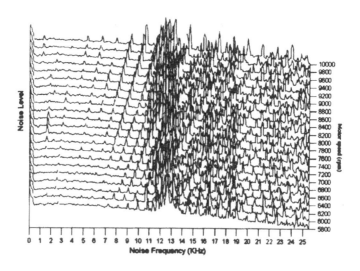

Figure 4.102: Spectra of acoustic noise of FDB spindle motor using CV-BLDC mode.

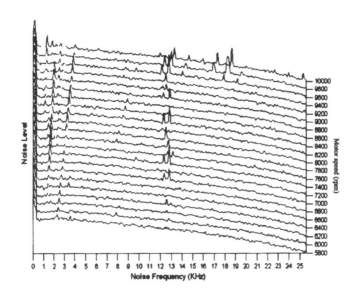

Figure 4.103: Spectra of acoustic noise of FDB spindle motor using PMSM mode.

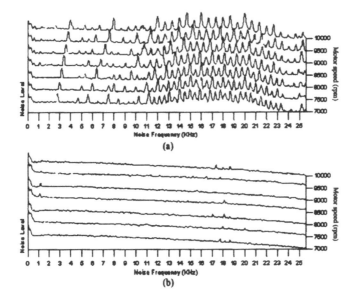

Figure 4.104: Spectra of acoustic noise of ADB spindle motor - (a) CV-BLDC drive mode, (b) PMSM drive mode.

Figure 4.104 shows experimental results of the acoustic noise tests performed on an ADB spindle motor driven by CV-BLDC drive mode and PMSM drive mode. Comparing these results with those of FDB spindle motors, the effects of the current-jumping can be disclosed more clearly. These results show that reasonable drive mode is more important for reduction of acoustic noise in the ADB spindle motors.

Chapter 5

Servo Track Writer

5.1 Introduction

Servo track writing (STW) is the process that defines tracks and sectors on the disk platters by creating the servo patterns so that the HDD servo system can later use them to identify tracks and sectors, and to measure the relative position of the read head relative to the center of the track. Accuracy and precision required in the HDD servomechanism have been emphasized, and the associated challenges have been explained in chapter 2 and chapter 3. Performance of the HDD servomechanism is affected by the precision at which the position feedback signal can be generated. The STW process, therefore, plays an important role in making the continuously growing trend of track density a reality.

The HDD servomechanism generates position feedback using the servo patterns as reference, which are written on the disks when no reference is present on the disks or inside the HDD. There are several mechanisms used to provide the reference while servo track writing is performed. Each of these methods has its own pros and cons in respect to factors such as cost, throughput, reliability etc. These methods will be discussed in this chapter. For the time being, we refer to servo track writing as a key stage in the process of manufacturing of HDD that employs precision mechatronics and control. Although there exists servo writing method that makes use of the mechanics available inside the HDD, the common practice so far is to use an ultra-precision equipment, known as *Servo Track Writer* (STW) or *Servo Writer*.

There are two critical control problems in the process of servo writing with desired accuracy:

1. All the patterns required to define the tracks and sectors must be placed in a concentric fashion. The STWs that are commonly in use controls the position of the write head with respect to the redial position on the

disk by controlling the position of the actuator arm with the help of an external mechanical push-pin or optical push-pin. Any disturbance and eccentricity present during this process will appear as writen-in RRO for the head positioning servomechanism of HDD. This written-in RRO increases TMR that must be compensated for by the servomechanism of an operational HDD. Naturally, the requirements on the accuracy in positioning the write heads is more stringent in the STW than in the HDD servo system. The magnitude of the written-in RRO is a yardstick used to measure the performance of the STW in creating servo patterns.

2. The servo sectors of any track must be precisely aligned with the servo sectors of adjacent tracks. In an STW, this is done with the help of a clock track written prior to the writing of servo tracks. A separate head is used to sense the transitions in the clock track, and these transitions are used as the reference timing marks for writing the transitions in the servo patterns. The misalignment between the servo patterns and timing information on the clock track can be measured by *timing jitter*. Excessive timing jitter causes distortion in read back waveform that the recording channel must compensate.

Figure 5.1 shows the functional block diagram of a servo track writer. The loop consisting of the optical position sensor and/or PES demodulation controls the position of the write head so that concentric tracks can be created, one track at a time. Unevenness of the track is exaggerated in this figure to illustrate that the proximity with which the tracks can be created depends on the vibrations in head-disk assembly. The clock signal generated from the transitions in a clock track controls the tangential position (along the track) of the transitions in servo patterns. A servo sector consists of a series of magnetic transitions. Since these patterns in one track must be aligned with those of the adjacent tracks, controlling their tangential positions precisely is an important requirement for servo writing. The quality of the servo-written tracks is assessed by measuring both the written-in RRO and the timing jitter.

The mechatronics and control system used in STW must be of ultra-high precision which makes the equipment expensive. As explained later in this chapter, the conventional servo writing process needs openings in the drive enclosure to make the heads, media and actuator of the HDD accessible to the mechanics of STW. It makes it necessary to carry out the process of servo writing in a very clean environment so that the space inside the HDD does not get contaminated with particles that can damage the head and media. Need for good clean room facility further increases the cost of manufacturing. The throughput of this process is an important factor to be considered.

Let

- ω be the RPM of the disks during the STW process,

- T be the desired track density (TPI) of the drive to be servo written,

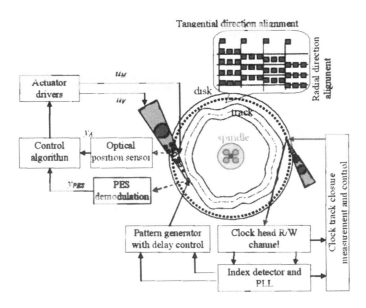

Figure 5.1: Block diagram of a servo track writer.

- L inch be range of radius of the disk over which servo tracks are to be created,

- s be the number of steps involved in writing one servo track with each step involving one rotation of the disk, and

- t_s be the seek-settle time (seconds) taken by the position control loop of the STW to move the head from one radial location to the adjacent one.

Then the time t required to servo write one disk is

$$t = LT(60/\omega + t_s)s. \tag{5.1}$$

For example, for a 7,200 RPM, 100 kTPI 3.5" HDD, we have around 1.5 inch radial length to be written with servo patterns. If the seek-settle time is $t_s = 2$ ms, and if creating one track requires to write two half-tracks, i.e., $s = 2$, then the time taken to complete servo writing this drive is, $t = 1.5 \times (60/7200 + 0.002) \times 2 = 3100$ seconds $= 51.66$ minutes. For this given example, seek-settle in positioning the write head to a new location takes up 10 minutes for the entire STW process, while 41.6 minutes is taken up by disk rotation without any involvement of head movement. It is obvious that use of faster actuators and better seek-settle control, higher spindle speed and less number of tracks all help to reduce servo writing time and hence increase STW throughput.

Because of the continuous growth in areal density (and hence the TPI) and desire for reduced manufacturing cost to survive in the highly competitive market, there has been a focused interest over the years in improving the precision of writing by STW, its throughput, and less usage of cleanroom space [155], [21], [191], [198], [222].

This chapter presents briefly the principles of operation and other important aspects of few schemes of servo track writing, namely,

1. the scheme with external positioner using a mechanical push pin,

2. the scheme using optical measurement based positioning system,

3. self servowriting, and

4. multidisk servo writing and its variants.

The first two schemes are HDA-based (Head Disk Assembly) servo writing, i.e., they use heads and actuators of the HDD to create the servo patterns on its disks. But they use external mechanism to control both radial and tangential positions of written patterns. The self servo writing scheme also uses the HDD's heads and actuator, but no external mechanism is required. This scheme writes one track and use the signal from this track as the reference for creating an adjacent track. Multidisk servo writing uses a dedicated STW to write servo patterns on a stack of disks. These servo written disks are later assembled in different HDDs. Besides these 4 major schemes, we also introduce briefly the process of creating servo patterns using printing. Through these discussions, we explain how the position and timing accuracies are controlled to achieve the desirable servo writing quality at an acceptable clean room throughput.

5.2 HDA Servowriting

The HDA servowriting employs the components of the assembled HDA to create servo patterns on the disks with the help of external sensors for radial and tangential positions, and the control loops. The external sensors are mounted on vibration-free platform using rigid fixtures. The patterns are magnetically written on the disk using the heads of the HDA, and the disks are spun by the HDA's own spindle motor. However, the position of the actuator arm is identified and displacement is measured using external mechanism - either a mechanical fixture called push-pin or an optical push-pin. The location of the pattern along the track is determined by another mechanism external to the HDA.

5.2.1 Control of Tangential Position using Clock Head

In order to align the servo patterns on different tracks, the servo patterns must be placed at appropriate locations along the track and, therefore, the process requires precise reference along the disk's circumference, i.e., the reference for tangential position. This reference is created by first writing a *clock track*, which contains pre-defined number of regularly spaced transitions. Writing of the clock track and reading its transitions later are done with a separate head other than the heads of the HDA. This head, known as the *clock head*, is external to the HDA and is inserted inside at the time of STW. Creation of the clock track involves few steps:

1. decide the number of clock cycles required per revolution,

2. write the clock track with precisely the desired number of transitions in it, and

3. verify the quality of the clock track.

Since a separate head is used for the clock track, the clock head signal is available throughout the entire process of writing servo patterns. A *phase lock loop* (PLL) circuit takes the signal from the clock head as its reference and the VCO (*Voltage Controlled Oscillator*) of the PLL produces a clock signal synchronized to the clock track. The clock track also include an INDEX pattern that defines the beginning of the track. Using the INDEX mark from the clock track and the synchronized clock signal from the PLL, a timing generator circuit produces reference for every sector and the fields within a sector. These references and the clock signal are then used to trigger the process of writing servo patterns on different tracks.

Calculation of the Number of Clock Cycles per Revolution

Care must be taken in deciding the number of cycles to be written on the clock track. If the disk rotates at N_{rpm} RPM and if the frequency of the patterns to be written is f_0, then a nominal, integer number of clock cycles (N) can be determined by $N = f_0 \times \frac{N_{rpm}}{60} = 60 f_0 / N_{rpm}$.

The rotational speed varies in any practical system. To accommodate such variations in speed (e.g.: $\sigma = 0.01\%$), the clock track is written in few attempts. Every time after writing it, quality of the clock track is verified. If the written clock track does not meet the specifications, the process is repeated.

Following the notation of [21], let Δ second be the interval between sampling the servo sectors in HDD, N_{rpm} be the spindle speed in RPM, f_{clock} be the clock track frequency, and the integer m_{sector} be the number of sectors per revolution. Then we have,

$$\frac{1}{\Delta} = \frac{m_{sector} N_{rpm}}{60},$$

which determines how fast the PES will be sampled by the HDD head positioning servomechanism. This parameter affects the performance of the head positioning servo controller.

Since the number of sectors is divisible by the motor pole number and there must be integer number n_{sector} of clock cycles for each servo sector, N must be integer multiple of $m_{sectors}$. The various numbers in the above equation must be adjusted to determine an appropriate N.

Writing and Verification of the Clock Track

An INDEX mark, a special unique pattern, is written first prior to writing the pre-programmed patterns for the clock track transitions. This unique pattern is used by the circuit to recognize the beginning of the track. The INDEX pattern, after it has been written, is sensed by the read head after one revolution of the disk. As soon as the INDEX is detected, the write electronics is triggered to start writing the transition patterns for the clock track. The writing of the transitions is done using a free running clock signal, and is stopped as soon as the INDEX mark is detected again.

It should be noted that the number of cycles written on the clock track (N) depends among many factors on the spindle speed and period of the free running clock. Any variation in either the spindle speed or the frequency of the clock causes the number of clock cycles written be different from pre-calculated N. Such variation within the span of one revolution makes the transitions in the clock track to be unequally spaced, which is not desired. It is also possible that the gap between the clock transition writen last and the first transition of the INDEX does not satisfy the specifications. These cases are known as imperfect *clock closure*, and for both cases, the clock track must be re-written. In case of re-writing the clock track, the speed of the spindle motor is dithered. The process is repeated until a good clock track is written.

Once the clock track is written, a PLL is allowed to lock to the signal from the clock track to produce a reference clock signal that is synchronized to the tangential positions of the disk. The synchronized clock output of the PLL divides the time of one revolution into N equal time slots. This clock signal can also be used to define the boundaries of the sectors.

Since the clock tracks written offer a more accurate tangential position signal than the spindle motor back EMF zero-crossing, the clock signal can be used as speed feedback for accurate speed control during the process of servo writing. For example, a few clock counts per revolution can be changed to achieve the spindle speed dithering.

During the servo writing process, the clock signals are written by inserting a thin film head into the HDD enclosure. This insertion is performed by a mechanical unit of the clock head module of STW. After the process of servo writing is over, the thin film head is extracted. Thus, there is a need for an opening in the drive enclosure to insert clock head. So the servo writing process must be carried out in a clean room environment to minimize the harmful effects caused by contaminations of the environment inside drive enclosure. The opening is covered with a seal at the end of the servo writing, before the drive is taken out of the clean room.

Writing of Servo Pattern

The patterns of transitions defining different fields of the servo sectors are stored in a shift register of N bytes, from where they are shifted out serially to the write circuit. The clock signal generated by the PLL locked to the clock track is used to shift the bits out of the shift register. The output of the shift register, being 0 or 1, is written as the servo pattern at the desirable tangential position determined by the clock count. The INDEX mark in the clock track is also used to synchronize the pattern generator. If the patterns for all servo sectors are the same then the required size of the shift register is shortened as the same pattern can be repeated.

5.2.2 Control of Radial Position using Mechanical Push-pin

Another very important task in the servo writing process is to ensure accurate and precise control of the radial position of the write head in compliance to the requirements of track density or TPI.

The traditional method of servo writing uses an external mechanical positioner whose position is controlled with ultra-high precision. The VCM of the HDA is biased with a DC current forcing the actuator arm to remain firmly in contact with a part of the mechanical positioner known as the *push-pin*. Since the actuator arm of the VCM is in contact with the mechanical positioner, controlling the position of the mechanical positioner is equivalent to controlling the position of the VCM actuator's arm. The external mechanical positioner provides an accurate radial reference position for servo writing process. The external positioner and the pushpin pushes the VCM and therefore the write head to the desired radial positions. Once the head is positioned at the desired location, the servo pattern shift register, triggered by the clock signal, sends out pattern to be written.

The push-pin, being part of a positioning system external to the HDD, must be inserted into the enclosure of the HDD so that the HDA's actuator can be biased to force the arm firmly against the push-pin. The insertion of the push-pin can be either from the side of the HDD enclosure or from the top of the enclosure.

A rotary actuator with a grating as position sensor can be used as the external positioner. A feedback controller is used to control with ultra-precision the position of the actuator of external positioner. One such device available in the market is MirocE micro positioner PA 4046 G whose resolution can be as high as 4.68 nanoradians with an accuracy of 35 nanoradians rms*.

It is possible to achieve a servo bandwidth greater than 1 kHz for the position control loop using the external positioner, which is a VCM actuator with optical feedback. However, when the push-pin pushes the actuator of the HDD,

---

*See webpage at http://www.microesys.com/

the total moving mass is increased. Besides, the contact between the two is not perfectly rigid and it causes significant resonances in the frequency response of the overall system. This resonance at around 800 Hz limits the achievable servo bandwidth to around 300 Hz [198]. Lower servo bandwidth means reduction in positioning accuracy and therefore increased written-in runout. It also results in longer seek-settle time and thus reduces the throughput of the servo track writing process.

This positioning system is external to the HDD enclosure, and is usually located at the bottom of the nest tray of STW which rigidly clamps the HDD enclosure. Insertion of the mechanical push-pin inside the enclosure to make it in contact with the HDA's actuator makes it necessary to leave an openning in the device enclosure. Thus, it is essential to perform the servo track writing using the above mentioned method inside a clean room so as to minimize any contamination to the disk platters.

The HDD and the external positioner should be securely and rigidly clamped in a nest or tray, typically made of marble, to minimize the vibrations between them. Furthermore, the nest or tray is placed on a vibration isolator which absorbs external vibration or shock and prevents them from being transmitted to the HDA and pusher during the servo writing process. However, using vibration isolator or using heavy mass as the platform can not eliminate the vibration contributed by the rotation of spindle motor and disks, vibration caused by the movement of the arm, and vibration of the suspension and slider excited by the air circulating at very high speed. These vibrations are still present at the time of servo track writing and they induce written-in errors.

5.2.3 Control of Radial Position using Optical Push-pin

The arm of the external VCM used as the mechanical push-pin contributes large share to the total moving mass during the servo track writing. Such heavy mass is not desirable as it increases response time and therefore reduces throughput. Various approaches have been tried to get rid of the external and heavy mechanical actuator. One of these methods uses a laser beam shining from the top on the E-block of the VCM actuator which carries a fine grating sheet scale [198]. The position of the actuator arm can be precisely determined using the laser beam and the grating sheet scale, and therefore, be controlled using closed loop feedback system. In this case, the plant to be controlled by the servomechanism is only the VCM actuator of the HDD and not any external mechanical structure. The bandwidth achievable is quite high and similar to that of the HDD servomechanism.

As reported in [198], the sheet scale pasted on the VCM actuator can be of printed diffraction grating of 1 μm pitch which produces an optical source signal of 0.25 μm. When using a grating of 10 μm pitch with an optical source of 5 μm pitch, a sensing resolution of 0.5 nm can be achieved. Moreover, this system is less sensitive to vertical vibration of the actuator arm compared to the case where the position of the arm is sensed from the side of the actuator. Similar to the scheme with mechanical push-pin, any vibration due to the spindle motor and disk rotation, arm and suspension excited by the disk and air rotation are still present.

Using the grating on the actuator arm solves the problem with heavy mass of the external mechanical positioner. This, however, does not solve the problem of keeping opennings in the HDA enclosure. Although there is no need for a physical contact between the external positioner and the internal actuator, it still requires a clock head to be inserted for creating tangential reference. One possible solution is to apply such grating on the spindle motor. With two gratings, one on the actuator arm and one on the spindle, there is no need for inserting an external push-pin or external clock head inside the HDA's enclosure, and the process of STW can be performed outside clean room environment. This will reduce the cost of STW process significantly, provided the gratings are available at cheap price. Current cost of these diffraction gratings makes this approach not suitable for commercial application. It should be noted that each HDD is required to be provided with a pair of diffraction gratings only for the sake of STW process and these gratings are never used in the life of the HDD.

5.3 Media STW

Two methods described earlier, the conventional method with mechanical push-pin and with optical push-pin, writes the servo patterns on the disks using the write heads of the HDD. These methods, also known as HDA servowriting, are carried out after the components have been assembled inside the HDD enclosure. The *Media STW*, on the other hand, creates the servo pattern on the disks prior to the assembly of the HDD. High precision, high rotating speed spindle motor and low vibration mechanics are used for writing multiple disks simultaneously at higher accuracy. Such servo writers which are independent of HDA and writes servo at recording media level were developed [155] [150] [191]. These equipments are known by different names such as *multidisk servo track writer*, *media level servo track writer* (MSTW), or *bulk servo writer*.

The servo writing process consists of two stages. The first stage writes the servo patterns on as many as ten disks or more simultaneously in a clean room using the MSTW machine. These servo written disks are later assembled in the HDD. If each HDD contains only two disks, an MSTW writing 10 disks simultaneously effectively writes servo patterns for five HDDs in one run. Writing multiple disks at a higher spindle speed means shorter writing time per disk leading to better throughput of the servo writing process. The MSTW uses active air bearing spindle motor with very low vibration, low vibration actuator mechanism, and high speed electronics which can not be used in an HDD due to cost constraints. Special techniques such as actuator arm with low vibration [154], air shroud around the disks [42], very well balanced disk-spindle pack by centering the disks, biasing the disks, or active control methods (see [87]-[86] and the references therein) can be used in the MSTW to enhance its performance by controlling the mechanical vibration, timing error, and position sensing error to a relatively lower level compared with those of HDA servo writing.

Once the patterns are written on disks using MSTW, these disks can be assembled in an HDD. An alternative method suggests assembling the HDD with one disk servowritten by MSTW and few other virgin disks with no servo patterns. Then the servo patterns from the master disk are replicated on other disk surfaces. In both cases, the patterns written on the master disk can be either the complete servo pattern or intermitently written servo patterns. For the case of complete servo pattern, all the servo tracks are written by the MSTW. On the other hand, the intermittent servo patterns can be of different types such as *spiral pattern* [190] or partial servo track writing [22]. In the first of these two cases, the write head is moved precisely from the outer edge of the disk towards the inner edge while the sector patterns are being written. This creates a pattern of sector marks spiralling from outer diameter to the inner diameter. For partial writing, few servo tracks are written at different radii of the disk leaving the reamining sections blank.

The processes of creating the intermitent servo patterns or servo tracks can be carried out in the HDA, instead of media level writing. During the second stage, the complete servo pattern is written on all disk surfaces by referring to the initial patterns and bank writing the remaining disks [94] or using self servowriting. Which process can be used in the second stage is determined by the type of initial pattern used.

When a disk with pre-written servo patterns is assembled in the HDD, the eccentricity of the disk and therefore of the patterns with reference to the centre of rotation of the spindle must be tightly controlled. Misalignment between the center of disk and the center of rotation introduces RRO. Even the force with which the disks are clamped causes deformation in the disk which increases the RRO further. These problems can be alleviated using one of the following methods:

1. Mark the disks during the MSTW process so that these marks can be used later during the assembly of the disks in the HDA to make them properly aligned,

2. Write the final servo pattern or user data by not following the eccentric master servo track centers but a virtual data track centers such as the zero acceleration path (ZAP) reported in [33],

3. Leave the ID region unwritten in the process of MSTW. Deformation of disk due to clamping is more severe in the inner region. The blank

ID region of the disk can be servo-written later using self servowriting technique. This technique is discussed later in section 5.4 of this chapter.

These methods can help to reduce RRO in the head positioning servomechanism of the assembled HDD, but RRO can not be completely eliminated if disks with pre-written patterns are assembled in an HDA. The RRO that remains after all these precautionary and corrective measures must be corrected by the head positioning servomechanism of HDD.

Application of air bearing spindle motors reduces the nonrepeatable radial error, which can be in the range of about 0.5 μin when carrying a load of 12 disks of 95 mm diameter [†]. At rotational speed of above 10,000 RPM, which is desired for fast servo writing, the vibration due to disk NRRO, suspension, as well as slider-disk interaction is no longer negligible. Such radial movements displace the center of a servo-written track, while the vibrations from slider and suspension displaces the write head causing distortions in the servo pattern and, therefore, increases the written-in RRO and track encroachment. Incremental reduction of vibration from various sources via improved electromagnetic design of the spindle motor and using bearings with higher stiffness and other components with low vibration is still necessary to improve the STW quality with increasing TPIs. If a self servowriting loop with capability of writing servo patterns using the mechanics and head/medium of the HDD is added, then it is possible to suppress vibrations via active control. This measure improves the quality of servowriting.

5.4 Self Servowriting

Conventional methods of servo track writing need external devices for positioning of the write heads of the HDD and writes the servo patterns using these heads. They face increasingly difficult challenges as the move towards ever increasing TPI continues in the industry. These challenges come from different factors. Firstly, conventional servo writing requires few revolutions of the spindle to create one servo track and, therefore, the total time required to servo write an HDD increases proportionately with increasing number of tracks per surface. Because of this decrease in the throughput of individual STW more servo-writers are required to meet the production target. This, in turn, requires more floor space in clean room since these STWs are operated in such environment. Both MSTW and HDA level STW suffers from this problem. Secondly, for drives of smaller form factors, the jigs and fixtures of the STW becomes smaller. It is challenging to design the pushpin mechanism that is small and yet sufficiently stiff. Finally, if HDA servo-writing is to be used then the drives must be assembled with expensive components of better quality as this servo-writing makes use of product level components, i.e., the

---

[†]see for example http://www.seagullsolutions.net/

components of the HDD. Usually relatively low cost (compared to those available as the state of the art technology) spindle motors with various imbalance conditions are used in HDDs to keep the cost of the final product competitive.

The demand for higher throughput without increasing the production cost was the primary motivation for developing an alternative method of writing servo tracks called the *self-servo track writing* (SSTW) which can be performed in an area outside expensive clean room as this method does not require any opening on the HDD's enclosure. Since the technique uses the mechanics of the HDD itself, there is no need to design mechanical pushpin and the form factor of the HDD is not an issue anymore. Mechanical jigs and fixtures are still required to clamp the HDD rigidly on a vibration-free platform when servowriting is carried out.

5.4.1   Basic Concept

The self-servo track writing (SSTW) refers to a process of servowriting that uses as references the radial (for servo) and tangential (for timing) positions regenerated from information previously written on the medium and sensed by the MR head of the HDD itself [38], [221]. The functional block diagram is similar to that of Figure 5.1 except for the additional role of the data head as the clock head. It can servo-write an HDD without any external push-pin or an optical position sensor. The process of self-servowriting is shown in Figure 5.2 and it generally involves the following steps [222]:

Step 1: Write some tracks or at least one track called the *seed track(s)* (the ith STW track). One possible way of creating seed track is to write servo patterns after biasing the actuator of the HDD firmly against the crash-stop [43]. Or the seed tracks can be pre-written using MSTW on disks before the disks are assembled in the HDD [22].

Step 2: With the seed track written on disks, the head slider is moved away from the center of the seed track using the VCM actuator in a controlled way such that the read sensor of the MR head can still sense the transitions in the servo wedges of the seed track. The readback signal generated by scanning the transitions $j, j+1, \cdots$ of the seed track is used as feedback for the closed loop system controlling the MR head. While the position of the slider is controlled, the write head writes an auxiliary servo pattern or the final servo pattern on the $j+1, j+2, \cdots$ wedge positions to create the $i+1^{th}$ STW track.

Step 3: If $i+1$ is less than the number predefined according to the ID-OD span and desired TPI, then set the $i+1^{th}$ track as the new seed track, and go to Step 2. Else all the tracks have been written, and stop the process.

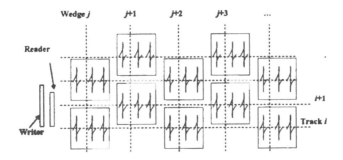

Figure 5.2: Illustration of SSTW process.

In the above mentioned process, the MR head reads previously written information so that the the the actuator's radial position can be controlled and the servo burst can be written at the next position. Therefore, the geometry of the read-write head is a crucial factor in the realization of this method. Its geometry should be such that for any radial position between the ID and the OD location, the read element can read at least one of the previously written tracks or timing information with reasonably good signal-to-noise ratio (SNR) when the write-head writes the next servo pattern or timing pattern or both. When the jth timing mark of the ith STW track is scanned by the read element placed off-track from the center of the track, it is used to determine the position of the $(j+1)^{th}$ timing mark of the $(i+1)^{th}$ STW track which muct be aligned with the $(j+1)^{th}$ timing mark of the ith STW track. To make this alignment possible, the error contributed by timing delays from the read and write electronics, and the physical separation between the read and the write elements must be precisely measured, calibrated, and controlled. In practice, the error cannot be reduced to zero. Moreover, the error in one track affects the position of the timing marks in the following track. If proper measure is not taken, the error continues to accumulate from the starting track till the end track. Such *clock error propagation* and *track error prorogation* must not be allowed to grow. Methods used to contain the propagation of these errors are explained next.

5.4.2 Track Propagation

Figure 5.3 shows a simplified block-diagram of a SSTW servo loop where $C(z)$ and $P(z)$ represent the transfer functions of the controller and the plant, respectively. The external signal $n(z)$ represents all torque disturbances, including D/A quantization noise, power amplifier noise, and any torque due to air-turbulence impinging upon the actuator, suspension and slider. Moreover, there are disturbances ($d(z)$) due to non-repeatable disk motions, and PES demodulation noise ($v(z)$) which includes electrical noise and A/D quantization

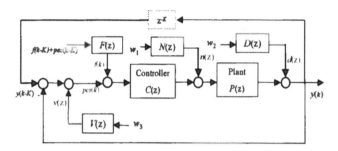

Figure 5.3: Noise and disturbances in a track propagation process.

noise. The position of the write head with respect to a perfectly circular track on the disk is $y(k)$ and the position error is $pes(k)$.

Let us denote the period of one revolution of spindle by T_p and the sampling interval by T_s. Let $K = T_p/T_s$, then $y(k - K)$ represents the track profile of the previous track. Similarly, $pes(k - K)$ represents the position error when writing with reference to the previous track. The read head follows on the track $y(k - K)$ which is the reference input for the SSTW servo system, i.e., $y(k)$ from one revolution becomes the reference for the next written track. From Figure 5.3, we see that

$$y(k) = \quad [\frac{CP}{1 + CP} + \frac{F}{1 + CP}]y(k - K) + \frac{1}{1 + CP}d(k) + \frac{P}{1 + CP}n(k)$$
$$+ \frac{CP}{1 + CP}v(k) - \frac{F}{1 + CP}d(k - K) - \frac{PF}{1 + CP}n(k - K). \quad (5.2)$$

The control objectives can be defined using the block diagram in Figure 5.4 as [45], which are,

a. to design a feedback controller $C(z)$ to achieve a low TMR, i.e., to minimize $\|\Phi_{yw}\|_2$, the H_2 norm of the transfer function from noise vector $w=[w_1, w_2 \ w_3]'$ to track profile y;

b. to design a feedforward compensator $F(z)$ to contain the error propagation, i.e., $\|\Phi_{y(k)/y(k-K)}\|_\infty < 1$, the H_∞ norm of the transfer function from $y(k - K)$ to $y(k)$ to be less than one;

While the H_2 optimal control is a standard one which has been discussed in Section 3.4, a simple solution to make the magnitude of the transfer function

$$\Phi_{y(k)/y(k-K)} = \frac{CP + F}{1 + CP}$$

less than unity is

$$F(z) = \Phi(1 + PC) - CP,$$

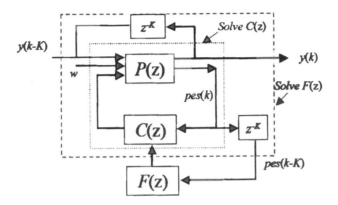

Figure 5.4: Block diagram for solving the control design problem.

where Φ is a weighting factor that can be chosen to be less than 1.

Instead of using the track profile $f(k - K)$ of the previous track for controlling the error through the use of $F(z)$, part of the $f(k - K)$ information can be used in its place for the same purpose [222]. For example, some repeatable components of $f(k - K)$ which represents the written-in runout signal can be used and gradually reduced in the writing process.

We can use the frequency domain properties of the PES signal to evaluate the performance of the self servowriting process. Figure 5.5 shows the frequency spectrum of PES in adjacent tracks for the case of self-servowriting without proper error containment algorithm used. There is a steady growth of the frequency component of PES at a frequency approximately 15 times the spindle frequency. It grows continuously until the process reaches a situation where the non-circularity written in the track becomes too large and the loop fails.

When the error containment algorithm is used, the error does not grow steadily as shown in the Figure 5.5. The use of the compensator $F(z)$ can effectively stop the error amplitude from growing and allows more tracks to be propagated using the self-servowriting process.

5.4.3   Clock Propagation

Accurate reference of tangential displacement is essential for writing the magnetic transitions of the new track aligned with the transitions of the previous track, i.e., the reference track. This is achieved with the help of a *phase lock loop* (PLL) that generates a clock signal precisely locked to the timing marks sensed from the reference track. The performance of the PLL is adversely affected by the noise entering the loop, and the generated clock does not remain perfectly in phase with the readback signal sensed from the reference track.

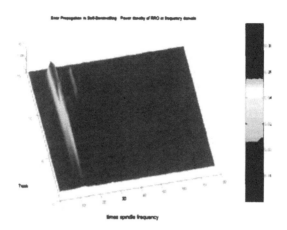

Figure 5.5: Frequency domain PES in adjacent tracks without error containment.

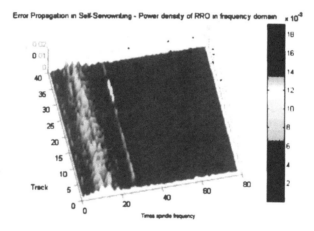

Figure 5.6: Frequency domain PES in adjacent tracks with error containment.

This affects the alignment of the transitions written on the new track with those of the reference track. The new track, which is not aligned with the reference track, is used as the reference for the next track. So the alignment of the transitions in the next track with those of the reference track is not expected. Success of self-servowriting depends on the magnitude of the track-to-track error as well as on the propagation of error from start to end.

Let $P(z)$ be the model representing PLL dynamics that is biased by a look up table and $C(z)$ be the PLL loop filter. Similar to the case of radial prorogation, we can define the noise and disturbance sources for the PLL. Let $v(z)$ be the time sensing (trigger) noise produced by a combination of head

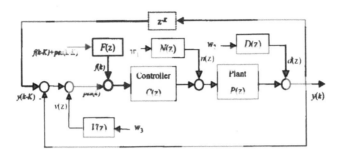

Figure 5.7: Noise and disturbances in a clock propagation process.

and electronic detection noise, $d(z)$ be the write noise which is the random difference between the position where a write is commanded and where it is actually written, $n(z)$ be the slow timing variation due to HDA geometry which depends on radial position, and electronics delay time drift $p(k)$ represents the delay at each step. A control strategy similar to that used for containment of radial error prorogation can be developed to stop the propagation of timing error.

It is reported in [171] that $v(z)$ is one of the dominant noise contributors for propagation of timing error. Use of a narrower di-bit generally reduces the trigger noise. $d(z)$ is due to a combination of head, media, and electronic noise. Approximately one quarter of the final alignment noise power in this system can be traced to this source. Similar to the track error propagation case, $F(z)$ in the timing loop can also be designed such that the timing error does not propagate. With suitably designed detector and reasonable system write noise, it is possible to achieve an alignment error with $\sigma \leq 1$.

5.4.4 Concluding Remarks

1. Self-servo track writing (SSTW) uses the amplitude of the read signal from a previously written track as a measurement for the distance of the head from the center of the previously written track. This is used as a feedback signal in a closed loop system that controls the position of the read/write head when a new track is created. Using such a control system structure, the disk and spindle vibrations are corrected by the servo loop. The transitions from the previous track are also used by a PLL to produce timing reference for placement of transitions in the new track.

2. SSTW radial direction step size needs to be calculated based on periodic calibration of geometrical error in the radial direction across the whole disk radius. Use of wrong step size may cause one track to erase the other tracks.

3. The step size should be calculated in the tangential direction also based on periodic calibration of geometrical error. Using wrong step size should be avoided, otherwise one track may erase other track's space.

4. In SSTW, we can use higher PES sampling frequency to achieve a more effective servo design. The write to read switching time, limited by switching off the inductive writing head, must be minimized because such delay reduces the sampling frequency.

5. Similar to HDD servo, missing trigger patterns must be detected. When the trigger pattern detector either triggers before the desired written trigger due to noise, or fails to trigger due to a defective or missing trigger pattern, the STW process must continue to robustly propagate and write servo and timing patterns. To achieve this, the detector can be enabled for only a short time (150 ns) before the trigger pattern is expected. This greatly limits noise and defect induced "extra" triggers. Alternatively, if a trigger pattern does not occur within a specified time after the expected time, a fake trigger is generated by the system hardware. Moreover, all trigger times can be compared in software against the expected time. If the trigger falls outside a specified duration the trigger is ignored for propagation purposes and all measurements and writing are referenced to the last valid trigger. The process has a specific threshold for consecutive invalid triggers which causes process termination. In practice such termination is very rare as the invalid trigger rate is typically less than 1e-5 [171].

To place narrower servo tracks closer without being limited by the spindle motor runout and head/disk assembly vibration, a hybrid STW (HSTW) system can be used which is a combination of the conventional STW and the SSTW.

5.5   A Laboratory-scale Example

The equipment used for servo track writing is a high precision mechatronics system with specification for error tolerance in positioning of the write-head in nanometer scale. An experimental setup, with the capability of servo-writing at a precision of 2.37 nm RMS is described in the following subsection. The setup is designed and developed in the Data Storage Institute, Singapore (DSI) [226]. This system is also capable of simultaneous servo-writing of multiple disks. It uses an active control loop for suppression of vibrations.

5.5.1   Configuration of the System

The hybrid servo track writing (HSTW) system is shown in Figure 5.8.

Figure 5.8: HSTW experimental setup.

In a typical multidisk servo track writer, a high speed active air bearing spindle is used where a pack of 10 to 20 disks can be easily mounted and removed. Since this setup is built to demonstrate the working principle and it is not meant for commercial use, a fluid bearing spindle motor is used instead of air bearing spindle for the simplicity of realization. It can carry up to five $2\frac{1}{2}$ inch disks. A *MicroE micro positioner*[‡] is used to rotate the actuator. Each arm of the actuator is embedded with a low voltage monolithic multi-layer piezo actuator chip PL033[§] of dimension $3mm \times 3mm \times 2mm$, as shown in Figure 5.9, so that these arms can be deflected and hence the head moved using the piezoelectric effect. With a driving voltage of 100 V, the PL033 PZT chip can generate 0.5 μm displacement. However, the mechanical amplification provided by the arm results in larger deflection of the slider. In the system developed at DSI, the displacement produced at the read/write head is about $\pm 3\mu m$ for an input driving voltage of 30V, indicating a factor of 20 mechanical amplification. After tuning and optimization of the design, such as using balanced arm design having the actuator arm's center of gravity aligned to the center of its rotation, the first minor and the first major resonances of the system are observed at 5.8 kHz and 7 kHz, respectively. These resonances limit the achievable servo bandwidth to about 2 kHz.

We note hat the configuration in Figure 5.9 is effectively a PZT "actuated arm" driving a commercial suspension that has a higher mass compared with that of the actuated suspension or slider discussed in section 3.6. To achieve higher servo bandwidth and better positioning accuracy, actuated suspension

[‡]See webpage at http://www.microesys.com/

[§]Product of Physik Instrumente (PI) GmbH & Co. KG, Karlsruhe/Palmbach, Germany, http://www.physikinstrumente.de/products/prdetail.php?secid=1-14

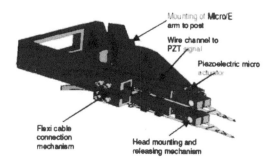

Figure 5.9: Actuator arm design with embedded PZT chip.

or actuated slider or actuated head is preferred to the "actuated arm". Moreover because of the additional height of the PZT element, it requires more spacing between disks and thus limits the access to only two disk surfaces instead of four surfaces in the design STW designed. Nevertheless, by using such fixture, the R/W head can move in wider range by controlling the PZT only without any rotation of the MicroE. Because of the structure of actuated arm used in this setup, the R/W head can be changed and the system can use state of the art R/W head that does not necessarily come with an actuated suspension commercially available.

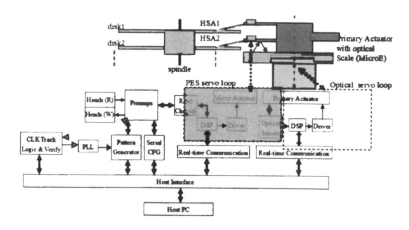

Figure 5.10: HSTW experimental system block diagram.

The block diagram of the HSTW system is shown in Figure 5.10. The gain of the clock preamp chip SR1581 can be set via programmable registers. One MR R/W head is attached to this amplifier to perform writing and reading of the clock track. Clock heads used in typical STW systems are wider than the MR heads used in this setup. Since the MR R/W head is also used as clock head, the clock track is propagated wider than the head's dimension so that the reader can sense the clock information reliably. The preamplifier is carried on a flex cable which provides write/read signal path. Signal tracks for differential read back signal are shielded with ground plane to reduce noise.

Two MR data heads are connected to the on board SR1581 preamp with programmable write current and MR bias current as well as head selection. There are four other terminals for connecting the piezoelectric micro actuators. The signal tracks for connecting PZT are separated from the R/W signals by a ground track.

Two on board SA5209 variable gain low noise radio frequency amplifiers serve as buffers for read back servo pattern and clock signals. The differential signals are transmitted to the R/W board via shielded coaxial cables. It also routes bi-directional serial configuration data to or from preamplifiers. On the R/W circuitry, two 32P4752 read channel IC's are used to recover clock pulse and servo burst. The servo bursts are digitized by AD7482 and then send to DSP module for servo control.

The pattern generation circuitry uses three Altera Flex 10KA CPLD devices. It has a PLL circuit to generate the write clock which is driven by the clock track read back signal. A clock track closure measurement circuit is also developed so that a desirable clock can be written after compensating the timing error in an initial trial clock track. The 4 MHz clock track signal is also stepped down to lower frequency for spindle motor speed control via the CPLD.

The pattern generator adaptor bridges the pattern generation board and the R/W electronics. It serves to extend the capability of the original pattern generation board. The Altera EPM9320 CPLD on the adaptor is programmed to receive configuration data from the pattern generation board and then sent them to preamplifiers, read channels on R/W board, spindle motor control unit and a delay generator accordingly. Registers for preamplifiers, read channels and motor control units can also be read by sending command via the pattern generator board.

5.5.2 Measurement and Reduction of Disk-Spindle Pack Imbalance

Since the imbalance of the disk-spindle pack generates radial force and thus excites vibrations in the spindle motor and disk, its value is typically controlled to be below certain amount in high speed HDDs and servo track writers.

To measure the disk-spindle pack imbalance, Figure 5.11 shows a disk spindle pack mounted on a flexure [86]. The disk-spindle pack eccentricity force drives the disk spindle pack to vibrate on the flexure bearing. A laser Doppler Vibrometer (LDV) is used to measure the vibration velocity signal whose amplitude is proportional to the eccentricity force, and hence the eccentricity. A bandpass filter whose center frequency is at the disk rotating speed can be used to filter out the noise in the amplitude noise.

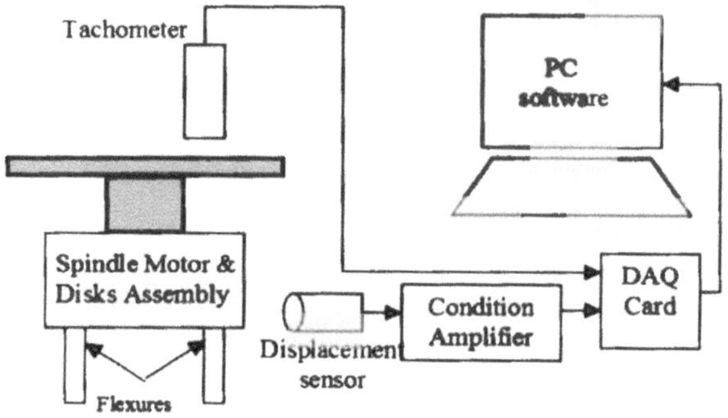

Figure 5.11: Schematic of an imbalance measurement machine with flexible bearing.

Let the output signal of the bandpass filter be represented by $V_f e^{j\delta}$, which means that it has an amplitude V_f and phase δ with reference to the motor index measured by a tachometer. To calibrate the amplitude and phase of the machine, first, the amplitude and phase of the disk-spindle pack imbalance is read, and its value is represented by $\vec{V}_1$. Next a known mass w gram is placed at 0 angle at radius R mm with respect to the motor index mark, the imbalance reading is represented by $\vec{V}_2$. The calibration vector

$$\vec{V}_3 = \vec{V}_2 - \vec{V}_1 = V_3 e^{j\gamma}$$

where V_3 and γ are the amplitude and angle of the calibration vector respectively.

The actual imbalance $\vec{I}_m$ will be

$$\vec{I}_m = V_f \frac{wR}{V_3} e^{\delta - \gamma}.$$

If we have the knowledge of the imbalance in disk-spindle pack, the imbalance can be corrected by adding or removing equivalent amount of mass. A spacer ring with 24 screws of known masses is used. By adjusting the screw's radial distance to the center of rotation, a fine change in center of gravity can be inflicted and hence the imbalance can be corrected. Interested readers may refer to [86], [87], [217] and the references therein for more discussions on the balancing techniques.

5.5.3   Control System

The control block diagram is shown in Figure 5.12. Let $P_1(s)$ be the MicroE arm model, $C_1(s)$ be the controller available in the MicroE, $K(s)$ be the model (can be assumed 1 for simplicity) of the optical sensing device, and n_1 be the noise in the sensing loop. The PZT actuator loop in the MicroE arm is modeled as $P_2(s)$, with its controller is represented as $C_2(s)$. Position of the R/W head is y_3. Let T_{n-1} be the $n-1^{th}$ track written on the disk, v_1 be the suspension and slider vibrations, v_2 be the vibrations in spindle and disk, y_{ref} be the track center of the $n-1$ track, y_4 be the measured difference between the track center of the $n-1$ track and the track to be written, and n_2 be the measurement noise contributed mainly by the R/W process. Two feed-forward signals, u_3 and u_4, are used where u_3 is the RRO measured from the previous track (i.e., T_{n-1}) with $C_2(s)$ cut open and u_4 is the filtered y_1.

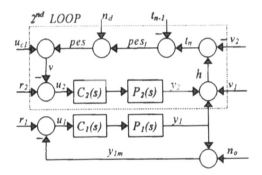

Figure 5.12: HSTW servo loop block diagram.

$C_2(s)$ is designed such that the PZT loop compensates for the vibrations between head and media, i.e., the head position $Y_2(s)$ follows $V_2(s) - V_1(s)$, to minimize the written-in error. The reading element is used as the sensor to measure the vibrations. We note that taking out T_{n-1} from the loop (by injecting u_3) relaxes the error propagation problem and reduces the written-in error similar to the self-servo track writing method. Referring to the Figure 5.12, we have

$$
\begin{aligned}
t_n &= h - v_2 \\
&= T_s(s)(r_2 - n_d) + S_2(s)(v_1 - v_2) \\
&\quad + S_2(s)T_1(s)(r_1 - n_0) + S_2(s)S_1(s)P_1(s)d_1,
\end{aligned}
\tag{5.3}
$$

where $T_1(s), T_2(s), S_1(s)$ and $S_2(s)$ are the complementary sensitivity functions and sensitivity functions for the optical servo loop and the second servo loop respectively. In this configuration, optical servo loop provides coarse movement of the head to the desired track and enables the second loop to control precisely the head with reference to the previous ideal track center by decoupling the previous written-in error T_{n-1} from PES for writing the current servo track. The PZT actuator can attenuate the PES induced by head and disk vibrations by reducing their effect via the feedback loop using the R/W head as the sensor as opposed to using the external optical sensor. Furthermore, the selection of controller $C_2(s)$ for vibration $V_2(s) - V_1(s)$ rejection and noise n_2 attenuation can follow the same way as in HDD servo design.

5.5.4   Test Results

In the experiment set up only two 1.27 mm thick glass disks were installed and they were spun at 5,400 RPM. After the servo writing process, the NRRO is measured with the microactuator feedback loop disabled, i.e., with only the MicroE optical loop. The measured NRRO is found to be at the level of $3\sigma = 17.9$ nm. This has been an improvement from the initial 21.5 nm when no air shroud around the disk was employed.

With an open loop cross-over frequency of 885 Hz when we close the servo loop in the servo writing process, vibrations below 800 Hz can be attenuated. Figure 5.13 shows the PES spectrum with a normal peak filter and with a phase lead peak filter. As shown in the figure, the PZT loop effectively rejected the vibrations at 650 Hz and below using the phase lead peak filter. The PES 3σ achieved is 6.4 nm. Assuming that a 10% track width is required of the 3σ, this STW configuration can easily support writing of 395 kTPI on the disks spun by fluid bearing motor. With better control design, actuator and sensor technologies, we can expect the low frequency vibration and also the peak at 3.8 kHz (contributing roughly 0.3 nm rms error to the PES) better attenuated, and achieve a 425 kTPI on such a platform.

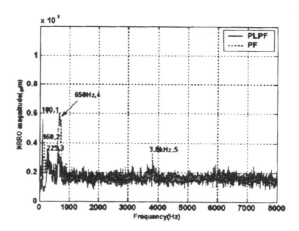

Figure 5.13: PES spectrum with normal peak filter and phase lead peak filter.

The example given above shows the feasibility of servo track writing with ultra high TPI. In this hybrid STW example, the optical position feedback loop determines the average track center whereas the previously written servo information determines where the next servo burst is laid. The control design philosophy for the PZT actuator loop follows that of the HDD servo control design. Although limited by the head width and the PZT actuator bandwidth, this setup is capable of achieving a PES 3σ of 6.4 nm. Use of bettwe sensor and actuator technologies can help to achieve better servo loop shaping and help to achieve servo track writing at even narrower track width despite the mechanical vibrations.

5.6 Printing the patterns

The processes discussed so far for creating servo tracks use either an HDA or a more precise equipment which is also built on the same principle used in HDD. However, since creating reference marks is the main objectives of servo writing, it would serve the same if these marks can be created by some other processes. Moreover, if these references are created by a batch process then the bottleneck of the servo writing stage can be greately reduced. Several methods have been explored to find a suitable, batch processing alternative methods for creating servo patterns. Two of these methods, nano-imprint and magnetic print, are elaborated next.

5.6.1 Nano-Imprint

The first of these nonconventional methods of creating servo reference is to use pre-embossed disks. In this approach, the disks are embossed with the servo patterns prior to applying the layer of magnetic coating on them. The first such disk was produced by Sony; one may refer to [215] and the references therein for more details. The required patterns of the servo sectors are embossed on the disk creating land-and-groove patterns. When such a disk is put in an HDD, the land-and-groove patterns cause head to disk separations to change and hence the amplitude of readback signal varies according to the embossed pattern. Such variations in signal amplitude can be used to encode track numbers as well as the servo bursts. In recent years, the *nanoimprint lithography* (NIL) technology makes it possible to stamp a rigid mold with land-and-groove patterns as small as 10 nm into a soft polymeric layer [35].

There are different methods of developing nano patterns into recordable bits. One of them has the following stages to be followed:

1. lithography of the substrate using the master designed for servo function,

2. sputter deposition and post process - magnetic layer is coated in this stage,

3. dc-erase the disk surface so that the entire disk surface is magnetized in the same polarization.

As an alternative, these steps can be performed in a slightly different order - the disk is coated first with magnetic film by sputtering followed by overcoating, then the lithography is performed on the media, followed by lubrication and other post-process techniques, and finally dc-erasing the patterned disk.

Land-and-groove patterns on the disk affect the flying of the head slider on the spinning disk. Patterns must be identified so that they have minimum effects on the flyability. Improving the magnetic properties, increasing the throughput, and manufacturability are some of the concerns associated with the imprint technique of creating servo sectors [165].

One advantage of the nano-imprint method arises from the fact that the magnetized segments of the disk are better isolated by the grooves. This permits much higher signal overwrite and, as a result, higher signal output can be achieved. However, making the nanostructure for each bit increases considerably the number of fine structures in 100 nanometer scale defined by the NIL and hence increase the manufacturing cost. Discrete track recording (DTR) medium that requires less dimensional control can be a bridge to the fully patterned media. This makes the NIL pattern on HDD substrate with sub-100 nm scale minimum features more cost effective [200].

Both DTR and patterned media results in higher SNR when data is read from these media. Such technologies will push magnetic recording density higher and will make the bits more square in shape. But such system requires servo control with higher precision and novel servo pattern design and PES generation scheme. The servowriting process, on the other hand, gives way to the fabrication of the nanostructures.

5.6.2   Magnetic Printing

This approach uses a lithographically (such as NIL) patterned master disk and copies the magnetization pattern from the master disk to slave disks. The master disk is placed face to face with a slave disk and an external magnetic field is applied to the master disk. Due to the master disk's shielding effects, the highly coercive medium of the slave disk can be *written in* with the servo pattern in one shot. The master disk, made of more permeable soft magnetic film with higher magnetic moment than the recording media, provides a physical shape (and not the magnetic pattern) to guide the external field. Hence, the information on the master disk is not erased when a large external field is applied [97], [98].

Critical issues in this scheme include (i) the design of the master disk pattern to have effective guidance of magnetic flux to switch on the magnetic media, and (ii) close contact of the master disk and slave disk for effective uniform printing. These issues must be resolved before successful magnetic printing can be a reality. Signal obtained from patterns created by magnetic printing has noise level higher than the signal obtained from patterns written by inductive head.

Although these methods have not beed successfully realized for commercial production, they are potential candidates for creating servo patterns that will bring about enormous increase in the throughput of servo writing. Servo patterns for an entire disk can be done in one shot and there is no need for servo writing on a rotating disk which is prone to vibration. Nevertheless, all these competing technologies have to show their capability of producing quality disks with desired features of given specifications to generate quality signal at competitive cost. Besides, it must be possible to scale the feature size so that future areal densities are supported by these techniques.

Bibliography

[1] Abramovitch, D.Y., Wang, F. and Franklin, G.F. Disk drive pivot non-linearity modeling, part 1: Frequency domain, *Proceedings of the 1994 American Control Conference*, June, 1994.

[2] Abramovitch, D.Y. Rejecting rotational disturbance on small disk drives using rotational accelerometers, *Proceedings of the 1996 IFAC World Congress*, San Francisco, CA, USA, July 1996.

[3] Abramovitch, D.Y. Customizable coherent servo demodulation for disk drives, *IEEE Transactions on Mechatronics*, Vol. 3, No. 3, pp. 184-193, September 1998.

[4] Abramovitch, D.Y., Hurst, T. and Henze, D. An overview of the PES pareto method for decomposing baseline noise sources in hard disk position error signals, *IEEE Transactions on Magnetics*, Vol. 34, No. 1, pp. 17-23, January 1998.

[5] Abramovitch, D.Y. and Franklin, G.F. A brief history of disk drive control, *IEEE Control Systems Magazine*, Vol. 22, No. 3, pp. 28-42, June 2002.

[6] Al-Mamun, A., Lee, T.H. and Low, T.S. Frequency domain identification of a hard disk driver actuator, *Mechatronics*, Vol. 12, No. 4, pp. 563-574, May 2002.

[7] Albrecht, T.R. and Sai, F. Load/Unload technology for disk drives, *IEEE Transactions on Magnetics*, Vol. 35, No. 2, pp. 857-862, March 1999.

[8] Anderson, B.D.O. and Moore, J.B. *Optimal Control - Linear Quadratic Method*, Prentice Hall, Englewood Cliffs, New Jersey, USA, 1990.

[9] Aruga, K. *et al.* High-speed orthogonal power effect actuator for recording at over 10,000 TPI, *IEEE Transactions on Magnetics*, Vol. 32, No. 3, pp. 1756-1761, May 1996.

[10] Atsumi, T. *et al.* Vibration servo control design for mechanical resonant modes of a hard disk drive actuator, *JSME, Series C*, Vol. 46, No. 3, pp. 819-827, September 2003.

[11] Athans, M. and Falb, P. *Optimal Control: An Introduction to the Theory and Its Applications*, McGraw-Hill Book Company, 1986.

[12] Becerra, R.C., Jahns, T.M. and Ehsani, M. Four-quadrant sensorless brushless ECM drive, *IEEE Applied Power Electronics Conference and Exposition*, pp. 202-209, March 1991.

[13] Berger, C.S. and Peduto, D. Robust digital control using multirate input sampling, *International Journal of Control*, Vol. 67, No. 5, pp. 813-824, 1997.

[14] Bi, C. and Liu, Z.J. Analysis of unbalanced pulls in hard disk drive spindle motors using a hybrid method, *IEEE Transactions on Magnetics*, Vol. 32, No. 5, pp. 4308-4310, September 1996.

[15] Bi, C. *et al.* Reduction of acoustic noise in FDB spindle motors by using drive technology, *IEEE Transactions on Magnetics*, Vol. 39, No. 2, pp. 800-805, March 2003.

[16] Bi, C., Jiang, Q. and Lin, S. Estimate the effects of unbalanced magnetic pull of PM synchronous motors, *Proceedings of ICEMS-05*, Nanjing, P.R. China, September 2005.

[17] Bi, C., Jiang, Q. and Lin, S. Unbalanced magnetic pull induced by the EM structure of PM spindle motor, *Proceedings of the ICEMS-05*, Nanjing, China, September 2005.

[18] Bobroff, N. Recent advances in displacement measuring interferometry, *Measurement Science and Technology*, Vol. 4, No. 9, pp. 907-926, September 1993.

[19] Bodson, M., Sacks, A. and Khosla, P. Harmonic generation in adaptive feedforward cancellation schemes, *IEEE Transactions on Automatic Control*, Vol. 39, No. 9, pp. 1939-1944, September 1994.

[20] Bolognani, S., Tubiana, L. and Zigliotto, M. Extended Kalman filter tuning in sensorless PMSM drives, *IEEE Transactions on Industry Applications*, Vol. 39, No. 6, pp. 1741-1747, November - December, 2003.

[21] Brown, D.H. and Sri-Jayantha, S.M. Development of a servowriter for magnetic disk storage applications, *IEEE Transactions on Instrumentation and Measurement*, Vol. 39, No. 2, pp. 409-415, April 1990.

[22] Brunnett, D. and Lloyd, L. *Partial Servo Write Fill In*, US Patent No US6714376.

[23] Bryson, A.E. and Ho, Y.C. *Applied Optimal Control*, Halsted Press, Washington DC, 1975.

[24] Chang, J.K. and Ho, H.T. LQG/LTR frequency loop shaping to improve TMR budegt, *IEEE Transactions on Magnetics*, Vol. 35, No. 5, pp. 2280-2282, September 1999.

[25] Chang, S.H. *et al.* Finite element analysis of a piezoelectrically actuated slider for magnetic recording, *IEEE Transactions on Magnetics*, Vol. 27, No. 1, pp. 698-704, January 1991.

[26] Chang, Y.B. *et al.* Prediction of track misregistration due to disk flutter in hard disk drive, *IEEE Transactions on Magnetics*, Vol. 38, No. 2, Part: 2, pp. 1441-1446, March 2002.

[27] Chen, B.M. H_∞ *control and its applications*, (Lecture Notes in Control and Information Sciences), Springer, New York, 1998.

[28] Chen, B.M. *et al.* An H_∞ almost disturbance decoupling robust controller design for a piezoelectric bimorph actuator with hysteresis, *IEEE Transactions on Control Systems Technology*, Vol. 7, No. 2, pp. 160-174, March 1999.

[29] Chen, B.M., Lee, T.H. and Venkataramanan, V. *Hard Disk Drive Servo Systems*, Advances in Industrial Control Series, Springer, New York, 2002.

[30] Chen, B.M. *et al.* Composite nonlinear feedback control for linear systems with input saturation: theory and an application, *IEEE Transactions on Automatic Control*, Vol. 48, No. 3, pp. 427-439, March 2003.

[31] Chen, S.X. *et al.* Design of fluid bearing spindle motors with controlled unbalanced magnetic forces, *IEEE Transactions on Magnetics*, Vol. 33, No. 5, pp. 2638-2640, September 1997.

[32] Chen, S.X., and Low, T.S. Analysis of spindle motor performance sensitivity to excitation schemes, *IEEE Transactions on Magnetics*, Vol. 34, No. 2, pp. 486-488, March 1998.

[33] Chen, Y. *et al. Repeatable Runout Compensation using a Learning Algorithm with Scheduled Parameters*, U. S. Patent No. 6,437,936, August 2002.

[34] Chew, K.K. and Tomizuka, M. Digital control of repetitive errors in disk drive systems, *Control Systems Magazine*, Vol. 10, No. 1, pp. 16-20, January 1990.

[35] Chou, S.Y. Patterned magnetic nanostructures and quantized magnetic disks, *Proceedings of the IEEE*, Vol. 85, No. 4, pp. 652-671, April 1997.

[36] Comstock, R.L. and Workman, M.L. Data storage on rigid disks, in *Magnetic Recording Handbook*, Mee, C.D. and Daniel, E.D, editors, pp. 655-771. McGraw Hill, 1990.

[37] Consoli, A., Scarcella, G. and Testa, A. Industry application of zero-speed sensorless control techniques for PM synchronous motors, *IEEE Transactions on Industry Applications*, Vol. 37, No. 2, pp. 513-521, March - April 2001.

[38] Cribbs, D.F., Ellenberger, M.L. and Hassler, J.W. Jr. *Self Servo-writing Disk Drive and Method*, U.S. Patent 5,448,429, 1995.

[39] D'Angelo III, C. and Mote Jr. C.D. Aerodynamically excited vibration and flutter of a thin disk rotating at supercritical speed, *Journal of Sound and Vibration*, Vol. 168, No. 1, pp. 15-30, 1993.

[40] Dahl, P.R. Solid friction damping of mechanical vibration, *AIAA Journal*, Vol. 14, No. 12, pp. 1675-1682, December 1976.

[41] de Oliveira, M.C., Geromel, J.C. and Bernussou, J. An LMI optimization approach to multiobjective and robust H_∞ controller design for discrete-time systems, *Proceedings of the 38th IEEE Conference on Decision and Control*, Vol. 4, pp. 3611-3616, 1999.

[42] Deeyiengyang, S. and Ono, K. Suppression of resonance amplitude of disk vibrations by squeeze air bearing plate, *IEEE Transactions on Magnetics*, Vol. 37, No. 2, Part: 1, pp. 820-825, March 2001.

[43] Deng, Y. and Guo, L. *Disk Drive with Self Sero Writing Capability*, U.S. Patent No. 6,798,610, September 2004.

[44] Du, C., Zhang, J. and Guo, G. Vibration analysis and control design comparison of HDDs using ball bearing and fluid bearing spindles, *Proceedings of the American Control Conference*, Anchorage, pp. 13781383, May 810, 2002.

[45] Du, C., Zhang, J. and Guo, G. Vibration modeling and control design for self servo-track writing, *IEEE/ASME Transactions on Mechatronics*, Vol. 10, No. 1, pp. 122-127, February 2005.

[46] Du, C. and Guo, G. Lowering the hump of sensitivity functions for discrete-time dual-stage systems, *IEEE Transactions on Control Systems Technology*, Vol. 13, No. 5, pp. 791-797, September 2005.

[47] Ehrlich, R. and Curran, D. Major HDD TMR sources and projected scaling with TPI, *IEEE Transactions on Magnetics*, Vol. 35, No. 21, pp. 885-891, March 1999.

[48] Ehrlich, R., Adler, J. and Hindi, H. Rejecting oscillatory, non-synchronous mechanical disturbances in hard disk drives, *IEEE Transactions on Magnetics*, Vol 37, No. 2, pp. 646-650, March 2001.

[49] Ertugrul, N. and Acarnley, P. A new algorithm for sensorless operation of permanent magnet motors, *IEEE Transactions on Industry Applications*, Vol. 30, No. 1, pp. 126-133, January - February, 1994.

[50] Evans, R., Griesbada, J. and Messner, W.C. Piezoelectric microactuator for dual-stage Control, *IEEE Transactions on Magnetics*, Vol. 35, No. 2, pp. 977-982, March 1999.

[51] Fan, L.S. *et al.* Magnetic recording head positioning at very high track densities using a microactuator-based two-stage servo system, *IEEE Transactions on Industrial Electronics*, Vol. 42, No. 3, pp. 222-233, June 1995.

[52] Fitzgerald, A.E., Kingsley, C. and Umans, S.D. *Electric Machinery*, McGraw-Hill Book Company, London, 1992.

[53] Francis, B.A. and Wonham, W.M. The internal model principle for linear multivariable regulators, *Applied Mathematics and Optimization*, Vol. 2, No. 2, pp. 170194, June 1975.

[54] Franklin, G.F., Powell, J.D. and Workman, M.L. *Digital Control of Dynamic Systems*, Addison-Wesley Publishing Company, Reading, MA, 3rd edition, 1997.

[55] Freudenberg, J., Middleton, R. and Stefanopoulou, A. A survey of inherent design limitations, *Proceedings of the American Control Conference*, Chicago, Illinois, USA, pp. 2987-3001, June 2000.

[56] Fujimoto, H., Hori, Y. and Kawamura, A., Perfect tracking control based on multirate feedforward control with generalized sampling periods, *IEEE Transactions on Industrial Electronics*, Vol. 48, No. 3, pp. 636-644, June 2001.

[57] Fujita, H. *et al.* A microactuator for head positioning system of hard disk drives, *IEEE Transactions on Magnetics*, Vol. 35, No. 2, pp. 1006-1010, March 1999.

[58] Gahinet, P. and Apkarian, P. Following control of a hard disk drive by using sampled-data H_∞ control, *Proceedings of IEEE International Conference on Control Applications*, Hawaii, USA, pp. 182-186, August 1999.

[59] Goh, T.B. *et al.* Design and implementation of a hard disk drive servo system using robust and perfect tracking approach, *The 38th Conference on Decision and Control*, Phoenix, AZ, USA, December 1999.

[60] Grochowski, E. and Halem, R.D. Technological impact of magnetic hard disk drives on storage systems, *IBM Systems Journal*, Vol. 42, No. 2, pp. 338-346, 2003.

[61] Guo, G. *et al.* A DSP based hard disk drive servo test stand, *IEEE Transactions on Magnetics*, Vol. 34, No. 2, pp. 480-482, March 1998.

[62] Guo, G., Hao, Q. and Low, T.S. A dual-stage control design for high track per inch hard disk drives, *IEEE Transactions on Magnetics*, Vol. 37, No. 2, Part 1, pp. 860-865, March 2001.

[63] Guo, G. and Zhang, J. Feedforward control for reducing disk-flutter-induced track misregistration, *IEEE Transactions on Magnetics*, Vol. 39, No. 4, pp. 2103-2108, July 2003.

[64] Guo, G. *et al. Active Control System and Method for Reducing Disk Flutter induced Track Misregistrations*, US patent 6,888,694, May 3, 2005.

[65] Guo, L., Martin, D. and Brunnett, D. Dual-stage actuator servo control for high density disk drives, *Proceedings of the 1999 IEEE/ASME International Conference on Advanced Intelligent Mechatronics*, pp. 132-137, September 1999.

[66] Guo, L. and Chen, Y. Disk flutter and its impact on HDD servo performance, *IEEE Transactions on Magnetics*, Vol. 37, No. 2, pp. 866-870, March 2001.

[67] Guo, W. *et al.* Dual stage actuators for high design rotating memory devices, *IEEE Transactions on Magnetics*, Vol. 34, No. 2, pp. 450-455, March 1998.

[68] Guo, W. *et al.* Linear quadratic optimal dual-stage servo control systems for hard disk drives, *Proceedings of the 24th Annual Conference of the IEEE Industrial Electronics Society*, IECON'98, Aachen, Germany, Vol. 3, pp. 1405-1410, August 31-September 4, 1998.

[69] Hamdi, E.S. *Design of Small Electrical Machines*, John Willey and Sons, 1994.

[70] Hanselman, D.C. *Brushless Permanent Magnet Motor Design*, McGraw-Hill, 1994.

[71] Hanselmann, H. and Mortix, W. High-bandwidth control of the head-positioning mechanism in a winchester disk drive, *IEEE Control Systems Magazine*, Vol. 7, No. 5, pp. 15-19, October 1987.

[72] Hao, Q. *et al.* A gradient based track-following controller optimization for hard disk drive, *IEEE Transactions on Industrial Electronics*, Vol. 50, No. 1, pp. 108-115, February 2003.

[73] Hara, S. *et al.* Repetitive control systems: a new type servo system for periodic exogenous signals, *IEEE Transactions on Automatic Control*, Vol. 33, No. 7, pp. 659668, July 1988.

[74] Harker, J.M. *et al.* A quarter century of disk file innovation *IBM Journal of Research and Development*, Vol. 25, No. 5, pp. 677-689, September 1981.

[75] Harnoy, A. *Bearing Design in Machinery*, Marcel Dekker, Inc, 2002.

[76] Heath, J. Simple approach to improve actuator bandwidth, *Data Storage*, pp. 48-60, September 2000.

[77] Hendershot, J.R. and Miller, T.J.E. *Design of Brushless Permanent Magnet Motors*, Magna Physics Publishing and Oxford University Press, 1994.

[78] Herrmann G. *et al.* Practical implementation of a novel anti-windup scheme in a HDD-dual-stage servo-system, *IEEE/ASME Transactions on Mechatronics*, Vol. 9, No. 3, pp. 580-592, September 2004.

[79] Heo, B. and Shen, I.Y. Taming disk and spindle rocking by damped laminated disks - an experimental study, *IEEE Transactions on Magnetics*, Vol. 35, No. 5, pp. 2304-2306, September 1999.

[80] Hindmarsh, J. *Electrical Machines and Their Applications*, Pergamon Press, 1984.

[81] Hipwell, R. *et al.* MEMS enhance hard disk drive performance, *http://www.waferprocessing.unaxis.com/en/download/mems.pdf*.

[82] Ho, H.T. Fast servo bang-bang seek control, *IEEE Transactions on Magnetics*, Vol. 33, No. 6, pp. 4522-4527, November 1997.

[83] Ho, H.T. Noise impact on servo TMR, *Proceedings of the American Control Conference*, New Mexico, Vol. 5, pp. 2906-2909, June 1997.

[84] Horowitz, R. and Li, B. Adaptive tracking-following servos for disk file actuators, *IEEE Transactions on Magnetics*, Vol. 32, No. 3, pp. 1779-1786, 1996.

[85] Horsley, D.A. *et al.* Design and fabrication of an angular microactuator for magnetic disk drives, *Journal of Microelectromechanical Systems*, Vol. 7, No. 2, pp. 141-148, June 1998.

[86] Hredzak, B., Guo, G. and Zhang, J. Investigation of the effect of hard disk drive unbalance on repeatable and non-repetable runout, *Proceedings of IEEE International Conference on Power Electronics & Drive Systems, PEDS*, pp. 1359-1363, Nov. 17-20, 2003.

[87] Hredzak, B. and Guo, G. New passive balancing algorithm for high density hard disk drives, *Journal of Mechanical Engineering Science, Proceedings of Institution of Mechanical Engineers of UK*, Vol. 218, No. 4, pp. 401-410, April 2004.

[88] Hu, X. *et al.* A TMR oriented optimization of suspension based milli-actuators, *Journal of Information Storage and Processing Systems*, Vol. 2, No. 3, pp. 163-168, February 2000.

[89] Huang, Y. *et al.* Design and analysis of a high bandwidth disk drive servo system using an instrumented suspension, *IEEE/ASME Transactions on Mechatronics*, Vol. 4, No. 2, pp. 196-206, June 1999.

[90] Huang, F.Y. *et al.* Active damping in HDD actuator, *IEEE Transactions on Magnetics*, Vol. 37, No. 2, Part: 1, pp. 847-849, March 2001.

[91] Humphrey, J. *et al.* Unobstructed and obstructed rotating disk flows: A summary review relevant to information storage system, in Bhusan, B., editor *Advances in Information Storage Systems*, pp. 79-110. ASME Press, NY, 1991.

[92] Hwang, C.C., Cheng, S.P. and Chang, C.M. Design of high-performance spindle motors with concentrated windings, *IEEE Transactions on Magnetics*, Vol. 41, No. 2, pp. 971-973, February 2005.

[93] Iizuka, K. *et al.* Microcomputer control for sensorless brushless motor, *IEEE Transaction on Industry Applications*, Vol. IA-21, No. 4, pp. 595-601, May-June 1985.

[94] Inoue, T. *et al.* Improvement of RRO using hybrid-type STW for hard disk drives, *IEEE Transactions on Magnetics*, Vol. 37, No. 2, pp. 969-973, March 2001.

[95] Ioannou, P.A., Xu, H. and Fidan, B., Servo control design for a hard disk drive based on estimated head position at high sampling rates, *Proceedings of the American Control Conference, 2003*, Vol 1, pp. 731-736, June 4-6, 2003.

[96] Ioannou, P.A., Kosmatopoulos, E.B. and Despain, A.M., Position error signal estimation at high sampling rates using data and servo sector measurements, *IEEE Transactions on Control Systems Technology*, Vol. 11, No. 3, pp. 325-334, May 2003.

[97] Ishida, T. *et al.* Printed media technology for an effective and inexpensive servo track writing of HDDs, *IEEE Transactions on Magnetics*, Vol. 37, No. 4, pp. 1875-1877, July 2001.

[98] Ishida, T. *et al.* Magnetic printing technology - application to HDD, *IEEE Transactions on Magnetics*, Vol. 39, No. 2, Part 1, pp. 628-632, March 2003.

[99] Ishikawa, J. and Tomizuka, M. Pivot friction compensation using an accelerometer and a disturbance observer for hard disk drives, *IEEE-ASME Transactions on Mechatronics*, Vol. 3, No. 3, pp. 194-201, September 1998.

[100] Jahns, T.M., Becerra, R.C. and Ehsani, M. Integrated current regulation for a brushless ECM drive, *IEEE Transactions on Power Electronics*, Vol. 6, No. 1, pp. 118-126, January 1991.

[101] Jang, J.S.R, Sun, C.T. and Mizutani, E. *Neuro-fuzzy and Soft Computing*, Prentice Hall, Reading, NJ, 1997.

[102] Jiang, Q., Bi, C. and Huang, R.Y. A new phase-delay free method to detect back EMF zero-crossing points, *Proceedings of the Asia Pacific Magnetic Recording Conference APMRC'04*, Seoul, Korea, August, 2004.

[103] Johansson, R. *System Modeling & Identification*, Prentice Hall Information and System Sciences Series, editor - Kailath, T., 1992.

[104] Johnson, J.P., Ehsani, M. and Guzelgunler. Y. Review of sensorless methods for brushless DC, *Conference Record of the 1999 IEEE IAS Annual Meeting*, Vol. 1, pp. 143 - 150, October 1999.

[105] Kalman, R.E. and Bertram, J.E. A unified approach to the theory of sampling systems, *Journal of Franklin Institute*, Vol. 267, pp. 405-436, May, 1959.

[106] Karaman, M. and Messner, W.C. Robust dual stage HDD track follow control systems design for hand-off shaping, *Digest of APMRC 2002*, CA5, 2002.

[107] Kempf, C. *et al.* Comparison of four discrete-time repetitive control algorithms, *Proceedings of the American Control Conference*, pp. 2700-2704, 1992.

[108] Kenjo, T. and Sugawara, A. *Stepping Motors and Their Microprocessor Control*, Oxford University Press, 2nd Edition, 1994.

[109] Kenjo, T. and Nagamori, S. *Brushless Motors*, Sogo Electronics Press, 2003.

[110] Kim, Y.H., OH, D.H. and Ho, S.L. PES reduction in magnetic disk drives via a Low-TMR HGA and servo loop shaping, *IEEE Transactions on Magnetics*, Vol. 41, No. 2, pp. 779-783, February 2005.

[111] Klaasen, E. *et al.* Silicon fusion bonding and deep reactive ion etching: a new technology for microstructures, *Sensors and Actuators: A*, Vol. 52, No. 1-3, pp. 132-139, 1996.

[112] Kobayashi, M. *et al.* Multi-sensing servo with carriage-acceleration feedback for magnetic disk drives, *Proceedings of the American Control Conference, 1998*, Vol. 5, pp. 3038-3042, 1998.

[113] Kobayashi, M. *et al.* High-bandwidth servo control designs for magnetic disk drives, *Proceedings of IEEE/ASME International Conference on Advanced Intelligent Mechatronics 2001*, Vol. 2, pp. 1124 -1129, July 2001.

[114] Kobayashi, M., Nakagawa, S. and Nakamura, S. A phase-stabilized servo elontroller for dual-stage actuators in hard disk drives, *IEEE Transactions on Magnetics*, Vol. 39, No. 2, pp. 844-850, Part 1, March 2003.

[115] Kollar, I. Frequency domain system identification toolbox for use with Matlab, The Mathworks, 1994.

[116] Kranc, G.M. Input-Output analysis of multirate feedback systems, *IRE Transactions on Automatic Control*, Vol. 2, pp. 21-28, November 1957.

[117] Kulkarni, A.B. and Ehsani, M. A novel position sensor elimination technique for the interior permanent-magnet synchronous motor drive, *IEEE Transactions on Industry Applications*, Vol. 28, No. 1, pp. 144-150, January - February, 1992.

[118] Kurita, M. *et al.* An active-head slider with a piezoelectric actuator for controlling flying height, *IEEE Transactions on Magnetics*, Vol. 38, No. 5, pp. 2102-2104, Part 1, September 2002.

[119] Kurita, M. and Suzuki, K. Flying-Height adjustment technologies of magnetic head sliders, *IEEE Transactions on Magnetics*, Vol. 40, No. 1, pp. 332-336, Part 2, January 2004.

[120] Lam, B.H., Panda, S.K. and Xu, J.X. Reduction of periodic speed ripples in PM synchronous motors using iterative learning control, *Proceedings of the IEEE Industrial Electronics Confrence IECON-00*, Vol. 2, pp. 1406-1411, October 2000.

[121] Lee, H. Controller optimization for minimum position error signal of hard disk drives, *Proceedings of American Control Conference*, pp. 3081-3085, June 2000.

[122] Lee, L.C., Zheng, J. and Guo, G. In-phase optimization of hard disk drive actuator using genetic algorithms, *submitted to American Control Conference 2006*.

[123] Lee, S.H., Kim, Y.H. and Chung, C.C. Dual-stage actuator disk drives for improved servo performance: track follow, track seek, and settle, *IEEE Transactions on Magnetics*, Vol. 37, No. 4, Part: 1, pp. 1887-1890, July, 2001.

[124] Li, Y. and Tomizuka, M. Two-degree-of-freedom control with robust feedback control for hard disk servo systems, *IEEE/ASME Transactions on Mechatronics*, Vol. 4, No. 1, pp. 17-24, March 1999.

[125] Li, Y. and Horowitz, R. Mechatronics of electrostatic microactuators for computer disk drive dual-stage servo systems, *IEEE/ASME Transactions on Mechatronics*, Vol. 6, No. 2, pp. 111-121, June 2001.

[126] Li, Y. and Horowitz, R. Active suspension control with dual stage actuators in hard disk drives, *Proceedings of the 2001 American Control Conference*, pp. 2786-2791, Vol. 4, June 2001.

[127] Li, Y., Guo, G. and Wang, Y. Nonlinear control for fast settling in HDDs, *IEEE Transactions on Magnetics*, Vol. 40, No. 4, pp. 2086-2088, July 2004.

[128] Li, Z.M. *et al.* Optimal track-following design for toward highest tracks per inch in hard disk drives, *Journal of Information Storage and Processing System*, Vol. 3, pp. 27-42, 2001.

[129] Lin, R.L., Hu, M.T., Chen, S.C., and Lee, C.Y. Using phase-current sensing circuit as the position sensor for brushless DC motors without shaft position sensor, *Proceedings of the IEEE Conference on Industrial Electronic*, IECON-89, pp. 215-218, November 1989.

[130] Lin, S. *et al.* Effect of drive modes on the acoustic noise of fluid dynamic bearing spindle motors, *IEEE Transactions on Magnetics*, Vol. 39, No. 5, pp. 3277 - 3279, September 2003.

[131] Lin, S. *Advanced Control of micro PMSM and its Application in Hard Disk Drive*, PhD Thesis, National University of Singapore, 2005.

[132] Liu, B. and Yuan, Z.M. In-situ characterization of head-disk clearance, in *Proceedings of ASME/Tribology Division Symposium on Interface Tribology Towards 100 Gb/in2 and Beyond*, pp. 51-58, Seattle, USA, October 1-4, 2000.

[133] Liu, Z.J., Bi, C. and Low, T.S. Analysis of iron loss in hard disk drive spindle motors, *IEEE Transactions On Magnetics*, Vol. 33, No. 5, pp. 4089-4091, September 1997.

[134] Lou, Y. *et al.* Dual-stage servo with on-slider piezoelectric microactuator for hard disk drives, *IEEE Transactions on Magnetics*, Vol. 38, No. 5, pp. 2183-2185, September 2002,

[135] Low, T.S., Chen, S.X. and Gao, X. Robust torque optimization for BLDC spindle motors, *IEEE Transactions on Industrial Electronics*, Vol. 48, No. 3, pp. 656-663, June 2001.

[136] Lu, Y. *et al.* A MEMS-based electrostatic microactuator for hard disk drives, *submitted for publication.*

[137] Lucibello, P. Comments on "Nonlinear repetitive control", *IEEE Transactions on Automatic Control*, Vol. 48, No. 8, pp. 1470-1471, August 2003.

[138] Mahalik, N.P. *Mechatronics - Principles, Concepts and Applications*, Tata McGraw Hill, 2003.

[139] Matsui, N. Sensorless PM brushless DC motor drives, *IEEE Transaction on Industrial Electronics*, Vol. 43, No. 2, pp. 300-308, January/February 1996.

[140] McAllister, J.S. Characterization of disk vibrations on aluminum and alternate substrates, *IEEE Transactions on Magnetics*, Vol. 33, No. 1, pp. 968973, January 1997.

[141] McAllister, J.S. Disk flutter: causes and potential cures, *Data Storage*, pp. 29-34, May 1997.

[142] Mee, C.D. and Daniel, E.D. (editors), *Magnetic Recording*, Volume I and Volume II, McGraw-Hill, 1987.

[143] Messner, W. *et al.* A new adaptive learning rule, *IEEE Transactions on Automatic Control*, Vol. 36, No. 2, pp. 188-197, February 1991.

[144] Mohtadi, C. Bode's integral theorem for discrete-time systems, *IEE Proceedings - Control Theory and Applications*, Vol. 137, No. 2, pp. 57-66, March 1990.

[145] Moon, J.H., Lee, M.N. and Chung M.J. Repetitive control for the track-following servo system of an optical disk drive, *IEEE Transactions on Control Systems Technology*, Vol. 6, No. 5, pp. 663-670, September 1998.

[146] Moreira, J.C. Indirect sensing for rotor flux position of permanent magnet AC motors operation in a wide speed range, *Conference Record of IEEE IAS Annual Meeting*, pp. 401-407, October 1994.

[147] Morrison, C.I. and Stengel, R.F. Robust control system design using random search and genetic algorithm, *IEEE Transactions on Automatic Control*, Vol. 42, No. 6, pp. 835-839, 1997.

[148] Mou, J.Q. Design, fabrication and characterization of single crystal silicon microactuator for hard disk drives, *Journal of Micromechanism and Microengineering*, Vol. 14, pp. 1608-1613, 2004.

[149] Niu, Y. *et al.* A PZT micro-actuated suspension for high TPI hard disk servo systems, *IEEE Transactions on Magnetics*, Vol. 36, No. 5, pp. 2241-2243, September 2000.

[150] Noda, K. *et al.* Novel NRRO minimization algorithm for an ultra-high-density servotrack writer, *IEEE Transactions on Magnetics*, Vol. 33, No. 5, pp. 2626-2628, September 1997.

[151] Noyes, T. and Dickinson, W.E. The random access memory accounting machine II - The Magnetic-Disk Random-Access Memory, *IBM Journal*, pp. 72-75, January 1957.

[152] Ogata, K. *Modern control engineering*, Prentice Hall, Upper Saddle River, N.J., 4th edition, 2002.

[153] Ogasawara, S. and Akagi, H. An approach to position sensorless drive for brushless DC motors, *IEEE Transactions on Industry Applications*, Vol. 27, No. 5, pp. 928-933, September-October, 1991.

[154] Ong, E.H. *et al.* A low-turbulence high-bandwidth actuator for 3.5-inch hard disk drives, *IEEE Transactions on Magnetics*, Vol. 36, No. 5, pp. 2235-2237, September 2000.

[155] Ono, H. Architecture and performance of the ESPER-2 hard-disk drive servo writer, *IBM Journal of Research and Development*, Vol. 37, No. 1, pp. 311, 1993.

[156] Onuki, Y. and Ishioka, H. Compensation for repeatable tracking errors in hard drives using discrete-time repetitive controllers, *IEEE Transactions on Mechatronics*, pp. 132-136, 2001.

[157] Oswald, R.K. Design of a disk file head positioning servo, *IBM Journal of Research and Development*, Vol. 18, pp. 506-512, 1974.

[158] Pang, C.K. *et al.* Nano-position sensing and control in HDD dual-stage servo, *IEEE Conference on Control Applications*, Taipei, ROC, Sept 2004.

[159] Pang, C.K. *et al.* Suppressing sensitivity hump in HDD dual-stage servo systems, *Microsystem Technology*, Vol. 11, No. 8-10, pp. 653-662, August 2005.

[160] Pang, C.K. *et al.* NRRO rejection using online iterative control for high density storage, submitted for publication.

[161] Pannu, S. and Horowitz, R. Adaptive accelerometer feedforward servo for disk drives, *Proceedings of the 36th IEEE Conference on Decision and Control*, Vol. 5, pp. 4216-4218, 1997.

[162] Paskota, M. *et al.* Optimal simultaneous stabilization of linear single-input systems via linear state feedback control, *International Journal of Control*, Vol. 60, No. 4, pp. 483-498, 1994.

[163] Pintelon, R. *et al.* Parametric identification of transfer function in the frequency domain - a survey, *IEEE Transactions on Automatic Control*, Vol. 39, No. 11, pp. 2245-2260, November 1994.

[164] Reed, D.E. *et al.* Digital servo demodulation in a digital read channel, *IEEE Transactions on Magnetics*, Vol. 34, No. 1, Part 1, pp. 13-16, January 1998.

[165] Ross, C.A. Patterned magnetic recording media, *Annual Review on Material Research*, Vol. 31, pp. 203-235, 2001.

[166] Saberi, A., Sannuti, P. and Chen, B.M. H_2 *Optimal Control*, Prentice Hall, 1995.

[167] Saberi, A., Chen, B.M. and Sannuti, P. *Loop Transfer Recovery: Analysis and Design*, Springer, 1993.

[168] Sacks, A., Bodson, M. and Messner, W. Advanced methods for repeatable runout compensation, *IEEE Transactions on Magnetics*, Vol. 31, No. 2, pp. 1031-1037, March 1995.

[169] Sasaki, M. *et al.* Track-following control of a dual-stage hard disk drive using a neuro-control system, *Engineering Application of Artificial Intelligence*, Vol. 11, pp. 707-716, 1998.

[170] Schroeck, S.J. and Messner, W.C. On controller design for linear time-invariant dual-input single-output systems, *Proceedings of American Control Conference*, Vol. 6, pp. 4122-4126, San Diego, 1999.

[171] Schultz, M.D. *et al.* A self-servowrite clocking process, *IEEE Transactions on Magnetics*, Vol. 37, No. 4, pp. 1878-1880, July 2001.

[172] Schoukens, J. and Pintelon, R. *Identification of linear systems: A practical guide to accurate modelling*, Pergamon Press, London, 1991.

[173] Semba, T. *et al.* Dual-stage servo controller for HDD using MEMS actuator, *IEEE Transactions on Magnetics*, Vol. 35, No. 5, pp. 2271-2273, September 1999.

[174] Shao, J., Nolan, D. and Hopkins, T. A novel direct back EMF detection for sensorless brushless DC (BLDC) motor drives, *APEC 2002*, Vol. 1, pp. 33-37, 10-14 March 2002.

[175] Shen, I.Y. Recent vibration issues in computer hard disk drives, *Journal of Magnetism and Magnetic Materials*, Vol. 209, No. 1, pp. 6-9, February 2000.

[176] Singer, N.C. and Seering, W.P. Preshaping command input to reduce system vibration, *Journal of Dynamic Systems, Measurement and Control*, Vol. 112, No. 1, pp. 76-82, March 1990.

[177] Singh, B. *et al.* Performance analysis of fuzzy logic controlled permanent magnet synchronous motor drive, *Proceedings of the IEEE Conference on Industrial Electronics*, IECON-95, pp. 399 - 405, November 1995.

[178] Sivadasan, K.K. and Guo, G. Head suspension assembly with piezoelectric beam micro actuator, US Patent No. 6,522,050, February 2003.

[179] Soeno, Y. *et al.* Piezoelectric piggy-back microatuator for hard disk drive, *IEEE Transactions on Magnetics*, Vol. 35, No. 2, Part 1, pp. 983-987, March 1999.

[180] Solsona, J., Valla, M.I. and Muravchik, C. A nonlinear reduced order observer for permanent magnet synchronous motors, *IEEE Transactions on Industrial Electronics*, Vol. 43, No. 4, pp. 492-497, August 1996.

[181] Sommargren, G.E. A new laser measurement system for precision metrology, *Precision Engineering*, Vol. 9, No. 4, pp. 179-184, 1987.

[182] Song, D., Schnur, D. and Boutaghou, Z.E. Discharge mechanism for electrostatic fly control, *IEEE Transactions on Magnetics*, Vol. 40, No. 4, pp. 3162-3164, Part 2, July 2004.

[183] Sri-Jayantha, S.M. *et al.* $TrueTrack^{TM}$ servo technology for high TPI disk drives, *IEEE Transactions on Magnetics*, Vol. 37, No. 2, Part 1, pp. 871-876, March 2001.

[184] Sri-Jayantha, S.M. *et al.* Repeatable runout free servo architecture in direct access storage device, *US patent No. 6,097,565*, August 2000.

[185] Srikrishna, P. and Kasetty, K. Predicting track misregistration (TMR) from disk vibration of alternate substrate materials, *IEEE Transactions Magnetics*, Vol. 36, No. 1, pp. 171-176, January 2000.

[186] Stevens, L.D. The evolution of magnetic storage, *IBM Journal of Research and Development*, Vol. 25, No. 5, pp. 663-675, September 1981.

[187] Storagereview, Spindle Speed, *http://www.storagereview.com/guide2000/ref/hdd/*.

[188] Suzuki, K. *et al.* An active head slider using a piezoelectric cantilever for in situ flying-height control, *IEEE Transactions on Magnetics*, Vol. 39, No. 2, pp. 826-831, Part 1, March 2003.

[189] Suzuki, K. and Kurita, M. A MEMS-based active-head slider for flying height control in magnetic recording, *JSME International Journal, Series B - Fluids and Thermal Engineering*, Vol. 47, No. 3, pp. 453-458, August 2004.

[190] Swearingen, P.A. and Shepherd, S.H. *System for self-servowriting a disk drive*, US patent No. 5,668,679, September 1997.

[191] Takaishi, K. HDD servo technology for media level servo track writing, *IEEE Transactions on Magnetics*, Vol. 39, No. 2, Part 1, pp. 851-856, March 2003.

[192] Tagawa, N. *et al.* Development of novel PZT thin films for active sliders based on head load/unload on demand systems, *Microsystem Technology*, Vol. 8, No. 2-3, pp. 133-138, May 2002.

[193] Tagawa, N., Kitamtra, K. and Mori, A. Design and fabrication of MEMS-based active slider using double-layered composite PZT thin film in hard disk drives, *IEEE Transactions on Magnetics*, Vol. 39, No. 2, pp. 926-931, Part 1, March 2003.

[194] Tang, W.C., Lim, M.G. and Howe, R.T. Electrostatic comb drive levitation and control method, *Journal of Micro-Electromechanical Systems*, Vol. 1, No. 4, pp. 170178, December 1992.

[195] Tang, Y., Chen, S.X. and Low, T.S. Micro elecctrostatic actuators in dual-stage disk drives with high track density, *IEEE Transactions on Magnetics*, Vol. 32, No. 5, pp. 3851-3853, September 1996.

[196] Tokuyama, M. *et al.* Development of a Phi-shaped actuated suspension for 100-kTPI hard disk drives, *IEEE Transactions on Magnetics*, Vol. 37, No. 4, Part 1, pp. 1884-1886, July 2001.

[197] Toshiyoshi, H., Mita, M. and Fujita, H. A MEMS piggyback actuator for hard-disk drives, *Journal of Micro-Electromechanical Systems*, Vol. 11, No. 6, pp. 648-654, December 2002.

[198] Uematsu, Y., Fukushi, M. and Taniguchi, K. Development of the pushpin free STW, *IEEE Transactions on Magnetics*, Vol. 37, No. 2, Part 1, pp. 964-968, March 2001.

[199] Vidyasagar, M. Statistical learning theory and randomized algorithms for control, *IEEE Control Systems Magazine*, Vol. 18, No. 6, pp. 69-85, December 1998.

[200] Wachenschwanz, D. *et al.* Design of a manufacturable discrete track recording medium, *IEEE Transactions on Magnetics*, Vol. 41, No. 2, pp. 670-675, February 2005.

[201] Wang, F. *et al.* Disk drive pivot nonlinearity modeling part ii: Time domain, in *Proceedings of the 1994 American Control Conference*, pp. 2604-2607, June 1994.

[202] Wang, S.X. and Taratorin, A.M. *Magnetic Information Storage Technology*, McGraw Hill, 1999.

[203] Wang, Y., Tan, Y.L. and Guo, G. Robust nonlinear co-ordinated excitation and TCSC control for power systems, *IEE Proceedings - Generation, Transmission and Distribution*, Vol. 149, No. 3, pp. 367-372, May 2002.

[204] Weaver, P.A. and Ehrlich, R. The use of multirate notch filters in embedded servo disk drives, in *Proceedings of American Control Conference*, pp. 4156-4160, June 1993.

[205] Weerasooriya, S. and Phan, D.T. Discrete-time LQG/LTR design and modeling of a disk drive actuator tracking servo system, *IEEE Transactions on Industrial Electronics*, Vol. 42, No. 3, pp. 240-247, June 1995.

[206] Weerasooriya, S., Zhang, J.L. and Low, T.S. Efficient implementation of adaptive feedforward runout cancellation in a disk drive, *IEEE Transactions on Magnetics*, Vol. 32, No. 5, pp. 3920-3922, September 1996.

[207] White, M. and Lu, W. Hard disk drive bandwidth limitations due to sampling frequency and computational delay, *Proceedings of IEEE/ASME International Conference on Advanced Intelligent Mechatronics*, pp. 120-125, 1999.

[208] Wong, W.E., et al., *Detection of track misregistration within user data channel*, US Patent No. 7,009,805 B2, March 2006.

[209] Workman, M.L. *Adaptive Proximate Time Optimal Servomechanism*, Stanford University, PhD thesis, 1987.

[210] Workman, M.L., Kosut, R.L. and Franklin, G.F. Adaptive proximate time-optimal control: Continuous time case, in *Proceedings of the American Control Conference*, pp. 589-594, 1987.

[211] Workman, M.L., Kosut, R.L., and Franklin, G.F. Adaptive proximate time-optimal control: Discrete time case, in *Proceedings of the Conference on Decision and Control*, pp. 1548-1553, December 1987.

[212] Wu, D., Guo, G. and Chong, T.C. Adaptive compensation of microactuator resonance in hard disk drives, *IEEE Transactions on Magnetics*, Vol. 36, No. 5, pp. 2247-2250, September 2000.

[213] Wu, D., Guo, G. and Chong, T.C. Mid-frequency disturbance suppression via micro-actuator in dual-stage HDDs, *IEEE Transactions on Magnetics*, Vol. 38, No. 5, pp. 2189-2191, Part 1, September 2002.

[214] Xu, J.X., Lee, T.H. and Zhang, H. Comparative studies on repeatable runout compensation using iterative learning control, in *Proceedings of the American Control Conference*, pp. 2834-2839, 2001.

[215] Yada, H. *et al.* High areal density recording using an MR/Inductive head and pre-embossed rigid magnetic disk *IEEE Transactions on Magnetics*, Vol. 30, No. 2, pp. 404-409, March 1994.

[216] Yada, H. and Takeda, T. A coherent maximum likelihood head position estimator for PERM disk drives, *IEEE Transactions on Magnetics*, Vol. 32, No. 3, Part 2, pp. 1867-1872, May 1996.

[217] Yang, Q. *et al.* Study on the influence of friction in an automatic ball balancing system, *Journal of Sound and Vibration*, Vol. 285, No. 1-2, pp. 73-99, July 2005.

[218] Yamaguchi, T. *et al.* Mode switching control design with initial value compensation and its application to head positioning control on magnetic disk drives, *IEEE Transactions on Industrial Electronics*, Vol. 43, No. 1, pp. 65-73, February 1996.

[219] Yamaguchi, T., Numasato, H. and and Hirai, H. A mode-switching control for motion control and its application to disk drives: design of optimal mode-switching conditions, *IEEE/ASME Transactions on Mechatronics*, Vol. 3, No. 3, pp. 202-209, September 1998.

[220] Yamaguchi, T. *et al.* Servo bandwidth and positioning accuracy design for high track density disk drives, *IEEE Transactions on Magnetics*, Vol. 36, No. 5, pp. 2216-2218, September 2000.

[221] Yarmchuk, E. J. *et al.* Radial self-propagation pattern generation for disk file servo writing, U.S. Patent 5,907,447, 1999.

[222] Ye, H. *et al.* Radial error propagation issues in self servo track writing technology, *IEEE Transaction on Magnetics*, Vol. 38, No. 5, pp. 2180-2182, September 2002.

[223] Yeack-Scranton, C.E. *et al.* An active slider for practical contact recording, *IEEE Transactions on Magnetics*, Vol. 26, No. 5, pp. 2478-2483, September 1990.

[224] Zhang, J.L. *et al.* Modified AFC runout cancellation and its influence on track following, *Proceedings of the 23rd International Conference of the IEEE Industrial Electronics Society*, IECON'97, Vol. 1, No. 9-14, pp. 35-40, November 1997.

[225] Zhang, J.L. *et al.* Modified adaptive feedforward runout compensation for dual stage servo system, *IEEE Transactions on Magnetics*, Vol. 36, Vol. 5, Part 1, pp. 35813584, September 2000.

[226] Zhang, J.L. *et al.* *7nm PES 3 Sigma @ March 2005*, Project Report, Data Storage Institute, May 2005.

[227] Zhang, Q.D. *et al.* Design of high-speed magnetic fluid bearing spindle motor, *IEEE Transactions on Magnetics*, Vol. 37, No. 4, pp. 2647-2650, July 2001.

[228] Zhang, X.J. *et al.* A new method to minimize the commutation torque ripple in trapezoidal BLDC motor with sensorless drive, *Conference Proceedings of PIEMC*, Vol. 2, pp. 607-611, August 2000.

[229] Zheng, J., Guo, G. and Wang, Y. Identification and decentralized control of a dual-actuator hard disk drive system, *IEEE Transactions on Magnetics*, Vol. 41, No. 9, pp. 2515-2521, September 2005.

[230] Zheng, J., Guo, G. and Wang, Y. Nonlinear PID control of linear plants for improved disturbance rejection, *Proceedings of the 16th IFAC World Congress*, Czech Republic, 4-8 July, 2005.

[231] Zhu, Z.Q. and Howe, D. Electromagnetic noise radiated by brushless permanent magnet DC drives, *The 6th International Conference on Electrical Machines and Drives*, pp. 606-611, September 1993.

Index

T - #0163 - 101024 - C0 - 229/152/22 [24] - CB - 9780849372537 - Gloss Lamination